T0214289

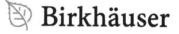

Applied and Numerical Harmonic Analysis

Series Editor
John J. Benedetto
University of Maryland
College Park, MD, USA

Editorial Advisory Board

Akram Aldroubi
Vanderbilt University
Nashville, TN, USA

Douglas Cochran
Arizona State University
Phoenix, AZ, USA

Hans G. Feichtinger
University of Vienna
Vienna, Austria

Christopher Heil
Georgia Institute of Technology
Atlanta, GA, USA

Stéphane Jaffard
University of Paris XII
Paris, France

Jelena Kovačević
Carnegie Mellon University
Pittsburgh, PA, USA

Gitta Kutyniok
Technische Universität Berlin
Berlin, Germany

Mauro Maggioni
Duke University
Durham, NC, USA

Zuowei Shen
National University of Singapore
Singapore, Singapore

Thomas Strohmer
University of California
Davis, CA, USA

Yang Wang
Michigan State University
East Lansing, MI, USA

More information about this series at http://www.springer.com/series/4968

Christopher Heil

Metrics, Norms, Inner Products, and Operator Theory

Christopher Heil
School of Mathematics
Georgia Institute of Technology
Atlanta, GA, USA

ISSN 2296-5009   ISSN 2296-5017   (electronic)
Applied and Numerical Harmonic Analysis
ISBN 978-3-030-09737-0   ISBN 978-3-319-65322-8   (eBook)
https://doi.org/10.1007/978-3-319-65322-8

This book is published under the imprint Birkhäuser, www.birkhauser-science.com, by the registered
company Springer Nature Switzerland AG part of Springer Nature
The registered company address is: Gewerbestrasse 11, 6330 Cham, Switzerland

In memory, Christopher Andrew Heil, 1920–2011

ANHA Series Preface

The *Applied and Numerical Harmonic Analysis (ANHA)* book series aims to provide the engineering, mathematical, and scientific communities with significant developments in harmonic analysis, ranging from abstract harmonic analysis to basic applications. The title of the series reflects the importance of applications and numerical implementation, but richness and relevance of applications and implementation depend fundamentally on the structure and depth of theoretical underpinnings. Thus, from our point of view, the interleaving of theory and applications and their creative symbiotic evolution is axiomatic.

Harmonic analysis is a wellspring of ideas and applicability that has flourished, developed, and deepened over time within many disciplines and by means of creative cross-fertilization with diverse areas. The intricate and fundamental relationship between harmonic analysis and fields such as signal processing, partial differential equations (PDEs), and image processing is reflected in our state-of-the-art *ANHA* series.

Our vision of modern harmonic analysis includes mathematical areas such as wavelet theory, Banach algebras, classical Fourier analysis, time-frequency analysis, and fractal geometry, as well as the diverse topics that impinge on them.

For example, wavelet theory can be considered an appropriate tool to deal with some basic problems in digital signal processing, speech and image processing, geophysics, pattern recognition, biomedical engineering, and turbulence. These areas implement the latest technology from sampling methods on surfaces to fast algorithms and computer vision methods. The underlying mathematics of wavelet theory depends not only on classical Fourier analysis, but also on ideas from abstract harmonic analysis, including von Neumann algebras and the affine group. This leads to a study of the Heisenberg group and its relationship to Gabor systems, and of the metaplectic group for a meaningful interaction of signal decomposition methods. The unifying influence of wavelet theory in the aforementioned topics illustrates the justification for providing a means for centralizing and disseminating information from the broader, but still focused, area of harmonic analysis. This will be a key role of *ANHA*. We intend to publish with the scope and interaction that such a host of issues demands.

Along with our commitment to publish mathematically significant works at the frontiers of harmonic analysis, we have a comparably strong commitment to publish major advances in the following applicable topics in which harmonic analysis plays a substantial role:

Antenna theory	*Prediction theory*
Biomedical signal processing	*Radar applications*
Digital signal processing	*Sampling theory*
Fast algorithms	*Spectral estimation*
Gabor theory and applications	*Speech processing*
Image processing	*Time-frequency and*
Numerical partial differential equations	*time-scale analysis*
	Wavelet theory

The above point of view for the *ANHA* book series is inspired by the history of Fourier analysis itself, whose tentacles reach into so many fields.

In the last two centuries Fourier analysis has had a major impact on the development of mathematics, on the understanding of many engineering and scientific phenomena, and on the solution of some of the most important problems in mathematics and the sciences. Historically, Fourier series were developed in the analysis of some of the classical PDEs of mathematical physics; these series were used to solve such equations. In order to understand Fourier series and the kinds of solutions they could represent, some of the most basic notions of analysis were defined, e.g., the concept of "function." Since the coefficients of Fourier series are integrals, it is no surprise that Riemann integrals were conceived to deal with uniqueness properties of trigonometric series. Cantor's set theory was also developed because of such uniqueness questions.

A basic problem in Fourier analysis is to show how complicated phenomena, such as sound waves, can be described in terms of elementary harmonics. There are two aspects of this problem: first, to find, or even define properly, the harmonics or spectrum of a given phenomenon, e.g., the spectroscopy problem in optics; second, to determine which phenomena can be constructed from given classes of harmonics, as done, for example, by the mechanical synthesizers in tidal analysis.

Fourier analysis is also the natural setting for many other problems in engineering, mathematics, and the sciences. For example, Wiener's Tauberian theorem in Fourier analysis not only characterizes the behavior of the prime numbers, but also provides the proper notion of spectrum for phenomena such as white light; this latter process leads to the Fourier analysis associated with correlation functions in filtering and prediction problems, and these problems, in turn, deal naturally with Hardy spaces in the theory of complex variables.

Nowadays, some of the theory of PDEs has given way to the study of Fourier integral operators. Problems in antenna theory are studied in terms of unimodular trigonometric polynomials. Applications of Fourier analysis abound in signal processing, whether with the fast Fourier transform (FFT), or filter design, or the adaptive modeling inherent in time-frequency-scale methods such as wavelet theory. The coherent states of mathematical physics are translated and modulated Fourier transforms, and these are used, in conjunction with the uncertainty principle, for dealing with signal reconstruction in communications theory. We are back to the raison d'être of the *ANHA* series!

University of Maryland
College Park

John J. Benedetto
Series Editor

Contents

Preface

Mathematics is a very large subject, and many mathematicians consider some particular area to be their "main research field." Algebra, topology, probability, and discrete mathematics are only a few examples of areas of mathematics. My own field is *analysis*,[1] which (broadly speaking) is the science of functions. *Real analysis* tends to deal with functions whose domain is a set of real numbers or a set of vectors in $\mathbb{R}^d$, while *complex analysis* deals with functions of one or more complex variables. Sometimes we work with individual functions, but even more insight can often be gained by considering *sets* or *spaces* of functions that have some property in common (e.g., the space of all differentiable functions). Functions do not exist in isolation, and by looking at their properties as a family we gain new understanding. The same is true for other objects that we encounter in mathematics—we often learn more by studying spaces of objects than by just focusing on the objects themselves. Equally important is how objects in one space can be transformed by some operation into objects in another space, i.e., we want to study *operators* on spaces—functions that map one space (possibly itself a space of functions) into another space. For example, the two premier operators of calculus are differentiation and integration. Integration converts objects (functions) that lie in some appropriate space of integrable functions into objects (functions) that belong to a space of differentiable functions, while differentiation does the reverse. Each of these two *operators* maps one space of functions into another space of functions.

This text is an introduction to the main types of spaces that pervade analysis. These are *metric spaces*, *normed spaces*, and *inner product spaces* (all of which are particular types of an even more general class of sets known as *topological spaces*). Essentially, a metric space is a set on which we can define a notion of "distance," or "metric," between points in the set. Normed spaces are metric spaces, but the concept of a normed space is more restrictive, in

---

[1] Hence I am an *analyst*, and I like to say that the main thing I do in my job is to listen to functions as they tell me their problems.

two respects. First, a normed space must be a vector space (and hence there is a way to add vectors together, and to multiply a vector by a scalar). Second, the existence of a norm not only provides us with a way to measure the distance between vectors, but also gives each individual vector in the space a "length." Inner product spaces have an even more restrictive definition, because each pair of vectors must be assigned an inner product, which in essence tells us the "angle" between these vectors. The Euclidean space $\mathbb{R}^d$ and its complex analogue $\mathbb{C}^d$ are familiar examples of inner product spaces. We will introduce and study metrics, norms, and inner products in this text, as well as operators that transform one such space into another space.

Audience

This text is aimed at students who have some basic knowledge of undergraduate real analysis, and who have previous experience reading and writing mathematical proofs. The intended reader of this text is a motivated student who is ready to take an upper-level, proof-based undergraduate mathematics course. No knowledge of measure theory or advanced real analysis is required (although a student who has taken some upper-level real analysis will of course be better prepared to handle this material). A brief review of the needed background material is presented in Chapter 1, and can be summarized as follows.

- Real and complex numbers.
- Functions, including one-to-one (or injective) functions and onto (or surjective) functions, inverse functions, direct and inverse images.
- Countable and uncountable sets.
- The definition and properties of the supremum and infimum of a set of real numbers.
- Sequences and series of real numbers, as covered in an undergraduate calculus course.
- Differentiation and integration, as covered in an undergraduate calculus course.
- The definition and basic properties of vector spaces, including independence, spans, and bases.

Aside from the above background material, this text is essentially self-contained, and can be used either as the text for an upper-level undergraduate mathematics course or as a text for independent study. The pace is fast and a considerable amount of material is covered. Most major proofs are included in detail, while certain other proofs are assigned as problems, and references are provided for any proofs that are omitted. Many exercises are included, and the reader should work as many as possible. A solutions manual for instructors is available upon request; instructions for obtaining a copy are given on the Birkhäuser website for this text.

Outline

Chapter 1 is a short quick-reference guide to notation, terminology, and background information that is assumed throughout the remainder of the text.

In Chapter 2 we introduce our first important class of spaces. These are *metric spaces*, which are sets on which we can define a notion of *distance* between points in the space. We show how the existence of a distance function, or *metric*, allows us to define the limit of a sequence of points, and then show that metric spaces have many properties that are similar to those that we are familiar with from $\mathbb{R}^d$. For example, we can define open and closed sets in a metric space. From this we proceed to define compact sets, and then to study the notion of *continuity* for functions on metric spaces. We end the chapter with two important theorems for metric spaces, namely, Urysohn's Lemma and the Baire Category Theorem.

In Chapters 3 and 4 we focus on a more restrictive class of spaces, the *normed spaces*. These are vector spaces on which we can define a *norm*, or *length function*. Every normed space is a metric space, but not all metric spaces are normed spaces. The more stringent requirements of a normed space mean that these spaces have a "richer structure" that allows us to prove more refined results than we could for generic metric spaces. In particular, we can form *linear combinations* of vectors (because our space is a vector space), but more importantly we can also form *infinite series*, because we can both add vectors and take limits. This gives us powerful tools, even when our space is infinite-dimensional. In particular, we study *Schauder bases*, which generalize the idea of a basis for a finite-dimensional vector space to something that is appropriate for infinite-dimensional spaces. We analyze a number of different specific examples, including the sequence spaces ℓ^p and spaces of continuous or differentiable functions. Several important (but optional) results are presented at the end of Chapter 4, including the Weierstrass and Stone–Weierstrass Theorems, the Arzelà–Ascoli Theorem, and the Tietze Extension Theorem.

We impose even more restriction in Chapter 5 by requiring that our space have an *inner product*. Every inner product space is a normed space, but not every normed space is an inner product space. This imparts even more structure to our space. In particular, in an inner product space we can define what it means for two vectors to be *orthogonal*, or perpendicular. This means that we have an analogue of the Pythagorean Theorem, and this in turn leads to the fundamental notion of an *orthogonal projection*, which we can use to find closest points and solve important existence and representation problems. We prove the existence of *orthonormal bases*, which are orthogonal systems that are also Schauder bases. These yield especially simple, and robust, representations of vectors. We construct several important examples of orthonormal bases, including both real and complex versions of the *trigonometric system*, which yields *Fourier series* representations of square-integrable functions.

Issues of *convergence* underlie much of what we do in Chapters 2–5. Closely related to the definition of a *convergent sequence* is the notion of a *Cauchy sequence*. Every convergent sequence is a Cauchy sequence, but a Cauchy sequence need not be convergent. A critical question is whether our space is *complete*, i.e., whether we have the good fortune to be in a space where every Cauchy sequence actually is convergent. A normed space that is complete is called a *Banach space*, and an inner product space that is complete is called a *Hilbert space* (there is no corresponding special name for metric spaces—a metric space that is complete is simply called a *complete metric space*).

Starting with Chapter 6 we focus on *operators*, which are functions (usually linear functions) that map one space into another space. We especially focus on operators that map one normed space into another normed space. For example, if our space X is a set of differentiable functions, then we might consider the differentiation operator, which transforms a function f in X to its derivative f', which may belong to another space Y. Because our spaces may be infinite-dimensional, there are subtleties that we simply do not encounter in finite-dimensional linear algebra. We prove that continuity of linear operators on normed spaces can be equivalently reformulated in terms of a property called *boundedness* that is much easier to analyze. This leads us to consider "meta-spaces" of bounded operators. That is, not only do we have a normed space X as a domain and a normed space Y as a codomain, but we also have a new space $\mathcal{B}(X, Y)$ that contains all of the bounded linear operators that map X into Y. This itself turns out to be a normed space! We analyze the properties of this space, and consider several particularly important examples of *dual spaces* (which are "meta-spaces" $\mathcal{B}(X, Y)$ where the codomain Y is the real line $\mathbb{R}$ or the complex plane $\mathbb{C}$).

Finally, in Chapter 7 we focus on operators that map one Hilbert space into another Hilbert space. The fact that the domain and codomain of these operators both have a rich structure allows us to prove even more detailed (and beautiful) results. We define the *adjoint* of an operator on a Hilbert space, study *compact operators* and *self-adjoint operators*, and conclude with the fundamental *Spectral Theorem* for compact self-adjoint operators. This theorem gives us a remarkable representation of every compact self-adjoint operator and leads to the *Singular Value Decomposition* of compact operators and $m \times n$ matrices.

Some Special Features

We list a few goals and special features of the text.

- Extensive exercises provide opportunities for students to learn the material by working problems. The problems themselves have a range of difficulties, so students can practice on basic concepts but also have the chance to think deeply in the more challenging exercises. The problems for each section are approximately ordered in terms of difficultly, with easier problems tending to occur earlier in the problem list and more difficult problems tending to be placed toward the end.

- While many texts present only the real-valued versions of results, in advanced settings and in applications the complex-valued versions are often needed (for example, in harmonic analysis in mathematics, quantum mechanics in physics, or signal processing in engineering). This text presents theorems for both the real and complex settings, but in a manner that easily allows the instructor or reader to focus solely on the real cases if they wish.

- The text presents several important topics that usually do not receive the coverage they deserve in books at this level. These include, for example, unconditional convergence of series, Schauder bases for Banach spaces, the dual of ℓ^p, topological isomorphisms, the Baire Category Theorem, the Uniform Boundedness Principle, the Spectral Theorem, and the Singular Value Decomposition for operators and matrices.

- The results presented in the text do not rely on measure theory, but an optional online *Chapter 8* covers related results and extensions to Lebesgue spaces that require measure theory.

Course Outlines

This text is suitable for independent study or as the basis for an upper-level course. There are several options for building a course around this text, three of which are listed below.

Course 1: Metric, Banach, and Hilbert spaces. A one-semester course focusing on the most important aspects of metric, Banach, and Hilbert spaces could present the following material:

Chapter 1: Sections 1.1–1.9
Chapter 2: Sections 2.1–2.9
Chapter 3: Sections 3.1–3.7
Chapter 4: Sections 4.1–4.5
Chapter 5: Sections 5.1–5.10

For classes with better-prepared students, Chapter 1 can be presented as a quick review, while for less-prepared classes more attention and detail can be spent on this material. This one-semester course would not cover the operator theory material that appears in Chapters 6 and 7, but interested students who complete the semester would be prepared to go on to study that material on their own.

Course 2: Core material plus operator theory. A fast-paced one-semester course could cover the most important parts of the material on metric, Banach, and Hilbert spaces while still having time to cover selected material on operator theory. Here is one potential outline of such a course.

Chapter 2: Sections 2.1–2.4, 2.6–2.9
Chapter 3: Sections 3.1–3.7
Chapter 4: Sections 4.1–4.2
Chapter 5: Sections 5.1–5.10
Chapter 6: Sections 6.1–6.7
Chapter 7: Sections 7.1–7.8

Course 3: All topics. A two-semester course could cover the background material in Chapter 1, all of the foundations and details of metric, Banach, and Hilbert spaces as presented in Chapters 2–5, and the applications to operator theory that are given in Chapters 6 and 7. Such a course could cover most or all of the entire text in a two-semester sequence.

Further Reading

This text is an introduction to real analysis and operator theory. There are many possible directions for the reader who wishes to learn more, including those listed below.

- *Measure Theory*. We do not use measure theory, Lebesgue measure, or the Lebesgue integral in this text, except where a formal use of these ideas provides context or illumination (such as when we discuss Fourier series in Chapter 5). This does not affect the theory covered in the text and also keeps the text accessible to its intended audience. It does mean that we sometimes needed to be selective in choosing examples or mathematical applications. The reader who wishes to advance further in analysis must learn measure theory. There are a number of texts that take different approaches to the subject. For example, the beginning student may benefit from a treatment that starts with Lebesgue measure and integral (before proceeding to abstract measure theory). Some texts that take this approach include Stein and Shakarchi [SS05], Wheeden and Zygmund [WZ77], and the forthcoming text [Heil19]. Some classic comprehensive texts that cover abstract measure theory and more include Folland [Fol99] and Rudin [Rud87].

- *Operator Theory and Functional Analysis.* Chapters 6 and 7 provide an introduction to operator theory and functional analysis. More detailed and extensive development of these topics is available in texts such as those by Conway [Con90], [Con00], Folland [Fol99], Gohberg and Goldberg [GG01], Kreyszig [Kre78], and Rudin [Rud91].

- *Basis and Frame Theory.* A *Schauder basis* is a sequence of vectors in a Banach space that provides representations of arbitrary vectors in terms of unique "infinite linear combinations" of the basis elements. These are studied in some detail in this text. A *frame* is a basis-like system, but one that allows nonuniqueness and redundancy, which can be important in many applications. Frames are a fascinating subject, but are only briefly mentioned in this text. Some introductory texts suitable for readers who wish further study of these topics are the text [Heil11] and the texts by Christensen [Chr16], Han, Kornelson, Larsen, and Weber [HKLW07], and Walnut [Wal02].

- *Topology.* Metric, normed, and inner product spaces are the most common types of spaces encountered in analysis, but they are all special cases of *topological spaces.* The reader who wishes to explore abstract topological spaces in more detail can turn to texts such as Munkres [Mun75] and Singer and Thorpe [ST76].

Additional Material

Additional resources related to this text are available at the author's website,

http://people.math.gatech.edu/~heil/

In particular, an optional Chapter 8 is posted online that includes extensions of the results of this text to settings that require measure theory, covering in particular integral operators on Lebesgue spaces and Hilbert–Schmidt operators.

Acknowledgments

It is a pleasure to thank Lily Hu, Jordy van Velthoven, and Rafael de la Llave for their comments and feedback on drafts of this manuscript, and Shannon Bishop and Ramazan Tinaztepe for their help in finding typographical and other errors. We also thank Ben Levitt of Birkhäuser for his invaluable help in the editorial process.

June 17, 2017   *Christopher Heil*

Chapter 1
Notation and Preliminaries

This preliminary chapter is a quick-reference guide to the notation, terminology, and background information that will be assumed throughout the volume. Unlike the rest of the text, in this preliminary chapter we may state results without proof or without the motivation and discussion that is provided throughout the later chapters. Proofs of the facts reviewed in Sections 1.1–1.9 can be found in most calculus texts (such as [HHW18]), or in undergraduate analysis texts (such as [Rud76]). Proofs of the results in Sections 1.10–1.11 can be found in most linear algebra texts (such as [Axl15]).

We use the symbol □ to denote the end of a proof, and the symbol ◇ to denote the end of a definition, remark, example, or exercise. We also use ◇ to indicate the end of the statement of a theorem whose proof will be omitted. A detailed index of symbols employed in the text can be found at the end of the volume.

1.1 Numbers

The set of natural numbers is denoted by $\mathbb{N} = \{1, 2, 3, \dots\}$. The set of integers is $\mathbb{Z} = \{\dots, -1, 0, 1, \dots\}$, the set of rational numbers is $\mathbb{Q} = \{m/n : m, n \in \mathbb{Z}, n \neq 0\}$, $\mathbb{R}$ is the set of real numbers, and $\mathbb{C}$ is the set of complex numbers. As we will discuss in more detail in Notation 1.10.2, we often let the symbol $\mathbb{F}$ represent a choice of the real line or the complex plane according to context, i.e., $\mathbb{F}$ can be either $\mathbb{R}$ or $\mathbb{C}$. An element of $\mathbb{F}$ is called a *scalar*, i.e., a scalar is either a real or complex number according to context.

The *real part* of a complex number $z = a + ib$ (where $a, b \in \mathbb{R}$) is $\mathrm{Re}(z) = a$, and its *imaginary part* is $\mathrm{Im}(z) = b$. We say that z is *rational* if both its real and imaginary parts are rational numbers. The *complex conjugate* of z is $\bar{z} = a - ib$. The *modulus*, or *absolute value*, of z is

$$|z| = \sqrt{z\bar{z}} = \sqrt{a^2 + b^2}.$$

© Springer International Publishing AG, part of Springer Nature 2018
C. Heil, *Metrics, Norms, Inner Products, and Operator Theory*, Applied and
Numerical Harmonic Analysis, https://doi.org/10.1007/978-3-319-65322-8_1

If $z \neq 0$, then its *polar form* is $z = re^{i\theta}$ where $r = |z| > 0$ and $\theta \in [0, 2\pi)$. In this case the *argument* of z is $\arg(z) = \theta$.

Some useful identities are

$$z + \overline{z} = 2\operatorname{Re}(z), \qquad z - \overline{z} = 2i\operatorname{Im}(z),$$

and

$$|z + w|^2 = (z + w)(\overline{z} + \overline{w}) = |z|^2 + 2\operatorname{Re}(z\overline{w}) + |w|^2.$$

1.2 Sets

Given a set X, we often use lowercase letters such as x, y, z to denote elements of X. Below we give some terminology and notation for sets.

- If every element of a set A is also an element of a set B, then A is a *subset* of B, and in this case we write $A \subseteq B$ (note that this includes the possibility that A might equal B). A *proper subset* of a set B is a set $A \subseteq B$ such that $A \neq B$. We indicate this by writing $A \subsetneq B$.

- The *empty set* is denoted by $\varnothing$. The empty set is a subset of every set.

- The notation $X = \{x : x \text{ has property P}\}$ means that X is the set of all x that satisfy property P. For example, the union of a collection of sets $\{X_i\}_{i \in I}$ is

$$\bigcup_{i \in I} X_i = \{x : x \in X_i \text{ for some } i \in I\},$$

and their intersection is

$$\bigcap_{i \in I} X_i = \{x : x \in X_i \text{ for every } i \in I\}.$$

- If S is a subset of a set X, then the *complement* of S is

$$X \backslash S = \{x \in X : x \notin S\}.$$

We sometimes abbreviate $X \backslash S$ as S^C if the set X is understood.

- *De Morgan's Laws* state that

$$X \backslash \bigcup_{i \in I} X_i = \bigcap_{i \in I} (X \backslash X_i) \quad \text{and} \quad X \backslash \bigcap_{i \in I} X_i = \bigcup_{i \in I} (X \backslash X_i).$$

- The *Cartesian product* of sets X and Y is the set of all ordered pairs of elements of X and Y, i.e.,

$$X \times Y = \{(x, y) : x \in X, y \in Y\}.$$

- A collection of sets $\{X_i\}_{i \in I}$ is *disjoint* if $X_i \cap X_j = \varnothing$ whenever $i \neq j$. In particular, two sets A and B are disjoint if $A \cap B = \varnothing$.

- An *open interval* in $\mathbb{R}$ is any one of the following sets:

$$
\begin{aligned}
(a, b) &= \{x \in \mathbb{R} : a < x < b\}, & -\infty < a < b < \infty, \\
(a, \infty) &= \{x \in \mathbb{R} : x > a\}, & a \in \mathbb{R}, \\
(-\infty, b) &= \{x \in \mathbb{R} : x < b\}, & b \in \mathbb{R}, \\
(-\infty, \infty) &= \mathbb{R}.
\end{aligned}
$$

- A *closed interval* in $\mathbb{R}$ is any one of the following sets:

$$
\begin{aligned}
[a, b] &= \{x \in \mathbb{R} : a \leq x \leq b\}, & -\infty < a < b < \infty, \\
[a, \infty) &= \{x \in \mathbb{R} : x \geq a\}, & a \in \mathbb{R}, \\
(-\infty, b] &= \{x \in \mathbb{R} : x \leq b\}, & b \in \mathbb{R}, \\
(-\infty, \infty) &= \mathbb{R}.
\end{aligned}
$$

We refer to $[a, b]$ as a *bounded closed interval* or a *finite closed interval*.

- An *interval* in $\mathbb{R}$ is a set that is either an open interval, a closed interval, or one of the following sets:

$$
\begin{aligned}
(a, b] &= \{x \in \mathbb{R} : a < x \leq b\}, & -\infty < a < b < \infty, \\
[a, b) &= \{x \in \mathbb{R} : a \leq x < b\}, & -\infty < a < b < \infty.
\end{aligned}
$$

The sets $(a, b]$ and $[a, b)$ are sometimes called "half-open intervals," although they are neither open nor closed.

Problems

1.2.1. Prove De Morgan's Laws.

1.2.2. Let $\{E_k\}_{k \in \mathbb{N}}$ be a sequence of subsets of $\mathbb{R}^d$, and define

$$
\limsup E_k = \bigcap_{j=1}^{\infty} \left(\bigcup_{k=j}^{\infty} E_k \right), \qquad \liminf E_k = \bigcup_{j=1}^{\infty} \left(\bigcap_{k=j}^{\infty} E_k \right).
$$

Show that $\limsup E_k$ consists of those points $x \in \mathbb{R}^d$ that belong to infinitely many E_k, while $\liminf E_k$ consists of those x that belong to all but finitely many E_k (i.e., there exists some $k_0 \in \mathbb{N}$ such that $x \in E_k$ for all $k \geq k_0$).

1.3 Functions

Let X, Y, and Z be sets. We write $f\colon X \to Y$ to mean that f is a function whose *domain* is X and *codomain* (or *target*) is Y. Here is some terminology that we use to describe various properties of such a function f (also see Problems 1.3.2 and 1.3.3 for some facts about direct and inverse images).

- The *direct image* of a set $A \subseteq X$ under f is

$$f(A) = \{f(t) : t \in A\}.$$

- The *inverse image* of a set $B \subseteq Y$ under f is

$$f^{-1}(B) = \{t \in X : f(t) \in B\}. \tag{1.1}$$

- The *range* of f is the direct image of its domain X, i.e.,

$$\operatorname{range}(f) = f(X) = \{f(t) : t \in X\}.$$

- f is *surjective*, or *onto*, if $\operatorname{range}(f) = Y$.

- f is *injective*, or *one-to-one*, if $f(a) = f(b)$ implies $a = b$.

- f is a *bijection* if it is both injective and surjective.

- A bijection $f\colon X \to Y$ has an *inverse function* $f^{-1}\colon Y \to X$, defined by the rule $f^{-1}(y) = x$ if $f(x) = y$. The inverse function f^{-1} is also a bijection. Despite the similar notation, an inverse function should not be confused with the *inverse image* defined in equation (1.1). Only a bijection has an inverse function, but the inverse image $f^{-1}(B)$ is well-defined for every function f and set $B \subseteq Y$.

- The *composition* of a function $g\colon Y \to Z$ with $f\colon X \to Y$ is the function $g \circ f\colon X \to Z$ defined by $(g \circ f)(t) = g(f(t))$ for $t \in X$.

- If $Y = \mathbb{R}$ then we say that f is *real-valued*, and if $Y = \mathbb{C}$ then we say f is *complex-valued*.

Problems

1.3.1. Given a function $f\colon X \to Y$, prove the following statements.

(a) f is surjective if and only if for every $y \in Y$ there exists some $t \in X$ such that $f(t) = y$.

(b) f is injective if and only if for every $y \in \operatorname{range}(f)$ there exists a unique $t \in X$ such that $f(t) = y$.

(c) f is a bijection if and only if for every $y \in Y$ there exists a unique $t \in X$ such that $f(t) = y$.

1.3.2. Let $f\colon X \to Y$ be a function, let B be a subset of Y, and let $\{B_i\}_{i \in I}$ be a family of subsets of Y. Prove that

$$f^{-1}\left(\bigcup_{i \in I} B_i\right) = \bigcup_{i \in I} f^{-1}(B_i), \qquad f^{-1}\left(\bigcap_{i \in I} B_i\right) = \bigcap_{i \in I} f^{-1}(B_i),$$

and $f^{-1}(B^C) = (f^{-1}(B))^C$. Also prove that $f(f^{-1}(B)) \subseteq B$, and if f is surjective then equality holds. Show by example that equality need not hold if f is not surjective.

1.3.3. Let $f\colon X \to Y$ be a function, let A be a subset of X, and let $\{A_i\}_{i \in I}$ be a family of subsets of X. Prove that

$$f\left(\bigcup_{i \in I} A_i\right) = \bigcup_{i \in I} f(A_i).$$

Also prove that

$$f\left(\bigcap_{i \in I} A_i\right) \subseteq \bigcap_{i \in I} f(A_i), \qquad f(X) \backslash f(A) \subseteq f(A^C), \qquad A \subseteq f^{-1}(f(A)),$$

and if f is injective then equality holds in each of these inclusions. Show by example that equality need not hold if f is not injective.

1.4 Cardinality

We say that two sets A and B *have the same cardinality* if there exists a bijection f that maps A onto B, i.e., if there is a function $f\colon A \to B$ that is both injective and surjective. Such a function f pairs each element of A with a unique element of B and vice versa, and therefore is sometimes called a *one-to-one correspondence*.

Example 1.4.1. (a) The function $f\colon [0, 2] \to [0, 1]$ defined by $f(t) = t/2$ for $0 \le t \le 2$ is a bijection, so the intervals $[0, 2]$ and $[0, 1]$ have the same cardinality. This shows that a proper subset of a set can have the same cardinality as the set itself (although this is impossible for *finite* sets).

(b) The function $f\colon \mathbb{N} \to \{2, 3, 4, \dots\}$ defined by $f(n) = n + 1$ for $n \in \mathbb{N}$ is a bijection, so the set of natural numbers $\mathbb{N} = \{1, 2, 3, \dots\}$ has the same cardinality as its proper subset $\{2, 3, 4, \dots\}$.

(c) The function $f\colon \mathbb{N} \to \mathbb{Z}$ defined by

$$f(n) = \begin{cases} \frac{n}{2}, & \text{if } n \text{ is even,} \\ -\frac{n-1}{2}, & \text{if } n \text{ is odd,} \end{cases}$$

is a bijection of $\mathbb{N}$ onto $\mathbb{Z}$, so the set of integers $\mathbb{Z}$ has the same cardinality as the set of natural numbers $\mathbb{N}$.

(d) If n is a finite positive integer, then there is no way to define a function $f : \{1, \ldots, n\} \to \mathbb{N}$ that is a bijection. Hence $\{1, \ldots, n\}$ and $\mathbb{N}$ do not have the same cardinality. Likewise, if $m \neq n$ are distinct positive integers, then $\{1, \ldots, m\}$ and $\{1, \ldots, n\}$ do not have the same cardinality. $\diamond$

We use cardinality to define finite sets and infinite sets, as follows.

Definition 1.4.2 (Finite and Infinite Sets). Let X be a set. We say that X is *finite* if X is either empty or there exists an integer $n > 0$ such that X has the same cardinality as the set $\{1, \ldots, n\}$. That is, a nonempty set X is finite if there is a bijection of the form

$$f : \{1, \ldots, n\} \to X.$$

In this case we say that X *has n elements*.

We say that X is *infinite* if it is not finite. $\diamond$

We use the following terminology to further distinguish among sets based on cardinality.

Definition 1.4.3 (Countable and Uncountable Sets). We say that a set X is:

- *denumerable* or *countably infinite* if it has the same cardinality as the natural numbers, i.e., if there exists a bijection $f : \mathbb{N} \to X$,

- *countable* if X is *either* finite *or* countably infinite,

- *uncountable* if X is not countable. $\diamond$

Every finite set is countable by definition, and parts (b) and (c) of Example 1.4.1 show that $\mathbb{N}$, $\mathbb{Z}$, and $\{2, 3, 4, \ldots\}$ are countable. Here is another countable set.

Example 1.4.4. Consider $\mathbb{N}^2 = \mathbb{N} \times \mathbb{N} = \{(j, k) : j, k \in \mathbb{N}\}$, the set of all ordered pairs of positive integers.

We depict $\mathbb{N}^2$ in table format in Figure 1.1, with additional arrows that we will shortly explain. Every ordered pair (j, k) of positive integers appears somewhere in the table. In particular, the first line of pairs in the table includes all ordered pairs whose first component is $j = 1$, the second line of pairs lists those whose first component is $j = 2$, and so forth.

We define a bijection $f : \mathbb{N} \to \mathbb{N}^2$ by following the arrows in Figure 1.1:

$$f(1) = (1, 1),$$
$$f(2) = (1, 2),$$
$$f(3) = (2, 1),$$

$$f(4) = (3,1),$$
$$f(5) = (2,2),$$
$$f(6) = (1,3),$$
$$\vdots$$

In other words, once $f(n)$ has been defined to be a particular ordered pair (j,k), then we let $f(n+1)$ be the ordered pair that (j,k) points to next. In this way the outputs $f(1), f(2), f(3), \ldots$ give us a list of every ordered pair in $\mathbb{N}^2$. Thus $\mathbb{N}$ and $\mathbb{N}^2$ have the same cardinality, so $\mathbb{N}^2$ is denumerable and hence countable. ◇

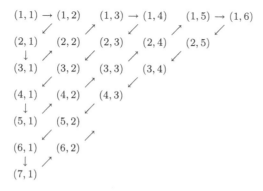

Fig. 1.1 $\mathbb{N}^2$ in table format, with the pattern of arrows for defining a bijection of $\mathbb{N}$ onto $\mathbb{N}^2$.

If X is a countable set, then we can create a *list* of the elements of X. There are two possibilities. First, a countable X set might be finite, in which case there exists a bijection $f: \{1, 2, \ldots, n\} \to X$ for some positive integer n. Since f is surjective, we therefore have

$$X = \text{range}(f) = \{f(1),\ f(2),\ \ldots,\ f(n)\}.$$

Thus the function f provides us with a way to list the n elements of X in some order. On the other hand, if X is countably infinite then there is a bijection $f: \mathbb{N} \to X$, and hence

$$X = \text{range}(f) = \{f(1),\ f(2),\ f(3),\ \ldots\}.$$

Thus the elements of X have again been listed in some order. For example, Example 1.4.4 shows that we can list the elements of $\mathbb{N}^2$ in the following way:

$$\mathbb{N}^2 = \{(1,1),\ (1,2),\ (2,1),\ (3,1),\ (2,2),\ (1,3),\ (1,4),\ (2,3),\ \ldots\}.$$

Although it may seem more natural to depict $\mathbb{N}^2$ as a "two-dimensional" table, as shown in Figure 1.1, because $\mathbb{N}^2$ is countable it is possible to make a "one-dimensional" *list* of all of the elements of $\mathbb{N}^2$.

Now we show that there exist infinite sets that are not countable.

Example 1.4.5. Let

$$S = (0,1) = \{x \in \mathbb{R} : 0 < x < 1\}$$

be the set of all real numbers that lie strictly between zero and one. We will use an argument by contradiction to prove that S is not countable. First we recall that every real number can be written in decimal form. In particular, if $0 < x < 1$ then we can write

$$x = 0.d_1 d_2 d_3 \ldots = \sum_{k=1}^{\infty} \frac{d_k}{10^k},$$

where each digit d_k is an integer between 0 and 9. Some numbers have two decimal representations, for example,

$$\frac{1}{2} = 0.5000\ldots = \frac{5}{10} + \sum_{k=2}^{\infty} \frac{0}{10^k},$$

but also

$$\frac{1}{2} = 0.4999\ldots = \frac{4}{10} + \sum_{k=2}^{\infty} \frac{9}{10^k}. \tag{1.2}$$

Any number whose decimal representation ends in infinitely many zeros also has a decimal representation that ends in infinitely many nines, but all other real numbers have a unique decimal representation.

Suppose that S is countable. In this case there is a bijection $f : \mathbb{N} \to S$, and therefore we can make a list of all the elements of S. If we let $x_n = f(n)$, then we can write S as follows:

$$S = \text{range}(f) = \{f(1),\ f(2),\ f(3),\ \ldots\} = \{x_1,\ x_2,\ x_3,\ \ldots\}.$$

This is a list of every real number between 0 and 1. Further, we can write each x_n in decimal form, say,

$$x_n = 0.d_1^n d_2^n d_3^n \ldots,$$

where each digit d_k^n is an integer between 0 and 9.

Now we will create another sequence of digits between 0 and 9. In fact, in order to avoid difficulties arising from the fact that some numbers have two decimal representations, we will always choose digits that are between 1 and 8. To start, let e_1 be any integer between 1 and 8 that does not equal d_1^1 (the first digit of the first number x_1). For example, if the decimal representation of x_1 happens to be $x_1 = 0.72839172\ldots$, then we let e_1 be any digit other than 0, 7, or 9 (so we might take $e_1 = 5$ in this case). Then we let e_2 be any integer between 1 and 8 that does not equal d_2^2 (the second digit of the second number x_2), and so forth. This gives us digits $e_1, e_2, \ldots$, and we let x be the real number whose decimal expansion has exactly those digits:

$$x = 0.e_1e_2e_3\ldots = \sum_{k=1}^{\infty} \frac{e_k}{10^k}. \tag{1.3}$$

Then x is a real number, and since none of the decimal digits e_k is 0 or 9, equation (1.3) is the unique decimal representation of x. Since x is a real number between 0 and 1, it must be one of the real numbers in the set S. Yet $x \neq x_1$, because the first digit of x (which is e_1) is not equal to the first digit of x_1 (why not—what if x_1 has two decimal representations?). Similarly $x \neq x_2$, because their second digits are different, and so forth. Hence x does not equal any element of S, which is a contradiction. Therefore S cannot be a countable set. ◇

Here are some properties of countable and uncountable sets (the proof is assigned as Problem 1.4.11).

Lemma 1.4.6. *Let X and Y be sets.*
(a) *If X is countable and $Y \subseteq X$, then Y is countable.*
(b) *If X is uncountable and $Y \supseteq X$, then Y is uncountable.*
(c) *If X is countable and there exists an injection $f \colon Y \to X$, then Y is countable.*
(d) *If X is uncountable and there exists an injection $f \colon X \to Y$, then Y is uncountable.* ◇

Example 1.4.7. (a) Let $\mathbb{Q}^+ = \{r \in \mathbb{Q} : r > 0\}$ be the set of all positive rational numbers. Given $r \in \mathbb{Q}^+$, there is a unique way to write r as a fraction in lowest terms. That is, $r = m/n$ for a unique choice of positive integers m and n that have no common factors. Therefore, by setting $f(r) = (m, n)$ we can define an injective map of $\mathbb{Q}^+$ into $\mathbb{N}^2$. Since $\mathbb{N}^2$ is countable and f is injective, we apply Lemma 1.4.6(c) and conclude that $\mathbb{Q}^+$ is countable.

A similar argument shows that $\mathbb{Q}^-$, the set of negative rational numbers, is countable. Problem 1.4.12 tells us that a union of finitely many (or countably many) countable sets is countable, so it follows that $\mathbb{Q} = \mathbb{Q}^+ \cup \mathbb{Q}^- \cup \{0\}$ is countable.

(b) We saw in Example 1.4.5 that the interval $(0, 1)$ is uncountable. Since $\mathbb{R}$ contains $(0, 1)$, Lemma 1.4.6(b) implies that $\mathbb{R}$ is uncountable. Also, since every real number is a complex number we have $\mathbb{R} \subseteq \mathbb{C}$, and therefore $\mathbb{C}$ is uncountable as well.

(c) Let $I = \mathbb{R} \setminus \mathbb{Q}$ be the set of irrational real numbers. Since $\mathbb{R} = I \cup \mathbb{Q}$, if I were countable then $\mathbb{R}$ would be the union of two countable sets, which is countable by Problem 1.4.12. This is a contradiction, so the set of irrationals must be uncountable.

Thus $\mathbb{Q}$ is countable, while I is uncountable. This may seem counterintuitive, since between any two rational numbers there is an irrational number, and between any two irrational numbers there is a rational number! $\diamond$

Problems

1.4.8. Prove equation (1.2).

1.4.9. Given sets A, B, and C, prove the following statements.

(a) A has the same cardinality as A.

(b) If A has the same cardinality as B, then B has the same cardinality as A.

(c) If A has the same cardinality as B and B has the same cardinality as C, then A has the same cardinality as C.

1.4.10. Prove that the closed interval $[0, 1]$ and the open interval $(0, 1)$ have the same cardinality by exhibiting a bijection $f \colon [0, 1] \to (0, 1)$.
  Hint: Do not try to create a *continuous* function f.

1.4.11. Prove Lemma 1.4.6.

1.4.12. (a) Show that if X and Y are countable sets, then their union $X \cup Y$ is countable.

(b) Prove that the union of finitely many countable sets $X_1, \ldots, X_n$ is countable.

(c) Suppose that $X_1, X_2, \ldots$ are countably many sets, each of which is countable. Prove that $\bigcup_{k=1}^{\infty} X_k = X_1 \cup X_2 \cup \cdots$ is countable. Thus the union of countably many countable sets is countable.
  Hint: Consider Figure 1.4.4.

1.4.13. Let F be the set of all functions $f \colon \mathbb{R} \to \mathbb{R}$, i.e., F is the set of all functions that map real numbers to real numbers. Prove that F is uncountable, and F does not have the same cardinality as the real line $\mathbb{R}$.

1.5 Sequences

Let J be a fixed set. Given a set X and points $x_j \in X$ for $j \in J$, we write $\{x_j\}_{j \in J}$ to denote the sequence of elements x_i indexed by the set J. We call J an *index set* in this context, and we refer to x_j as the *jth component* of the sequence $\{x_j\}_{j \in J}$. If we know that the x_j are real or complex *numbers*, then we often write a sequence as $(x_j)_{j \in J}$ instead of $\{x_j\}_{j \in J}$. If the index set J is understood, then we may write $\{x_j\}$, $\{x_j\}_j$, (x_j), or $(x_j)_j$.

Technically, a sequence $\{x_j\}_{j \in J}$ is shorthand for the function $x \colon J \to X$ whose rule is

$$x(j) = x_j, \qquad j \in J.$$

Consequently, the components x_j of a sequence need not be distinct, i.e., it is possible that we might have $x_i = x_j$ for some $i \neq j$.

Often the index set J is countable. The two most common situations are the finite index set $J = \{1, \dots, d\}$ and the countably infinite index set $J = \mathbb{N}$. If $J = \{1, \dots, d\}$, then we often write a sequence in list form as

$$\{x_n\}_{n=1}^d = \{x_1, \dots, x_d\} \qquad \text{or} \qquad (x_n)_{n=1}^d = (x_1, \dots, x_d).$$

Similarly, if $J = \mathbb{N} = \{1, 2, \dots\}$, then we often write

$$\{x_n\}_{n \in \mathbb{N}} = \{x_1, x_2, \dots\} \qquad \text{or} \qquad (x_n)_{n \in \mathbb{N}} = (x_1, x_2, \dots).$$

A *subsequence* of a countable sequence $\{x_n\}_{n \in \mathbb{N}} = \{x_1, x_2, \dots\}$ is a sequence of the form

$$\{x_{n_k}\}_{n \in \mathbb{N}} = \{x_{n_1}, x_{n_2}, \dots\}, \qquad \text{where } n_1 < n_2 < \cdots.$$

For example,

$$\{x_2, x_3, x_5, x_7, x_{11}, \dots\}$$

is a subsequence of $\{x_1, x_2, \dots\}$.

On occasion it is convenient to have an alternative way to write the components of an infinite sequence. As we discussed above, one way to denote the components of an infinite sequence x is with subscripts, as follows:

$$x = \{x_k\}_{k \in \mathbb{N}} = \{x_1, x_2, \dots\}.$$

Using this notation, the symbol x_k represents the kth component of the sequence x. However, it is sometimes preferable to write

$$x = \{x(k)\}_{k \in \mathbb{N}} = \{x(1), x(2), \dots\}. \tag{1.4}$$

Using this notation, the symbol $x(k)$ represents the kth component of x. In any particular situation we will use whichever of these two notations is more convenient.

Problems

1.5.1. Prove that the following sets are uncountable.

(a) A is the set of all sequences $x = (x_1, x_2, \dots)$, where each component x_k is an integer between 0 and 9.

(b) B is the set of all sequences $x = (x_1, x_2, \dots)$, where each component x_k is an integer between 0 and 4.

(c) C is the set of all sequences $x = (x_1, x_2, \dots)$, where each component x_k is either 0 or 1.

1.6 Suprema and Infima

Let S be a set of real numbers.

- S is *bounded above* if there exists a real number M such that $s \leq M$ for every $s \in S$. Any such number M is called an *upper bound* for S.

- S is *bounded below* if there exists a real number m such that $m \leq s$ for every $s \in S$. Any such number m is called a *lower bound* for S.

- S is *bounded* if it is bounded both above and below. Equivalently, S is bounded if and only if there is a real number $M \geq 0$ such that $|s| \leq M$ for all $s \in S$.

- v is a *maximum element* of S if $v \in S$ and $s \leq v$ for every $s \in S$.

- v is a *minimum element* of S if $v \in S$ and $s \geq v$ for every $s \in S$.

Not every set of real numbers S has a maximum or minimum element, even if it is bounded. For example, the open interval $I = (0, 1)$ has no maximum or minimum element. Often, a more useful notion than a maximum or minimum element is the supremum or infimum of a set S. The supremum is defined as follows.

Definition 1.6.1 (Supremum). Let S be a nonempty set of real numbers. We say that $u \in \mathbb{R}$ is the *supremum*, or *least upper bound*, of S if the following two statements hold.

(a) u is an upper bound for S, i.e., $s \leq u$ for every $s \in S$.

(b) If v is any upper bound for S, then $u \leq v$. That is, if $v \in \mathbb{R}$ and $s \leq v$ for every $s \in S$, then $u \leq v$.

We denote the supremum of S, if one exists, by $u = \sup(S)$. $\diamond$

For example, the supremum of the open interval $S = (0, 1)$ is $\sup(S) = 1$. Note that the supremum of a set need not belong to the set.

It is not obvious that every set that is bounded above has a supremum. We take the existence of suprema as an axiom, as follows.

Axiom 1.6.2 (Supremum Property of $\mathbb{R}$). Let S be a nonempty subset of $\mathbb{R}$. If S is bounded above, then there exists a real number $u = \sup(S)$ that is the supremum of S.   ◇

Here is an immediate but useful fact about the supremum of a set.

Lemma 1.6.3. *Let S be a nonempty subset of $\mathbb{R}$ that is bounded above. Then for each $\varepsilon > 0$ there exists some $x \in S$ such that*

$$\sup(S) - \varepsilon < x \leq \sup(S).$$

Proof. Since the set S is bounded above, its supremum $u = \sup(S)$ is a finite real number. Since u is the least upper bound for S, the number $u - \varepsilon$ is not an upper bound for S. Therefore there must exist some $x \in S$ such that $u - \varepsilon < x$. On the other hand, u is an upper bound for S, so we must also have $x \leq u$. Therefore $u - \varepsilon < x \leq u$.   □

We extend the definition of supremum to sets that are not bounded above by *declaring* that $\sup(S) = \infty$ if S is not bounded above. We also declare that $\sup(\varnothing) = -\infty$. Using these conventions, every set $S \subseteq \mathbb{R}$ has a supremum (although it might be $\pm\infty$). If $S = (x_n)_{n \in \mathbb{N}}$ is countable, then we often write $\sup_n x_n$ or $\sup x_n$ to denote the supremum instead of $\sup(S)$.

The *infimum*, or *greatest lower bound*, of S is defined in an entirely analogous manner, and is denoted by $\inf(S)$. Statements analogous to those made for suprema hold for infima.

To illustrate the use of suprema, we prove the following result. Further results about suprema and infima (including facts about unbounded sets of numbers) are listed in the problems below.

Lemma 1.6.4. *If $(x_n)_{n \in \mathbb{N}}$ and $(y_n)_{n \in \mathbb{N}}$ are two bounded sequences of real numbers, then*

$$\sup_{n \in \mathbb{N}} (x_n + y_n) \leq \sup_{n \in \mathbb{N}} x_n + \sup_{n \in \mathbb{N}} y_n.$$

Proof. For simplicity of notation, set $u = \sup x_n$ and $v = \sup y_n$. Since $(x_n)_{n \in \mathbb{N}}$ and $(y_n)_{n \in \mathbb{N}}$ are bounded, we know that u and v are finite real numbers. Since u is an upper bound for the x_n, we have $x_n \leq u$ for every n. Similarly, $y_n \leq v$ for every n. Therefore $x_n + y_n \leq u + v$ for every n. Hence $u + v$ is an upper bound for the sequence $(x_n + y_n)_{n \in \mathbb{N}}$, so this sequence is bounded and therefore has a finite supremum, say $w = \sup(x_n + y_n)$. By definition, w is the *least* upper bound for $(x_n + y_n)_{n \in \mathbb{N}}$. Since we saw above that $u + v$ is another upper bound for $(x_n + y_n)_{n \in \mathbb{N}}$, we must have $w \leq u + v$, which is exactly what we wanted to prove.   □

Problems

1.6.5. Given a nonempty set $S \subseteq \mathbb{R}$, prove the following statements about suprema. Also formulate and prove analogous results for infima.

Hint: We are not assuming that S is bounded, so first prove these results assuming S is bounded, and then separately consider the case of an unbounded set S.

(a) If S has a maximum element x, then $x = \sup(S)$.

(b) If $t \in \mathbb{R}$, then $\sup(S + t) = \sup(S) + t$, where $S + t = \{x + t : x \in S\}$.

(c) If $c \geq 0$, then $\sup(cS) = c \sup(S)$, where $cS = \{cx : x \in S\}$.

(d) If $a_n \leq b_n$ for every n, then $\sup a_n \leq \sup b_n$.

1.6.6. Let $(x_n)_{n \in \mathbb{N}}$ and $(y_n)_{n \in \mathbb{N}}$ be two sequences of real numbers (not necessarily bounded).

(a) Show that if $c > 0$, then

$$\sup_{n \in \mathbb{N}} c x_n = c \sup_{n \in \mathbb{N}} x_n \quad \text{and} \quad \sup_{n \in \mathbb{N}} (-c x_n) = -c \inf_{n \in \mathbb{N}} x_n,$$

where we declare that $c \cdot (\pm\infty) = \pm\infty$ and $-c \cdot (\pm\infty) = \mp\infty$.

(b) Prove that

$$\inf_{n \in \mathbb{N}} x_n + \inf_{n \in \mathbb{N}} y_n \leq \inf_{n \in \mathbb{N}} (x_n + y_n) \leq \sup_{n \in \mathbb{N}} (x_n + y_n) \leq \sup_{n \in \mathbb{N}} x_n + \sup_{n \in \mathbb{N}} y_n.$$

Show by example that each of the inequalities on the preceding line can be strict.

1.6.7. Given nonempty sets $A, B \subseteq \mathbb{R}$, prove that

$$\sup(A + B) = \sup(A) + \sup(B),$$

where $A + B = \{a + b : a \in A, b \in B\}$. Why does equality always hold here but not in Problem 1.6.6(b)?

1.6.8. Let S be a bounded, nonempty set of real numbers. Given a real number u, prove that $u = \sup(S)$ if and only if both of the following two statements hold.

(a) There does not exist any $s \in S$ such that $u < s$.

(b) If $v < u$, then there exists some $s \in S$ such that $v < s$.

1.7 Convergent Sequences of Numbers

We recall the definition of convergence for sequences of scalars.

Definition 1.7.1. Let $(x_n)_{n\in\mathbb{N}}$ be a sequence of real or complex numbers. We say that $(x_n)_{n\in\mathbb{N}}$ *converges* if there exists some real or complex number x such that for every $\varepsilon > 0$ there is an integer $N > 0$ such that

$$n \geq N \quad \Longrightarrow \quad |x - x_n| < \varepsilon.$$

In this case we say that x_n *converges to* x *as* $n \to \infty$ and write

$$x_n \to x \quad \text{or} \quad \lim_{n\to\infty} x_n = x \quad \text{or} \quad \lim x_n = x. \qquad \Diamond$$

Here are some basic properties of convergent sequences.

Lemma 1.7.2. *Let* $(x_n)_{n\in\mathbb{N}}$ *and* $(y_n)_{n\in\mathbb{N}}$ *be sequences of real or complex numbers.*

(a) *If* $(x_n)_{n\in\mathbb{N}}$ *converges, then it is bounded, i.e.,* $\sup|x_n| < \infty$.

(b) *If* $(x_n)_{n\in\mathbb{N}}$ *and* $(y_n)_{n\in\mathbb{N}}$ *both converge, then so does* $(x_n + y_n)_{n\in\mathbb{N}}$, *and*

$$\lim_{n\to\infty} (x_n + y_n) = \lim_{n\to\infty} x_n + \lim_{n\to\infty} y_n.$$

(c) *If* $(x_n)_{n\in\mathbb{N}}$ *converges and* c *is a real or complex number, then* $(cx_n)_{n\in\mathbb{N}}$ *converges, and*

$$\lim_{n\to\infty} cx_n = c \lim_{n\to\infty} x_n.$$

Proof. (a) Suppose that $x_n \to x$. Considering $\varepsilon = 1$, there must exist some $N > 0$ such that $|x - x_n| < 1$ for all $n \geq N$. Therefore, for $n \geq N$,

$$|x_n| = |x_n - x + x| \leq |x_n - x| + |x| \leq 1 + |x|. \tag{1.5}$$

Given an arbitrary $n \in \mathbb{N}$, either x_n is one of $x_1, \ldots, x_{N-1}$ or we have $n \geq N$ and therefore equation (1.5) holds. Hence in any case,

$$|x_n| \leq \max\{|x_1|, \ldots, |x_{N-1}|, |x| + 1\}.$$

Since the right-hand side of the line above is a constant independent of n, we see that the sequence $(x_n)_{n\in\mathbb{N}}$ is bounded.

We assign the proofs of parts (b) and (c) as Problem 1.7.4. $\square$

The following definition introduces some terminology for a sequence of real numbers that increases without bound. We do not say that such a sequence converges but instead say that it diverges to infinity.

Definition 1.7.3 (Divergence to Infinity). If $(x_n)_{n\in\mathbb{N}}$ is a sequence of real numbers, then we say that $(x_n)_{n\in\mathbb{N}}$ *diverges to* ∞ if given any $R > 0$ there is an $N > 0$ such that $x_n > R$ for all $n > N$. In this case we write

$$x_n \to \infty, \quad \lim_{n\to\infty} x_n = \infty, \quad \text{or} \quad \lim x_n = \infty.$$

We define *divergence to* $-\infty$ similarly. $\Diamond$

Problems

1.7.4. Prove parts (b) and (c) of Lemma 1.7.2, and use induction to extend part (b) of the lemma to finite sums of limits.

1.7.5. Assume that $(x_n)_{n\in\mathbb{N}}$ is a monotonically increasing sequence of real numbers, i.e., each x_n is real and $x_1 \leq x_2 \leq \cdots$. Prove that $(x_n)_{n\in\mathbb{N}}$ converges if and only if $(x_n)_{n\in\mathbb{N}}$ is bounded, and in this case we have

$$\lim_{n\to\infty} x_n = \sup_{n\in\mathbb{N}} x_n.$$

Formulate and prove an analogous result for monotonically decreasing sequences.

1.8 Infinite Series of Numbers

We say that a series $\sum_{n=1}^{\infty} c_n$ of real or complex numbers *converges* if there is a real or complex number s such that the *partial sums*

$$s_N = \sum_{n=1}^{N} c_n$$

converge to s as $N \to \infty$. In this case $\sum_{n=1}^{\infty} c_n$ is defined to be s, i.e.,

$$\sum_{n=1}^{\infty} c_n = \lim_{N\to\infty} s_N = \lim_{N\to\infty} \sum_{n=1}^{N} c_n = s.$$

If the series $\sum_{n=1}^{\infty} c_n$ does not converge, then we say that it *diverges*. We sometimes use the shorthand $\sum c_n$ or $\sum_n c_n$ to denote a series.

Here are two particular examples of infinite series.

Lemma 1.8.1. (a) *If z is a scalar with $|z| < 1$, then $\sum_{k=0}^{\infty} z^k$ converges and has the value*

$$\sum_{k=0}^{\infty} z^k = \frac{1}{1-z}.$$

Conversely, if $|z| \geq 1$, then $\sum_{k=0}^{\infty} z^k$ does not converge.

(b) *If z is any scalar, then $\sum_{k=0}^{\infty} z^k/k!$ converges and has the value*

$$\sum_{k=0}^{\infty} \frac{z^k}{k!} = e^z. \qquad \diamond$$

An important property of convergent series is given in the next lemma, whose proof is assigned as Problem 1.8.5.

Lemma 1.8.2. *If $\sum_{n=1}^{\infty} c_n$ is a convergent series of scalars, then*

$$\lim_{n \to \infty} c_n = 0. \qquad \Diamond$$

Example 1.8.3. The converse of Lemma 1.8.2 is false in general. For example, consider $c_n = 1/n$. Although the scalars $1/n$ converge to zero as $n \to \infty$, it is nonetheless the case (this is Problem 1.8.6) that

$$the\ series\ \sum_{n=1}^{\infty} \frac{1}{n}\ does\ not\ converge! \qquad \Diamond$$

We often deal with series in which every c_n is a *nonnegative real number*. In this case there are only the following two possibilities (see Problem 1.8.7):

- If $c_n \geq 0$ for every n and the sequence of partial sums $\{s_N\}_{N \in \mathbb{N}}$ is bounded above, then the series $\sum c_n$ converges to a nonnegative real number. In this case we write

$$\sum_{n=1}^{\infty} c_n < \infty.$$

- If $c_n \geq 0$ for every n and the sequence of partial sums $\{s_N\}_{N \in \mathbb{N}}$ is not bounded above, then s_N diverges to infinity. In this case, the series $\sum c_n$ diverges, and we say that $\sum c_n$ *diverges to infinity* and write

$$\sum_{n=1}^{\infty} c_n = \infty.$$

The following lemma regarding the "tails of a nonnegative series" is often useful.

Lemma 1.8.4 (Tails of Convergent Series). *If $c_n \geq 0$ for every n and $\sum c_n < \infty$, then*

$$\lim_{N \to \infty} \left(\sum_{n=N}^{\infty} c_n \right) = 0.$$

Proof. Assume that $x = \sum_{n=1}^{\infty} c_n$ exists. Given a positive integer N, the reader should show that the series $y_N = \sum_{n=N+1}^{\infty} c_n$ converges, and that $x - y_N = \sum_{n=1}^{N} c_n = s_N$, the Nth partial sum of the series that defines x. Since the partial sums converge to x, we see that

$$\left| \sum_{n=N}^{\infty} c_n \right| = |y_N| = |x - s_N| \to 0 \quad \text{as } N \to \infty. \qquad \square$$

Problems

1.8.5. Prove Lemma 1.8.2.

1.8.6. Prove that the *alternating harmonic series* $\sum (-1)^n \frac{1}{n}$ converges, but the *harmonic series* $\sum \frac{1}{n}$ diverges to infinity.

1.8.7. (a) Let a_n and b_n be real or complex numbers. Show that if $\sum a_n$ and $\sum b_n$ each converge, then $\sum (a_n + b_n)$ converges and has the value

$$\sum_{n=1}^{\infty} (a_n + b_n) = \sum_{n=1}^{\infty} a_n + \sum_{n=1}^{\infty} b_n.$$

(b) Exhibit real numbers a_n, b_n such that $\sum (a_n + b_n)$ converges but $\sum a_n$ and $\sum b_n$ do not converge.

1.8.8. Suppose that $c_n \geq 0$ for each $n \in \mathbb{N}$. Prove that $\sum c_n$ either converges or diverges to infinity.

1.9 Differentiation and the Riemann Integral

We assume that the reader is familiar with the basic concepts of single-variable undergraduate calculus. Some of these concepts, such as limits, were mentioned above, while others, such as continuity, will be discussed in detail in Chapter 2. Indeed, in Chapter 2 we will see how to formulate and understand convergence and continuity in the setting of *metric spaces,* and convergence and continuity will be recurring themes throughout Chapters 3–7.

In contrast, differentiation and integration play mostly supporting roles in this volume, rather than being the main topics of study. For example, we will typically make use of differentiation and integration only to formulate examples that illustrate other concepts. Differentiation and integration beyond the basic results of undergraduate calculus are the centerpieces of more advanced texts on measure theory; for more details we refer to texts such as [WZ77], [Rud87], [Fol99], [SS05], and [Heil19].

We will briefly review some facts and terminology connected with differentiation or integration that we will need in this volume.

Let $f \colon D \to \mathbb{F}$ be a real-valued or complex-valued function whose domain D is a subset of the real line $\mathbb{R}$. If $t \in D$ and there is some open interval I such that $t \in I \subseteq D$, then we say that f is *differentiable* at t if the limit

$$f'(t) = \lim_{h \to 0} \frac{f(t+h) - f(t)}{h}$$

exists. We call $f'(t)$ the derivative of f at the point t. If f is differentiable at t then it is continuous at t, but the converse need not hold. If f is differentiable at every point t in an open interval I, then we say that f *is differentiable on* I.

The Mean Value Theorem is one of most important results of differential calculus. A proof can be found in calculus texts such as [HHW18]. Note that this result holds only for *real-valued functions* (see Problem 1.9.2).

Theorem 1.9.1 (Mean Value Theorem). *Assume that $f: [a, b] \to \mathbb{R}$ is a continuous real-valued function, and f is differentiable at each point of (a, b). Then there exists some point $c \in (a, b)$ such that*

$$f'(c) = \frac{f(b) - f(a)}{b - a}. \qquad \diamondsuit$$

Now we turn to integration. Unless we specifically state otherwise, every integral that appears in this volume is a Riemann integral. If the Riemann integral of a real-valued or complex-valued function f on the bounded closed interval $[a, b]$ exists, then we say that f is *Riemann integrable* (see calculus texts such as [HHW18] for the precise definition of Riemann integrability). In this case we denote the value of its Riemann integral by

$$\int_a^b f(t)\, dt,$$

or sometimes by $\int_a^b f$ for short. Below are some facts about Riemann integrals.

- Every continuous real-valued or complex-valued function f whose domain is a closed interval $[a, b]$ is Riemann integrable.

- If f is a complex-valued function and we write $f = g + ih$, where g and h are real-valued, then f is Riemann integrable if and only if g and h are Riemann integrable, and in this case

$$\int_a^b f(t)\, dt = \int_a^b g(t)\, dt + i \int_a^b h(t)\, dt.$$

- Some discontinuous functions are Riemann integrable, but not all. For example, the function f that is 1 at every rational point and 0 at every irrational point is not Riemann integrable on any interval $[a, b]$.

- Every bounded *piecewise continuous* function on $[a, b]$ is Riemann integrable. A function f is piecewise continuous if we can partition the interval $[a, b]$ as $a = a_0 < a_1 < a_2 < \cdots < a_n = b$ and f is continuous on (a_{k-1}, a_k) for each $k = 1, \ldots, n$. A piecewise continuous function need not be continuous, but every bounded piecewise continuous function on $[a, b]$ is Riemann integrable.

Problems

1.9.2. This problem will show that the conclusion of the Mean Value Theorem can fail for complex-valued functions. Set $f(t) = e^{it}$ for $t \in [0, 2\pi]$. This function f is continuous on $[0, 2\pi]$ and is differentiable at every point of $(0, 2\pi)$. Prove that there is no point $c \in (0, 2\pi)$ such that

$$f'(c) = \frac{f(2\pi) - f(0)}{2\pi - 0}.$$

1.10 Vector Spaces

A vector space is a set V that is associated with a scalar field $\mathbb{F}$ (always either $\mathbb{F} = \mathbb{R}$ or $\mathbb{F} = \mathbb{C}$ in this volume) and two operations that allow us to add vectors together and to multiply a vector by a scalar. Here is the precise definition.

Definition 1.10.1 (Vector Space). A *vector space* over the scalar field $\mathbb{F}$ is a set V that satisfies the following conditions.

Closure Axioms

- (1) Vector addition: Given any two elements $x, y \in V$, there is a unique element $x + y$ in V, which we call the *sum* of x and y.
- (2) Scalar multiplication: Given $x \in V$ and a scalar $c \in \mathbb{F}$, there exists a unique element cx in V, which we call the *product* of c and x.

Addition Axioms

- (3) Commutativity: $x + y = y + x$ for all $x, y \in V$.
- (4) Associativity: $(x + y) + z = x + (y + z)$ for all $x, y \in V$.
- (5) Additive Identity: There exists an element $0 \in V$ that satisfies $x + 0 = x$ for all $x \in V$. We call this element 0 the *zero vector* of V.
- (6) Additive Inverses: For each element $x \in V$, there exists an element $(-x) \in V$ that satisfies $x + (-x) = 0$. We call $-x$ the *additive inverse* of x, and we declare that $x - y = x + (-y)$.

Multiplication Axioms

- (7) Associativity: $(ab)x = a(bx)$ for all $a, b \in \mathbb{F}$ and $x \in V$.
- (8) Multiplicative Identity: Scalar multiplication by the number 1 satisfies $1x = x$ for every $x \in V$.

Distributive Axioms

- (9) $c(x + y) = cx + cy$ for all $x, y \in V$ and $c \in \mathbb{F}$.
- (10) $(a + b)x = ax + bx$ for all $x \in V$ and $a, b \in \mathbb{F}$.   ◇

Another name for a vector space is *linear space*. We will call the elements of a vector space *vectors* (regardless of whether they are numbers, sequences, functions, operators, tensors, or other types of objects). The *trivial vector space* is $V = \{0\}$.

The scalar field associated with the vector spaces in this volume will always be either the real line $\mathbb{R}$ or the complex plane $\mathbb{C}$. In order to deal with both of these possibilities, we introduce the following notation.

Notation 1.10.2 (Scalar Field Notation and the Symbol $\mathbb{F}$). The *scalar field* for all vector spaces in this volume will be either the real line $\mathbb{R}$ or the complex plane $\mathbb{C}$. When we want to allow both possibilities we let the symbol $\mathbb{F}$ denote the scalar field, i.e., $\mathbb{F}$ can be either $\mathbb{R}$ or $\mathbb{C}$. $\Diamond$

The elements of the scalar field are often called *scalars*. Hence if $\mathbb{F} = \mathbb{R}$ then the unqualified term *scalar* means *real number*, while if $\mathbb{F} = \mathbb{C}$ then *scalar* means *complex number*.

Example 1.10.3. A real-valued function on a set X is a function that takes real values, and therefore has the form $f \colon X \to \mathbb{R}$. Likewise, a complex-valued function is a function of the form $f \colon X \to \mathbb{C}$. Since every real number is a complex number, every real-valued function is also complex-valued. However, a complex-valued function need not be real-valued. For example, $g(t) = |t|$ is both real-valued and complex-valued on the domain $X = \mathbb{R}$, while $f(t) = e^{it} = \cos t + i \sin t$ is complex-valued but not real-valued on $X = \mathbb{R}$.

Notation 1.10.2 allows us to consider both possibilities together. Specifically, a *scalar-valued function* means a function of the form $f \colon X \to \mathbb{F}$, where $\mathbb{F}$ stands for our choice of either $\mathbb{R}$ or $\mathbb{C}$. Such a function can be either real-valued or complex-valued, depending on whether we choose $\mathbb{F}$ to be $\mathbb{R}$ or $\mathbb{C}$. If $\mathbb{F} = \mathbb{R}$ then our functions can only be real-valued, while if $\mathbb{F} = \mathbb{C}$ then our functions are complex-valued, which includes the real-valued functions as special cases. $\Diamond$

The beginning student may simply want to assume that $\mathbb{F} = \mathbb{R}$ on a first reading of the volume. In this case, all numbers are real numbers, the word "scalar" means "real number," vectors in $\mathbb{F}^d = \mathbb{R}^d$ have real components, scalar-valued functions are real-valued functions, and so forth.

We will give some examples of vector spaces.

Example 1.10.4. d-dimensional *Euclidean space* $\mathbb{F}^d$ is the set of all d-tuples of scalars, i.e.,

$$\mathbb{F}^d = \big\{ (x_1, \dots, x_d) : x_1, \dots, x_d \in \mathbb{F} \big\}.$$

This is a vector space whose scalar field is $\mathbb{F}$. If $\mathbb{F} = \mathbb{R}$ then $\mathbb{F}^d$ is simply $\mathbb{R}^d$, while if $\mathbb{F} = \mathbb{C}$ then $\mathbb{F}^d = \mathbb{C}^d$. $\Diamond$

Here is another example of a vector space. At first glance it may seem that this space is entirely unlike the Euclidean space $\mathbb{F}^d$, but in fact there are many ways in which they are quite similar (see Problem 1.10.11).

Example 1.10.5. Let $\mathcal{F}[0,1]$ be the set of all scalar-valued functions whose domain is the closed interval $[0,1]$:

$$\mathcal{F}[0,1] \;=\; \{f : [0,1] \to \mathbb{F} : f \text{ is a function on } [0,1]\}.$$

If $\mathbb{F} = \mathbb{R}$ then $\mathcal{F}[0,1]$ is the set of all real-valued functions on $[0,1]$, while if $\mathbb{F} = \mathbb{C}$ then $\mathcal{F}[0,1]$ is the space of complex-valued functions on $[0,1]$. For example, taking $0 \le t \le 1$, each of

$$f(t) = |t|, \qquad g(t) = t^2, \qquad h(t) = e^{-|t|}, \qquad k(t) = e^{-t^2}, \qquad (1.6)$$

is a function on the domain $[0,1]$. These are all real-valued functions, and hence are also complex-valued, since every real number is a complex number. Therefore f, g, h, and k are all elements of $\mathcal{F}[0,1]$ regardless of whether we choose $\mathbb{F} = \mathbb{R}$ or $\mathbb{F} = \mathbb{C}$, and so we say that *f, g, h, and k are vectors in* $\mathcal{F}[0,1]$. On the other hand, the function

$$e(t) \;=\; e^{\pi i t} \;=\; \cos \pi t + i \sin \pi t, \qquad t \in [0,1],$$

is complex-valued, so e is a vector in $\mathcal{F}[0,1]$ if we choose $\mathbb{F} = \mathbb{C}$, but it is not an element of $\mathcal{F}[0,1]$ if $\mathbb{F} = \mathbb{R}$.

Not every element of $\mathcal{F}[0,1]$ is continuous. For example,

$$d(t) \;=\; \begin{cases} 1, & t \text{ is rational,} \\ 0, & t \text{ is irrational,} \end{cases}$$

is a real-valued function, so $d \in \mathcal{F}[0,1]$ regardless of whether $\mathbb{F} = \mathbb{R}$ or $\mathbb{F} = \mathbb{C}$. However, d is not continuous at any point.

Vector addition on $\mathcal{F}[0,1]$ is the usual addition of functions. That is, if f and g are two elements of $\mathcal{F}[0,1]$, then $f + g$ is the function defined by the rule

$$(f + g)(t) \;=\; f(t) + g(t), \qquad t \in [0,1].$$

For scalar multiplication, if $f \in \mathcal{F}[0,1]$ and $c \in \mathbb{F}$ is a scalar, then cf is the function given by the rule

$$(cf)(t) \;=\; cf(t), \qquad t \in [0,1].$$

The reader should verify that these operations satisfy all ten of the axioms of a vector space, and therefore $\mathcal{F}[0,1]$ is a vector space with respect to these two operations. The zero vector in this space is the *zero function* 0, i.e., the function that takes the value zero at every point. Note that we use the same symbol "0" to denote both the *zero function* and the *number zero*. It should usually be clear from context whether "0" denotes a function or a scalar. $\Diamond$

Once we know that a given set V is a vector space, we can easily check whether a subset Y is a vector space by applying the following lemma (the proof is Problem 1.10.10). In the statement of this lemma, we implicitly assume that the vector space operations on Y are the same operations that are used in V.

Lemma 1.10.6. *Let Y be a nonempty subset of a vector space V. If:*
(a) *Y is closed under vector addition, i.e.,*

$$x, y \in Y \implies x + y \in Y,$$

and

(b) *Y is closed under scalar multiplication, i.e.,*

$$x \in Y, \ c \in \mathbb{F} \implies cx \in Y,$$

then Y is itself a vector space with respect to the operations of vector addition and scalar multiplication that are defined on V. □

A subset Y of V that satisfies the conditions of Lemma 1.10.6 is called a *subspace* of V. Here is an example.

Lemma 1.10.7. *Let $C[0, 1]$ be the set of all continuous scalar-valued functions whose domain is $[0, 1]$:*

$$C[0, 1] = \{f : [0, 1] \to \mathbb{F} : f \text{ is continuous on } I\}.$$

Then $C[0, 1]$ is a subspace of $\mathcal{F}[0, 1]$, and hence is a vector space with respect to the operations of addition of functions and multiplication of a function by a scalar.

Proof. Since there are continuous functions on $[0, 1]$, we know that $C[0, 1]$ is a nonempty subset of $\mathcal{F}[0, 1]$. Now, if f and g are continuous on $[0, 1]$, then so is their sum, so $f + g \in C[0, 1]$. Likewise, if f is continuous, then so is cf for every scalar $c \in \mathbb{F}$. Lemma 1.10.6 therefore implies that $C[0, 1]$ is a subspace of $\mathcal{F}[0, 1]$. ◇

There is nothing special about the interval $[0, 1]$ in Example 1.10.5 or Lemma 1.10.7. If I is any interval in the real line, then the set $\mathcal{F}(I)$ that contains all scalar-valued functions with domain I is a vector space, and the subset $C(I)$ that consists of all continuous functions on I is also a vector space. To avoid multiplicities of brackets and parentheses, if $I = (a, b)$ then we usually write $C(a, b)$ instead of $C((a, b))$, and if $I = [a, b)$ then we usually write $C[a, b)$ instead of $C([a, b))$, and so forth.

If we restrict our attention to open intervals, then we can create further subspaces consisting of differentiable functions.

Example 1.10.8. Let I be an open interval in the real line, and let $C^1(I)$ be the set of all scalar-valued functions f that are differentiable on I and whose derivative f' is continuous on I:

$$C^1(I) = \{f : I \to \mathbb{F} : f \text{ is differentiable and } f' \text{ is continuous on } I\}.$$

If $\mathbb{F} = \mathbb{R}$ then functions in $C^1(I)$ are real-valued, while if $\mathbb{F} = \mathbb{C}$ then $C^1(I)$ includes both real-valued and complex-valued functions. Every differentiable function is continuous, so $C^1(I)$ is a subset of $C(I)$, but the examples below will show that it is a *proper* subset, i.e., $C^1(I) \subsetneq C(I)$.

To illustrate, take $I = \mathbb{R}$ and let $\mathbb{F}$ be either $\mathbb{R}$ or $\mathbb{C}$. The four functions f, g, h, k defined in equation (1.6) have the following properties.

- f is continuous but not differentiable on $\mathbb{R}$, so f belongs to $C(\mathbb{R})$ but does not belong to $C^1(\mathbb{R})$.

- g is differentiable and $g'(t) = 2t$ is continuous on $\mathbb{R}$ (in fact, g' is differentiable on $\mathbb{R}$), so $g \in C^1(\mathbb{R})$.

- h is continuous but not differentiable on $\mathbb{R}$, so $h \in C(\mathbb{R}) \backslash C^1(\mathbb{R})$.

- k is differentiable and $k'(t) = -2te^{-t^2}$ is continuous on $\mathbb{R}$ (in fact, k' is differentiable on $\mathbb{R}$), so $k \in C^1(\mathbb{R})$.

The reader should use Lemma 1.10.6 to prove that $C^1(I)$ is a subspace of $C(I)$. Since $C^1(I)$ is contained in but not equal to $C(I)$, we see that $C^1(I)$ is a *proper subspace* of $C(I)$, which is itself a proper subspace of $\mathcal{F}(I)$. $\Diamond$

The two functions g and k defined in equation (1.6) are actually *infinitely differentiable* on $\mathbb{R}$, i.e., $g', g'', g''', \ldots$ and $k', k'', k''', \ldots$ all exist and are differentiable at every point. Are there any functions that are differentiable on $\mathbb{R}$ but are not infinitely differentiable?

Example 1.10.9. Let $I = \mathbb{R}$ and define

$$w(t) = \begin{cases} t^2 \sin \frac{1}{t}, & t \neq 0, \\ 0, & t = 0. \end{cases} \tag{1.7}$$

The reader should check (this is Problem 1.10.13) that:

- w is continuous on $\mathbb{R}$, so $w \in C(\mathbb{R})$,
- w is differentiable *at every point of* $\mathbb{R}$, i.e., $w'(t)$ exists for each $t \in \mathbb{R}$, but
- w' is *not* continuous at every point of $\mathbb{R}$.

Therefore, although w is differentiable on $\mathbb{R}$, its derivative is not differentiable on $\mathbb{R}$ because it is not even continuous. We say that w is *once differentiable*, because $w'(t)$ exists for every t. However, w is not *twice differentiable*, because $w''(t)$ does not exist for every t. Because w is continuous it is an element of $C(\mathbb{R})$, but $w \notin C^1(\mathbb{R})$ because w' is not continuous. $\Diamond$

We can keep going and define $C^2(I)$ to be the space of all functions f such that both f and f' are differentiable on I and f'' is continuous on I. This is a proper subspace of $C^1(I)$. Then we can continue and define $C^3(I)$

and so forth, obtaining a nested decreasing sequence of spaces. The space $C^\infty(I)$ that consists of all infinitely differentiable functions is itself a proper subspace of each of these. Moreover, the set $\mathcal{P}$ consisting of all polynomial functions,

$$\mathcal{P} = \left\{ \sum_{k=0}^{N} c_k t^k : N \geq 0, c_k \in \mathbb{F} \right\},$$

is a proper subspace of $C^\infty(I)$. Thus we have the infinitely many distinct vector spaces

$$\mathcal{F}(I) \supsetneq C(I) \supsetneq C^1(I) \supsetneq C^2(I) \supsetneq \cdots \supsetneq C^\infty(I) \supsetneq \mathcal{P}.$$

Problems

1.10.10. Prove Lemma 1.10.6.

1.10.11. Let X be any set, and let $\mathcal{F}(X)$ be the set of all scalar-valued functions $f : X \to \mathbb{F}$.

(a) Prove that $\mathcal{F}(X)$ is a vector space.

(b) For this part we take $X = \{1, \ldots, d\}$. Explain why the Euclidean vector space $\mathbb{F}^d$ and the vector space $\mathcal{F}(\{1, \ldots, d\})$ are really the "same space," in the sense that each vector in $\mathbb{F}^d$ naturally corresponds to a function in $\mathcal{F}(\{1, \ldots, d\})$ and vice versa.

(c) Now let $X = \mathbb{N} = \{1, 2, 3, \ldots\}$, and let $\mathcal{S}$ be the set of all infinite sequences $x = (x_1, x_2, \ldots)$. In what sense are $\mathcal{F}(\mathbb{N})$ and $\mathcal{S}$ the "same" vector space?

1.10.12. Let I be an interval in the real line. We say that a function $f : I \to \mathbb{F}$ is *bounded* if $\sup_{t \in I} |f(t)| < \infty$.

(a) Let $\mathcal{F}_b(I)$ be the set of all bounded functions on I, and prove that $\mathcal{F}_b(I)$ is a proper subspace of $\mathcal{F}(I)$.

(b) Let $C_b(I)$ be the set of all bounded continuous functions on I. Prove that $C_b(I)$ is a subspace of $C(I)$. Show that if I is any type of interval *other than* a bounded closed interval $[c, d]$, then $C_b(I) \neq C(I)$.

Remark: If $I = [c, d]$ is a bounded closed interval, then every continuous function on I is bounded (we will prove this in Corollary 2.9.3). Therefore $C_b[c, d] = C[c, d]$.

1.10.13. Let w be the function defined in equation (1.7).

(a) Use the product rule to prove that w is differentiable at every point $t \neq 0$, and use the definition of the derivative to prove that w is differentiable at $t = 0$.

(b) Show that even though $w'(t)$ exists for every t, the derivative w' is not continuous at the origin.

1.11 Span and Independence

A *finite linear combination* (or simply a *linear combination*, for short) of vectors $x_1, \ldots, x_N$ in a vector space V is any vector that has the form

$$x = \sum_{k=1}^{N} c_k x_k = c_1 x_1 + \cdots + c_N x_N,$$

where $c_1, \ldots, c_N$ are scalars. We collect all of the linear combinations together to form the following set.

Definition 1.11.1 (Span). If A is a nonempty subset of a vector space V, then the *finite linear span* of A, denoted by $\mathrm{span}(A)$, is the set of all *finite linear combinations* of elements of A:

$$\mathrm{span}(A) = \left\{ \sum_{n=1}^{N} c_n x_n : N > 0, \ x_n \in A, \ c_n \in \mathbb{F} \right\}. \qquad (1.8)$$

We say that A *spans* V if $\mathrm{span}(A) = V$.

We *declare* that the span of the empty set is $\mathrm{span}(\varnothing) = \{0\}$. $\quad \Diamond$

We also refer to $\mathrm{span}(A)$ as the *finite span*, the *linear span*, or simply the *span* of A.

If $A = \{x_1, \ldots, x_n\}$ is a finite set, then we usually write $\mathrm{span}\{x_1, \ldots, x_n\}$ instead of $\mathrm{span}(\{x_1, \ldots, x_n\})$. In this case, equation (1.8) simplifies to

$$\mathrm{span}\{x_1, \ldots, x_n\} = \{c_1 x_1 + \cdots + c_n x_n : c_1, \ldots, c_n \in \mathbb{F}\}.$$

Similarly, if $A = \{x_n\}_{n \in \mathbb{N}}$ is a countable sequence, then we usually write $\mathrm{span}\{x_n\}_{n \in \mathbb{N}}$ instead of $\mathrm{span}(\{x_n\}_{n \in \mathbb{N}})$. In this case equation (1.8) simplifies to

$$\mathrm{span}\{x_n\}_{n \in \mathbb{N}} = \left\{ \sum_{n=1}^{N} c_n x_n : N > 0, \ c_n \in \mathbb{F} \right\}.$$

Example 1.11.2. A *polynomial* is a function of the form

$$p(t) = \sum_{k=0}^{n} c_k t^k, \qquad t \in \mathbb{R},$$

where n is a nonnegative integer and $c_0, \ldots, c_n \in \mathbb{F}$. A *monomial* is a polynomial that has only one nonzero term, i.e., it is a polynomial of the form $c t^k$, where c is a nonzero scalar and k is a nonnegative integer.

Let $\mathcal{P}$ be the set of all polynomials, and let $\mathcal{M}$ be the subset of $\mathcal{P}$ that contains all monomials with the specific form t^k,

$$\mathcal{M} = \{t^k\}_{k=0}^{\infty} = \{1, t, t^2, t^3, \dots\}.$$

Every polynomial p is a finite linear combination of elements of $\mathcal{M}$, so span$(\mathcal{M}) = \mathcal{P}$. Therefore $\mathcal{M}$ spans $\mathcal{P}$. $\Diamond$

By definition, if x is a vector in span(A), then x is some finite linear combination of elements of A. In general there could be many different linear combinations that equal x. Often we wish to ensure that x is a *unique* linear combination of elements of A. This issue is related to the following notion.

Definition 1.11.3 (Linear Independence). A nonempty subset A of a vector space V is *finitely linearly independent* if given any integer $N > 0$, any choice of finitely many distinct vectors $x_1, \dots, x_N \in A$, and any scalars $c_1, \dots, c_N \in \mathbb{F}$, we have

$$\sum_{n=1}^{N} c_n x_n = 0 \iff c_1 = \dots = c_N = 0.$$

Instead of saying that a set A is *finitely linearly independent*, we sometimes abbreviate this to A *is linearly independent*, or even just A *is independent*. We *declare* that the empty set $\varnothing$ is a linearly independent set. $\Diamond$

Often the set A is a finite set or a countable sequence. Rewriting the definition for these cases, we see that a finite set $\{x_1, \dots, x_n\}$ is independent if

$$\sum_{k=1}^{n} c_k x_k = 0 \iff c_1 = \dots = c_n = 0.$$

A countable sequence $\{x_n\}_{n \in \mathbb{N}}$ is independent if and only if

$$\forall N \in \mathbb{N}, \quad \sum_{n=1}^{N} c_n x_n = 0 \iff c_1 = \dots = c_N = 0.$$

Example 1.11.4. As in Example 1.11.2, let $\mathcal{P}$ be the set of all polynomials and let $\mathcal{M} = \{t^k\}_{k=0}^{\infty}$. We will show that $\mathcal{M}$ is a linearly independent subset of $\mathcal{P}$. To do this, choose any integer $N \geq 0$, and suppose that

$$\sum_{k=0}^{N} c_k t^k = 0 \tag{1.9}$$

for some scalars $c_0, c_1, \dots, c_N$. Note that the vector on the left-hand side of equation (1.9) is a function, the polynomial $p(t) = \sum_{k=0}^{N} c_k t^k$. The vector 0 on the right-hand side of equation (1.9) is the zero element of the vector space

$\mathcal{P}$, which is the zero function. Hence what we are assuming in equation (1.9) is that the polynomial p equals the zero function. This means that $p(t) = 0$ for *every* t (not just for *some* t).

We wish to show that each scalar c_k in equation (1.9) must be zero. There are many ways to do this. We ask for a *direct* proof in Problem 1.11.10 and spell out an *indirect* proof here. The *Fundamental Theorem of Algebra* tells us that a polynomial of degree N can have at most N roots, i.e., there can be at most N values of t for which $p(t) = 0$. If the scalar c_N is nonzero, then $p(t) = \sum_{k=0}^{N} c_k t^k$ has degree N, and therefore can have at most N roots. But we know that $p(t) = 0$ for every t, so p has infinitely many roots. This is a contradiction, so we must have $c_N = 0$.

Since $c_N = 0$, if c_{N-1} is nonzero then p has degree $N - 1$, which again leads to a contradiction, since $p(t) = 0$ for every t. Therefore we must have $c_{N-1} = 0$.

Continuing in this way, we see that $c_{N-2} = 0$, and so forth, down to $c_1 = 0$. So we are left with $p(t) = c_0$, i.e., p is a constant polynomial. The only way a constant polynomial can equal the zero function is if $c_0 = 0$. Therefore every c_k is zero, so we have shown that $\mathcal{M}$ is independent. $\diamond$

A set that is both linearly independent and spans a vector space V is usually called a *basis* for V. We will present some generalizations of this notion in Section 4.5, so to distinguish between these various types of "bases" we will use the following terminology for a set that both spans and is independent.

Definition 1.11.5 (Hamel Basis). Let V be a nontrivial vector space. A set of vectors $\mathcal{B}$ is a *Hamel basis, vector space basis*, or simply a *basis* for V if

$$\mathcal{B} \text{ is linearly independent} \qquad \text{and} \qquad \text{span}(\mathcal{B}) = V. \qquad \diamond$$

For example, since Example 1.11.2 shows that the set $\mathcal{M} = \{t^k\}_{k=0}^{\infty}$ spans the set of polynomials $\mathcal{P}$ and Example 1.11.4 shows that $\mathcal{M}$ is independent, it follows that $\mathcal{M}$ is a Hamel basis for $\mathcal{P}$.

It can be shown that any two Hamel bases for a given vector space V have the same cardinality. Thus, if V has a Hamel basis that consists of finitely many vectors, say $\mathcal{B} = \{x_1, \ldots, x_d\}$, then any other Hamel basis for V must also contain exactly d vectors. We call this number d the *dimension* of V and set

$$\dim(V) = d.$$

On the other hand, if V has a Hamel basis that consists of infinitely many vectors, then any other Hamel basis must also be infinite. In this case we say that V is *infinite-dimensional* and we set

$$\dim(V) = \infty.$$

For finite-dimensional vector spaces, we have the following characterization of Hamel bases. The proof is assigned as Problem 1.11.12, and a similar characterization for general vector spaces is given in Problem 1.11.13.

Theorem 1.11.6. *A set of vectors $\mathcal{B} = \{x_1, \ldots, x_d\}$ is a Hamel basis for a finite-dimensional vector space V if and only if each vector $x \in V$ can be written as*

$$x = \sum_{n=1}^{d} c_n(x)\, x_n$$

*for a **unique choice** of scalars $c_1(x), \ldots, c_d(x)$.*   ◇

The trivial vector space $\{0\}$ is a bit of an anomaly in this discussion, since it does not contain any linearly independent subsets and therefore does not contain a Hamel basis. To handle this case we *declare* that the empty set ∅ is a Hamel basis for the trivial vector space $\{0\}$, and we set

$$\dim(\{0\}) = 0.$$

Problems

1.11.7. Let $e_k = (0, \ldots, 0, 1, 0, \ldots, 0)$ be the vector in $\mathbb{F}^d$ that has a 1 in the kth component and zeros elsewhere. Prove that $\{e_1, \ldots, e_d\}$ is a Hamel basis for $\mathbb{F}^d$ (this is the *standard basis* for $\mathbb{F}^d$).

1.11.8. Let V be a vector space.

(a) Show that if $A \subseteq B \subseteq V$, then $\mathrm{span}(A) \subseteq \mathrm{span}(B)$.

(b) Show by example that $\mathrm{span}(A) = \mathrm{span}(B)$ is possible even if $A \neq B$. Can you find an example where A and B are both infinite?

1.11.9. Given a linearly independent set A in a vector space V, prove the following statements.

(a) If $B \subseteq A$, then B is linearly independent.

(b) If $x \in V$ but $x \notin \mathrm{span}(A)$, then $A \cup \{x\}$ is linearly independent.

1.11.10. Without invoking the Fundamental Theorem of Algebra, give a direct proof that the set of monomials $\{t^k\}_{k=0}^{\infty}$ is a linearly independent set in $C(\mathbb{R})$.

1.11.11. For each $k \in \mathbb{N}$, define $e_k(t) = e^{kt}$ for $t \in \mathbb{R}$.

(a) Prove that $\{e_k\}_{k \in \mathbb{N}}$ is a linearly independent set in $C(\mathbb{R})$, i.e., if $n > 0$ is a positive integer, $c_1, \ldots, c_n$ are scalars, and $c_1 e_1 + \cdots + c_n e_n = 0$ (the zero function), then $c_1 = \cdots = c_n = 0$.

(b) Let 1 be the constant function defined by $1(t) = 1$ for every t. Prove that 1 does not belong to $\text{span}\{e_k\}_{k\in\mathbb{N}}$.

(c) Find a function f that does not belong to $\text{span}\{e_k\}_{k\geq 0}$, where $e_0 = 1$.

1.11.12. Prove Theorem 1.11.6.

1.11.13. Let $\mathcal{B} = \{x_i\}_{i\in I}$ be a subset of a vector space V. Prove that $\mathcal{B}$ is a Hamel basis for V if and only if every nonzero vector $x \in V$ can be written as

$$x = \sum_{k=1}^{N} c_k(x)\, x_{i_k}$$

for a unique choice of $N \in \mathbb{N}$, indices $i_1, \ldots, i_N \in I$, and nonzero scalars $c_1(x), \ldots, c_N(x) \in \mathbb{F}$.

Chapter 2
Metric Spaces

Much of what we do in real analysis centers on issues of *convergence* or *approximation*. What does it mean for one object to be close to (or to approximate) another object? How can we define the *limit* of a sequence of objects that appear to be converging in some sense? If our objects are points in $\mathbb{R}^d$, then we can simply grab a ruler, embed ourselves into d-dimensional space, and measure the physical (or *Euclidean*) distance between the points. Two points are close if the distance between them is small, and a sequence of points x_n converges to a point x if the distance between x_n and x shrinks to zero in the limit as n increases to infinity. However, the objects we work with are often not points in $\mathbb{R}^d$ but instead are elements of some other set X (perhaps a set of sequences, or a set of functions, or some other abstract set). Even so, if we can find a way to measure the distance between elements of X, then we can still think about approximation or convergence. We simply say that two elements are close if the distance between them is small, and a sequence of elements x_n converges to an element x if the distance from x_n to x shrinks to zero. If the *properties* of the distance function on X are similar to those of Euclidean distance on $\mathbb{R}^d$, then we will be able to prove useful theorems about X and its elements. We will make these ideas precise in this chapter.

2.1 Metrics

A *metric* on a set X is a function that assigns a *distance* to each pair of elements of X. This distance function cannot be completely arbitrary—it must have certain properties similar to those of the physical distance between points in $\mathbb{R}^d$. For example, the distance between any two points x, y in $\mathbb{R}^d$ is nonnegative and finite, and it is zero only when the two points are identical. Further, the distance between x and y is the same as the distance between y and x, and if we have three points x, y, z, then the length of any one

© Springer International Publishing AG, part of Springer Nature 2018
C. Heil, *Metrics, Norms, Inner Products, and Operator Theory*, Applied and
Numerical Harmonic Analysis, https://doi.org/10.1007/978-3-319-65322-8_2

side of the triangle that they determine is less than or equal to the sum of the lengths of the other two sides (this is called the *Triangle Inequality*). A metric is a function, defined on pairs of elements of a set, that has similar properties. Here is the precise definition.

Definition 2.1.1 (Metric Space). Let X be a nonempty set. A *metric* on X is a function d: $X \times X \to \mathbb{R}$ such that for all x, y, $z \in X$ we have:

(a) Nonnegativity: $0 \le d(x,y) < \infty$,

(b) Uniqueness: $d(x,y) = 0$ if and only if $x = y$,

(c) Symmetry: $d(x,y) = d(y,x)$, and

(d) The Triangle Inequality: $d(x,z) \le d(x,y) + d(y,z)$.

If these conditions are satisfied, then X is a called a *metric space*. The number $d(x,y)$ is called the *distance* from x to y.   ◇

We often refer to the elements of a metric space X as "points," and we mostly use lowercase letters such as x, y, z to denote elements of the metric space. If our metric space happens to also be a vector space (which is the case for most of the metric spaces we will see in this text), then we often refer to the elements as either "points" or "vectors," depending on which terminology seems more appropriate in a given context. If the elements of our set X are functions (which is the case for many of the examples in this text), then we may refer to them as "points," "vectors," or "functions." Further, if we know that the elements of X are functions, then we usually denote them by letters such as f, g, h (instead of x, y, z).

Remark 2.1.2. If d is a metric on a set X and Y is a nonempty subset of X, then by restricting the metric to Y we obtain a metric on the subset Y. That is,

$$d(x,y), \qquad x, y \in Y,$$

is a metric on Y, called the *inherited metric* on Y. Thus, once we have a metric on a set X, we immediately obtain a metric on every nonempty subset of X.   ◇

The following example shows that we can always put at least one metric on any nonempty set.

Example 2.1.3 (The Discrete Metric). If X is a nonempty set, then the *discrete metric* on X is

$$d(x,y) = \begin{cases} 1, & \text{if } x \ne y, \\ 0, & \text{if } x = y. \end{cases}$$

The reader should check that this function satisfies the requirements of a metric (this is Problem 2.1.13). Using this metric, the distance between any two different points x and y in X is one unit.   ◇

In practice, we usually want to do more than just define a metric on a set X. Typically, our set X has some kind of interesting properties, and we seek a metric that somehow takes those properties into account. We will see many examples of such metrics as we progress through the text.

Notation 2.1.4 (Scalar Field Notation). Most of the metric spaces that we will encounter in this text will also be vector spaces. By definition, a vector space has a scalar field that is associated with it. In this text the scalar field will always be either the real line $\mathbb{R}$ or the complex plane $\mathbb{C}$. Almost all of the results that we will prove for real vector spaces are also true for complex vector spaces and vice versa. As we declared in Notation 1.10.2, in order to treat both of these cases simultaneously and concisely we let the symbol $\mathbb{F}$ stand for a choice of either $\mathbb{R}$ or $\mathbb{C}$. We will explicitly identify those few circumstances in which there is a difference between results over the two scalar fields.

Thus, when $\mathbb{F} = \mathbb{R}$ the word "scalar" means a *real number*, and when $\mathbb{F} = \mathbb{C}$ it means a *complex number*. The student interested only in the real case can simply take $\mathbb{F} = \mathbb{R}$ throughout the text; in this case "scalar" simply means a real number. We include the complex case both for completeness and because it is important in many areas in mathematics (e.g., harmonic analysis), science (e.g., quantum mechanics), and engineering (e.g., signal processing). $\diamondsuit$

Example 2.1.5 (The Standard Metric). Consider $X = \mathbb{F}$ (i.e., X is either the real line $\mathbb{R}$ or the complex plane $\mathbb{C}$, depending on what we choose for $\mathbb{F}$). There are many metrics that we can place on $\mathbb{F}$, but the *standard metric* on $\mathbb{F}$ is the ordinary distance given by the absolute value of the difference of two numbers:
$$\mathrm{d}(x, y) = |x - y|, \qquad x, y \in \mathbb{F}.$$
The standard metric is not the only metric on $\mathbb{F}$, but whenever we work with $\mathbb{F}$ we will always assume, unless we explicitly state otherwise, that we are using the standard metric. $\diamondsuit$

When we deal with multiple metrics on a given space, we may distinguish between them by using subscripts, as in the next example.

Example 2.1.6. Given vectors $x = (x_1, \ldots, x_d)$ and $y = (y_1, \ldots, y_d)$ in $\mathbb{F}^d$, define
$$\mathrm{d}_2(x, y) = \left(|x_1 - y_1|^2 + \cdots + |x_d - y_d|^2\right)^{1/2}. \tag{2.1}$$
The reader can check that d_2 is a metric on $\mathbb{F}^d$ (see Problem 2.1.16). We call this the *Euclidean metric* or the *standard metric* on $\mathbb{F}^d$. Whenever we work with $\mathbb{F}^d$ we will assume, unless we explicitly state otherwise, that our metric is d_2.

The Euclidean metric may seem to be the "obvious" choice of metric for $\mathbb{F}^d$ because it represents the actual physical distance between the points x

and y. However, there are many applications in which it is more convenient to use a different metric. Two other common metrics on $\mathbb{F}^d$ are

$$d_1(x, y) = |x_1 - y_1| + \cdots + |x_d - y_d| \tag{2.2}$$

and

$$d_\infty(x, y) = \max\{|x_1 - y_1|, \ldots, |x_d - y_d|\}. \tag{2.3}$$

The proof that these are metrics is assigned as Problem 2.1.16.

Using the Euclidean metric, the distance between the point $x = (1, 1, \ldots, 1)$ and the origin $0 = (0, 0, \ldots, 0)$ is

$$d_2(x, 0) = \left(|1 - 0|^2 + \cdots + |1 - 0|^2\right)^{1/2} = d^{1/2}.$$

However, if we use the metric d_1 then the distance from x to 0 is

$$d_1(x, 0) = |1 - 0| + \cdots + |1 - 0| = d,$$

while the distance as measured by the metric d_∞ is

$$d_\infty(x, 0) = \max\{|1 - 0|, \ldots, |1 - 0|\} = 1.$$

Although these three metrics d_1, d_2, and d_∞ measure distance differently, they are closely related and share certain properties. For example, no matter whether we use d_1, d_2, or d_∞, the distance from $7x$ to 0 is exactly 7 times the distance from x to 0. In contrast, the discrete metric from Example 2.1.3 does not "respect" scalar multiplication in this way. Using the discrete metric, if $x \neq 0$ then we have both $d(x, 0) = 1$ and $d(7x, 0) = 1$. This type of scaling issue will be important when we study *norms* in Chapter 3.   ◇

Each of the metrics discussed in Example 2.1.6 is a perfectly good metric on $\mathbb{F}^d$, and the answer to "which metric should we use?" depends on what context or problem we are dealing with. If our problem involves actual physical length, then it is probably best to use the Euclidean metric, but in other contexts it may be more advantageous to use a different metric.

In the next example we introduce an infinite-dimensional vector space known as ℓ^1 (pronounced "little ell one"). The elements of this vector space are infinite sequences of scalars that satisfy a certain "summability" property, and we will define a metric on ℓ^1 that is related to summability.

Example 2.1.7 (The Space ℓ^1). Let $x = (x_k)_{k \in \mathbb{N}} = (x_1, x_2, \ldots)$ be a sequence of scalars (if $\mathbb{F} = \mathbb{R}$ then x is a sequence of real numbers, while if $\mathbb{F} = \mathbb{C}$ then x is a sequence of complex numbers). We define the ℓ^1-*norm of x* to be

$$\|x\|_1 = \|(x_k)_{k \in \mathbb{N}}\|_1 = \sum_{k=1}^{\infty} |x_k|. \tag{2.4}$$

We will explore *norms* in detail in Chapter 3; for now we can simply regard the number $\|x\|_1$ as being a measure of the "size" of x in some sense. This size is finite for some sequences and infinite for others. For example, if $x = (1, 1, 1, \dots)$ then $\|x\|_1 = \infty$, but for the sequence $y = (1, \frac{1}{4}, \frac{1}{9}, \dots)$ we have

$$\|y\|_1 \;=\; \sum_{k=1}^{\infty} \frac{1}{k^2} \;=\; \frac{\pi^2}{6} \;<\; \infty.$$

(The equality $\sum \frac{1}{k^2} = \frac{\pi^2}{6}$, which we prove in Example 5.10.5, is known as *Euler's formula.*)

We say that a sequence $x = (x_k)_{k \in \mathbb{N}}$ is *absolutely summable*, or just *summable* for short, if $\|x\|_1 < \infty$. We let ℓ^1 denote the space of all summable sequences, i.e.,

$$\ell^1 \;=\; \left\{ x = (x_k)_{k \in \mathbb{N}} \;:\; \|x\|_1 = \sum_{k=1}^{\infty} |x_k| < \infty \right\}.$$

Here are some examples.

- $x = (1, 1, 1, \dots) \notin \ell^1$.
- $y = \left(\frac{1}{k^2}\right)_{k \in \mathbb{N}} = \left(1, \frac{1}{4}, \frac{1}{9}, \frac{1}{16}, \dots\right) \in \ell^1$.
- $q = \left(\frac{1}{k}\right)_{k \in \mathbb{N}} = \left(1, \frac{1}{2}, \frac{1}{3}, \frac{1}{4}, \dots\right) \notin \ell^1$.
- $s = (1, 0, -1, 0, 0, 1, 0, 0, 0, -1, , 0, 0, 0, 0, 1, \dots) \notin \ell^1$.
- $t = \left(-1, 0, \frac{1}{2}, 0, 0, -\frac{1}{3}, 0, 0, 0, \frac{1}{4}, 0, 0, 0, 0, -\frac{1}{5}, \dots\right) \notin \ell^1$.
- If $p > 1$, then $u = \left(\frac{(-1)^k}{k^p}\right)_{k \in \mathbb{N}} \in \ell^1$.
- $v = (2^{-k})_{k \in \mathbb{N}} = \left(\frac{1}{2}, \frac{1}{4}, \frac{1}{8}, \dots\right) \in \ell^1$.
- If z is a scalar with absolute value $|z| < 1$, then $w = \left(z^k\right)_{k \in \mathbb{N}} \in \ell^1$.

Observe that when we determine whether a sequence $x = (x_k)_{k \in \mathbb{N}}$ belongs to ℓ^1, only the absolute values of the components x_k are important. Precisely, a sequence $x = (x_k)_{k \in \mathbb{N}}$ belongs to ℓ^1 if and only if the sequence of absolute values $y = (|x_k|)_{k \in \mathbb{N}}$ belongs to ℓ^1.

If $x = (x_1, x_2, \dots)$ and $y = (y_1, y_2, \dots)$ are two elements of ℓ^1, then their sum

$$x + y \;=\; (x_k + y_k)_{k \in \mathbb{N}} \;=\; (x_1 + y_1, x_2 + y_2, \dots)$$

also belongs to ℓ^1 (why?). Likewise, if $c \in \mathbb{F}$ is scalar, then $cx = (cx_1, cx_2, \dots)$ is an element of ℓ^1. Thus ℓ^1 is closed under addition of vectors and under scalar multiplication, and the reader should verify that it follows from this that ℓ^1 is a vector space. Therefore we often call a sequence $x = (x_k)_{k \in \mathbb{N}}$ in ℓ^1 a "vector." Although x is a sequence of infinitely many numbers x_k, it is just one object in ℓ^1, and so we also often say that x is a "point" in ℓ^1.

We will define a distance between points in ℓ^1 that is analogous to the distance between points in $\mathbb{F}^d$ given by the metric defined in equation (2.2). Specifically, we will show that

$$d_1(x, y) = \|x - y\|_1 = \sum_{k=1}^{\infty} |x_k - y_k|, \qquad x, y \in \ell^1, \qquad (2.5)$$

is a metric on ℓ^1. By definition, $d_1(x, y)$ is a nonnegative finite number when x and y belong to ℓ^1, and the symmetry condition $d_1(x, y) = d_1(y, x)$ is immediate. To prove the uniqueness condition, suppose that $d_1(x, y) = 0$ for some sequences $x = (x_k)_{k \in \mathbb{N}}$ and $y = (y_k)_{k \in \mathbb{N}}$ in ℓ^1. Then

$$d_1(x, y) = \sum_{k=1}^{\infty} |x_k - y_k| = 0.$$

Since every term in this series is nonnegative, the only way that the series can equal zero is if $|x_k - y_k| = 0$ for every k. Hence $x_k = y_k$ for every k, so x and y are the same sequence. Finally, if x, y, and z are three sequences in ℓ^1, then

$$
\begin{aligned}
d_1(x, z) &= \sum_{k=1}^{\infty} |x_k - z_k| \\
&= \sum_{k=1}^{\infty} |(x_k - y_k) + (y_k - z_k)| \\
&\leq \sum_{k=1}^{\infty} \left(|x_k - y_k| + |y_k - z_k)| \right) \\
&= \sum_{k=1}^{\infty} |x_k - y_k| + \sum_{k=1}^{\infty} |y_k - z_k| \qquad \text{(because the series converge)} \\
&= d_1(x, y) + d_1(y, z).
\end{aligned}
$$

This shows that the Triangle Inequality holds, and therefore d_1 satisfies the properties of a metric on ℓ^1.

There are many metrics that we can define on ℓ^1, but unless we specifically state otherwise, we will always assume that the metric on ℓ^1 is the ℓ^1-*metric* given by equation (2.5).

Using the ℓ^1-metric, the distance from the point $v = (2^{-k})_{k \in \mathbb{N}}$ to the origin $0 = (0, 0, \dots)$ is

$$d_1(v, 0) = \sum_{k=1}^{\infty} |2^{-k} - 0| = 1.$$

How does the distance from $7v$ to 0 relate to the distance from v to 0 when we measure distance with the metric d_1? $\diamond$

The following particular elements of ℓ^1 appear so often that we introduce a name for them.

Notation 2.1.8 (The Standard Basis Vectors). Given an integer $n \in \mathbb{N}$, we let δ_n denote the sequence

$$\delta_n = (0, \ldots, 0, 1, 0, 0, \ldots),$$

where the 1 is in the nth component and the other components are all zero. We call δ_n the *nth standard basis vector*, and we refer to the family $\{\delta_n\}_{n \in \mathbb{N}}$ as the *sequence of standard basis vectors*, or simply the *standard basis*.

We have infinitely many different standard basis vectors $\delta_1, \delta_2, \ldots$, and each δ_n is itself an infinite sequence of scalars. For clarity of notation, we therefore use the alternative notation for sequences that was presented in equation (1.4). Specifically, we denote the kth component of δ_n by

$$\delta_n(k) = \begin{cases} 1, & \text{if } k = n, \\ 0, & \text{if } k \neq n, \end{cases}$$

and therefore write

$$\delta_n = \big(\delta_n(k)\big)_{k \in \mathbb{N}}. \qquad \Diamond$$

The word "basis" in Notation 2.1.8 should be regarded as just a name for now. In Section 4.5 we will consider the issue of in what sense $\{\delta_n\}_{n \in \mathbb{N}}$ is or is not a "basis" for ℓ^1 or other spaces.

Here is a different infinite-dimensional vector space whose elements are also infinite sequences of scalars.

Example 2.1.9 (The Space ℓ^∞). Given a sequence of scalars

$$x = (x_k)_{k \in \mathbb{N}} = (x_1, x_2, \ldots),$$

we define the *sup-norm* or *ℓ^∞-norm of x* to be

$$\|x\|_\infty = \|(x_k)_{k \in \mathbb{N}}\|_\infty = \sup_{k \in \mathbb{N}} |x_k|. \tag{2.6}$$

Note that $\|x\|_\infty$ could be infinite; in fact, x is a bounded sequence if and only if $\|x\|_\infty < \infty$. We let ℓ^∞ denote the space of all bounded sequences, i.e.,

$$\ell^\infty = \Big\{x = (x_k)_{k \in \mathbb{N}} : \|x\|_\infty = \sup_{k \in \mathbb{N}} |x_k| < \infty\Big\}.$$

This is a vector space, and the reader should show that

$$d_\infty(x, y) = \|x - y\|_\infty = \sup_{k \in \mathbb{N}} |x_k - y_k|, \qquad x, y \in \ell^\infty, \tag{2.7}$$

defines a metric on ℓ^∞ (see Problem 2.1.18). Unless we state otherwise, we always assume that this *ℓ^∞-metric* is the metric that we are using on ℓ^∞.

The two sets ℓ^1 and ℓ^∞ do not contain exactly the same vectors. For example, some of the vectors x, y, q, s, t, u, v, w discussed in Example 2.1.7 belong to ℓ^∞ but do not belong to ℓ^1 (which ones?), and therefore $\ell^\infty \neq \ell^1$. On the other hand, the reader should prove that every vector in ℓ^1 also belongs to ℓ^∞, and therefore $\ell^1 \subsetneq \ell^\infty$. In fact, ℓ^1 is a subspace and not just a subset of ℓ^∞, so we say that ℓ^1 is a *proper subspace* of ℓ^∞.   ◇

We have to use a supremum rather than a maximum in equation (2.7) because it is not true that every bounded sequence has a maximum. For example, if we define $x = (x_k)_{k \in \mathbb{N}}$ by $x_k = k/(k+1)$, then x is a bounded sequence and hence is a vector in ℓ^∞, but it has no largest component x_k.

Next we give another example of an infinite-dimensional normed space. In this example the elements of the vector space are *functions* rather than sequences. We will use the interval $[0,1]$ as the domain of the functions, but this is only for convenience—we could use any interval I as our domain in this example.

Example 2.1.10. We introduced the space $\mathcal{F}[0,1]$ in Example 1.10.5. This is the set of all scalar-valued functions whose domain is the closed interval $[0,1]$:

$$\mathcal{F}[0,1] = \{f : [0,1] \to \mathbb{F} : f \text{ is a function on } [0,1]\}.$$

A "point" or "vector" in $\mathcal{F}[0,1]$ is a function that maps $[0,1]$ into $\mathbb{F}$. For example, the function h defined by

$$h(t) = \begin{cases} t, & \text{if } t \text{ is rational,} \\ 0, & \text{if } t \text{ is irrational,} \end{cases} \tag{2.8}$$

is one point or vector in $\mathcal{F}[0,1]$. The zero vector in $\mathcal{F}[0,1]$ is the *zero function* 0, i.e., the function that takes the value zero at every point. We denote this function by the symbol 0. That is, 0 is the function defined by the rule $0(t) = 0$ for every $t \in [0,1]$.

Since $\mathcal{F}[0,1]$ is a set, it is possible to define a metric on it. For example, the discrete metric is one metric on $\mathcal{F}[0,1]$. However, we usually seek a metric that incorporates some information about the elements of the set that is useful for the application we have in mind. Often, this requires us to restrict our attention from all possible elements of a space to some appropriate subspace. For example, consider the subspace of $\mathcal{F}[0,1]$ that consists of the bounded functions:

$$\mathcal{F}_b[0,1] = \{f \in \mathcal{F}[0,1] : f \text{ is bounded}\}.$$

By definition, a function f is bounded if and only if there is some finite number M such that $|f(t)| \leq M$ for all t. Given any function f (not necessarily bounded), we call

$$\|f\|_u = \sup_{t \in [0,1]} |f(t)| \tag{2.9}$$

the *uniform norm* of f. The bounded functions are precisely those functions whose uniform norm is finite:

$$f \in \mathcal{F}_b[0,1] \quad \Longleftrightarrow \quad \|f\|_u < \infty.$$

Restricting our attention to just the bounded functions, we define the *uniform distance* between two bounded functions f and g to be

$$d_u(f,g) = \|f-g\|_u = \sup_{0 \leq t \leq 1} |f(t) - g(t)|. \qquad (2.10)$$

According to Problem 2.1.20, d_u is a metric on $\mathcal{F}_b[0,1]$. We call d_u the *uniform metric* on $\mathcal{F}_b[0,1]$. In some sense, $d_u(f,g)$ is the "maximum deviation" between $f(t)$ and $g(t)$ over all t. However, it is important to note that the supremum in equation (2.10) need not be achieved, so there need not be an actual maximum to the values of $|f(t) - g(t)|$ (if f and g are continuous then it is true that there is a maximum to the values of $|f(t) - g(t)|$ over the finite closed interval $[a,b]$, but this need not be the case if f or g is discontinuous).

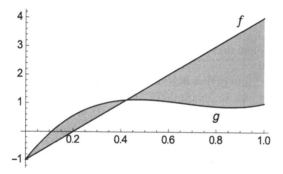

Fig. 2.1 Graphs of the functions $f(t) = 5t - 1$ and $g(t) = 9t^3 - 18t^2 + 11t - 1$. The region between the graphs is shaded.

To illustrate, consider the two continuous functions $f(t) = 5t - 1$ and $g(t) = 9t^3 - 18t^2 + 11t - 1$, whose graphs over the domain $[0,1]$ are shown in Figure 2.1. The distance between f and g, as measured by the uniform metric, is the supremum of $|f(t) - g(t)|$ over all values of $t \in [0,1]$. For these two specific functions f and g this supremum is achieved at the point $t = 1$, so

$$d_u(f,g) = \sup_{0 \leq t \leq 1} |f(t) - g(t)| = |f(1) - g(1)| = 4 - 1 = 3.$$

Thus, using this metric, the points f and g are 3 units apart in $\mathcal{F}_b[0,1]$. $\diamondsuit$

Here is a different metric on a space of functions. In order to define this metric we need to restrict our attention to functions that can be integrated, such as continuous functions.

Example 2.1.11. Let $C[0,1]$ denote the space of all continuous functions $f \colon [0,1] \to \mathbb{F}$. The sum of two continuous functions is continuous, and if f is continuous then so is cf for any scalar c. Therefore Lemma 1.10.6 implies that $C[0,1]$ is a subspace of $\mathcal{F}[0,1]$. In fact, since every continuous function on $[0,1]$ is bounded, $C[0,1]$ is a subspace of $\mathcal{F}_b[0,1]$. Consequently, if we restrict the uniform metric d_u to just the subspace $C[0,1]$ then we obtain a metric on $C[0,1]$.

However, d_u is not the only metric that we can define on $C[0,1]$. Recall that every continuous function f on $[0,1]$ is Riemann integrable, i.e., the Riemann integral $\int_0^1 f(t)\, dt$ exists for every continuous function f. The L^1-*norm* of f is the integral of its absolute value:

$$\|f\|_1 = \int_0^1 |f(t)|\, dt.$$

We use this to define the following important metric, which we call the L^1-*metric* on $C[0,1]$:

$$d_1(f,g) = \|f - g\|_1 = \int_0^1 |f(t) - g(t)|\, dt, \qquad f, g \in C[0,1]. \qquad (2.11)$$

The proof that d_1 satisfies the requirements of a metric on $C[0,1]$ is assigned as Problem 2.1.20.

Let $f(t) = 5t-1$ and $g(t) = 9t^3 - 18t^2 + 11t - 1$ be the two functions pictured in Figure 2.1. We saw in Example 2.1.10 that $d_u(f,g) = 3$. This is because the maximum value of $|f(t) - g(t)|$ over $t \in [0,1]$ is 3 units. The L^1-metric measures the distance between these functions in a different way. Instead of basing the distance on just the *supremum* of $|f(t) - g(t)|$, the L^1-metric takes all values of $|f(t) - g(t)|$ into account by computing the *integral* of $|f(t) - g(t)|$ over all t. For these two specific functions f and g, the L^1-metric is the *area* of the shaded region depicted in Figure 2.1:

$$
\begin{aligned}
d_1(f,g) &= \int_0^1 |f(t) - g(t)|\, dt \\
&= \int_0^1 |-9t^3 + 18t^2 - 6t|\, dt \\
&= \int_0^{1-\frac{\sqrt{3}}{3}} (9t^3 - 18t^2 + 6t)\, dt + \int_{1-\frac{\sqrt{3}}{3}}^1 (-9t^3 + 18t^2 - 6t)\, dt \\
&= \frac{16\sqrt{3} - 15}{12} \approx 1.059401\ldots.
\end{aligned}
$$

Thus f and g are fairly close as measured by the L^1-metric, being only a little over one unit apart. In contrast, f and g are a fairly distant 3 units apart as measured by the uniform metric. Neither of these values is any more "correct" than the other, but depending on our application, one of these ways of measuring may potentially be more *useful* than the other. ◇

Since the interval $[0, 1]$ is closed and finite, any continuous function on $[0, 1]$ must be bounded. Hence $C[0, 1]$ is a subset of $\mathcal{F}_b[0, 1]$, and so the uniform norm of every function $f \in C[0, 1]$ is finite. We could replace the interval $[0, 1]$ in Example 2.1.11 with any bounded closed interval $[a, b]$, but if we want to use any other type of interval I then we need to replace $C[0, 1]$ with the space $C_b(I)$ whose elements are *bounded* continuous functions $f : I \to \mathbb{F}$.

Problems

2.1.12. Determine which of the following is a metric on $\mathbb{R}$.
   (a) $d(x, y) = x - y$.
   (b) $d(x, y) = (x - y)^2$.
   (c) $d(x, y) = |x - y|^{1/2}$.
   (d) $d(x, y) = |x^2 - y^2|$.

2.1.13. Given a nonempty set X, show that the *discrete metric* defined in Example 2.1.3 is a metric on X.

2.1.14. Let X be a metric space. Prove that the following inequality (called the *Reverse Triangle Inequality*) holds for all x, y, $z \in X$:

$$\left| d(x, z) - d(y, z) \right| \leq d(x, y).$$

2.1.15. Determine whether the following sequences are elements of ℓ^1 or ℓ^∞:

$$x = \left(\frac{1}{k \ln(k+1)} \right)_{k \in \mathbb{N}}, \quad y = \left(\frac{1}{k \ln^2(k+1)} \right)_{k \in \mathbb{N}}, \quad z = \left(\frac{\ln^2(k+1)}{k^2} \right)_{k \in \mathbb{N}}.$$

2.1.16. Prove the following statements.
   (a) The functions d_1, d_2, and d_∞ from Example 2.1.6 are metrics on $\mathbb{F}^d$.
   (b) If $a, b \geq 0$, then $(a + b)^{1/2} \leq a^{1/2} + b^{1/2}$.
   (c) The following is a metric on $\mathbb{F}^d$:

$$d_{1/2}(x, y) = |x_1 - y_1|^{1/2} + \cdots + |x_d - y_d|^{1/2}, \qquad x, y \in \mathbb{F}^d.$$

   (d) If p is 1, 2, or ∞, then $d_p(cx, 0) = |c| \, d_p(x, 0)$ for all vectors $x \in \mathbb{F}^d$ and all scalars $c \in \mathbb{F}$, but this equality can fail when $p = 1/2$.

2.1.17. Given a fixed finite dimension $d \geq 1$, prove that there exist finite, positive numbers A_d, B_d such that the following inequality holds simultaneously for every $x \in \mathbb{F}^d$:

$$A_d \|x\|_\infty \leq \|x\|_1 \leq B_d \|x\|_\infty. \tag{2.12}$$

The numbers A_d and B_d can depend on the dimension, but they must be independent of the choice of vector x. What are the *optimal values* for A_d and B_d, i.e., what are the largest number A_d and the smallest number B_d such that equation (2.12) holds for every $x \in \mathbb{F}^d$?

2.1.18. Prove the following statements.

(a) The function d_∞ defined in equation (2.7) is a metric on ℓ^∞.

(b) $\|x\|_\infty \leq \|x\|_1$ for every $x \in \ell^1$, where d_1 is as defined in equation (2.4).

(c) There does not exist a finite constant $B > 0$ such that the inequality $\|x\|_1 \leq B \|x\|_\infty$ holds simultaneously for *every* $x \in \ell^1$.

2.1.19. Define a space of infinite sequences ℓ^2 and a corresponding metric d_2 that is analogous to the Euclidean metric on $\mathbb{F}^d$ defined in equation (2.1). Determine which of the sequences x, y, q, s, t, u, v, w from Example 2.1.7 belong to ℓ^2. Is $\ell^1 \subseteq \ell^2$? Is $\ell^2 \subseteq \ell^1$? Is $\ell^\infty \subseteq \ell^2$? Is $\ell^2 \subseteq \ell^\infty$?

2.1.20. Prove the following statements.

(a) The function d_u defined in equation (2.10) is a metric on $\mathcal{F}_b[0,1]$, and is therefore also a metric on the subspace $C[0,1]$.

(b) The function d_1 defined in equation (2.11) is a metric on $C[0,1]$.

(c) $\|f\|_1 \leq \|f\|_u$ for every $f \in C[0,1]$.

(d) There is a function $f \in C[0,1]$ such that $\|f\|_u = 1000$ yet $\|f\|_1 = 1$.

(e) There does not exist a finite constant $B > 0$ such that the inequality $\|f\|_u \leq B \|f\|_1$ holds simultaneously for *every* $f \in C[0,1]$.

2.2 Convergence and Completeness

If d is a metric on a set X, then the number $d(x, y)$ represents the distance from the point x to the point y with respect to this metric. We will say that points x_n are *converging* to a point x if the distance from x_n to x shrinks to zero as n increases. This is made precise in the following definition.

Definition 2.2.1 (Convergent Sequence). Let $\{x_n\}_{n \in \mathbb{N}}$ be a sequence of points in a metric space X. We say that $\{x_n\}_{n \in \mathbb{N}}$ is a *convergent sequence* if there is a point $x \in X$ such that

$$\lim_{n\to\infty} d(x_n, x) = 0.$$

That is, for every $\varepsilon > 0$ there must exist some integer $N > 0$ such that

$$n \geq N \qquad \Longrightarrow \qquad d(x_n, x) < \varepsilon.$$

In this case, we say that the points x_n *converge to* x or that x *is the limit of the points* x_n, and we denote this by writing $x_n \to x$ as $n \to \infty$, or simply $x_n \to x$ for short. ◇

Convergence implicitly depends on the choice of metric for X, so if we want to emphasize that we are using a particular metric, we may write $x_n \to x$ *with respect to the metric* d.

Example 2.2.2. Let $X = C[0, 1]$, the space of continuous functions on the domain $[0, 1]$. One metric on $C[0, 1]$ is the uniform metric defined in equation (2.10), and another is the L^1-metric defined in equation (2.11). For each integer $n \geq 0$, let $p_n(t) = t^n$. With respect to the L^1-metric, the distance from p_n to the zero function is the area between the graphs of these two functions, which is

$$d_1(p_n, 0) = \|p_n - 0\|_1 = \int_0^1 |t^n - 0|\, dt = \int_0^1 t^n\, dt = \frac{1}{n+1}.$$

This distance decreases to 0 as n increases, so p_n *converges to* 0 *with respect to the L^1-metric* (consider Figure 2.2). However, if we change the metric then the meaning of distance and convergence changes. For example, if we instead measure distance using the uniform metric, then for every n we have

$$d_u(p_n, 0) = \|p_n - 0\|_u = \sup_{0 \leq t \leq 1} |t^n - 0| = \sup_{0 \leq t \leq 1} t^n = 1.$$

Using this distance function, the two vectors p_n and 0 are 1 unit apart, no matter how large we choose n. The distance between p_n and 0 does not decrease with n, so p_n *does not* converge to 0 with respect to the uniform metric. ◇

Closely related to *convergence* is the idea of a *Cauchy sequence*, which is a sequence of points $\{x_n\}_{n\in\mathbb{N}}$ such that the distance $d(x_m, x_n)$ between two points x_m and x_n decreases to zero as m and n increase.

Definition 2.2.3 (Cauchy Sequence). Let X be a metric space. A sequence of points $\{x_n\}_{n\in\mathbb{N}}$ in X is a *Cauchy sequence* if for every $\varepsilon > 0$ there exists an integer $N > 0$ such that

$$m, n \geq N \qquad \Longrightarrow \qquad d(x_m, x_n) < \varepsilon. \qquad ◇$$

Thus, if $\{x_n\}_{n\in\mathbb{N}}$ is a *convergent* sequence, then there exists a point x such that x_n gets closer and closer to x as n increases, while if $\{x_n\}_{n\in\mathbb{N}}$

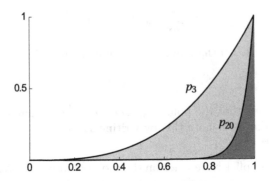

Fig. 2.2 Graphs of the functions $p_3(t) = t^3$ and $p_{20}(t) = t^{20}$. The area of the region under the graph of p_3 is $1/4$, while the area of the region under the graph of p_{20} is $1/21$.

is a *Cauchy* sequence, then the elements x_m, x_n of the sequence get closer and closer to each other as m and n increase. For simplicity, we often just write that "$\{x_n\}_{n \in \mathbb{N}}$ is convergent" if there exists a vector $x \in X$ such that $x_n \to x$, and we likewise use the phrase "$\{x_n\}_{n \in \mathbb{N}}$ is Cauchy" as a shorthand for "$\{x_n\}_{n \in \mathbb{N}}$ is a Cauchy sequence."

By applying the Triangle Inequality, we see that *every convergent sequence is Cauchy*.

Lemma 2.2.4 (Convergent Implies Cauchy). *If $\{x_n\}_{n \in \mathbb{N}}$ is a convergent sequence in a metric space X, then $\{x_n\}_{n \in \mathbb{N}}$ is a Cauchy sequence in X.*

Proof. Assume that $x_n \to x$. If we fix an $\varepsilon > 0$, then, by the definition of a convergent sequence, there exists some $N > 0$ such that $d(x, x_n) < \varepsilon/2$ for all $n \geq N$. Consequently, if $m, n \geq N$, then the Triangle Inequality implies that

$$d(x_m, x_n) \leq d(x_m, x) + d(x, x_n) < \frac{\varepsilon}{2} + \frac{\varepsilon}{2} = \varepsilon.$$

Therefore $\{x_n\}_{n \in \mathbb{N}}$ is Cauchy. $\square$

Example 2.2.5. Lemma 2.2.4 tells us that *every convergent sequence must be Cauchy*. The logically equivalent contrapositive formulation of this statement is that *a sequence that is not Cauchy cannot converge*. To illustrate, consider the sequence of standard basis vectors $\{\delta_n\}_{n \in \mathbb{N}}$ in the space ℓ^1. If $m < n$, then

$$\delta_m - \delta_n = (0, \ldots, 0, 1, 0, \ldots, 0, -1, 0, 0, \ldots),$$

where the 1 is in the mth component and the -1 is in the nth component. Using the metric for ℓ^1 defined in equation (2.5), we compute that

$$
\begin{aligned}
d_1(\delta_m, \delta_n) &= \|\delta_m - \delta_n\|_1 \\
&= 0 + \cdots + 0 + 1 + 0 + \cdots + 0 + |-1| + 0 + 0 + \cdots \\
&= 2.
\end{aligned}
$$

Thus the distance between δ_m and δ_n is 2 whenever $m \neq n$. Consequently, if $\varepsilon < 2$, then we can *never* have $d_1(\delta_m, \delta_n) < \varepsilon$, no matter how large we take m and n. Hence $\{\delta_n\}_{n \in \mathbb{N}}$ is not a Cauchy sequence in ℓ^1, and therefore it *cannot* converge to *any* point in ℓ^1. That is, there is no vector $x \in \ell^1$ such that $d_1(x, \delta_n) = \|x - \delta_n\|_1 \to 0$ as $n \to \infty$ (compare Problem 2.2.18, which asks for a *direct* proof that there is no vector $x \in \ell^1$ that δ_n converges to as $n \to \infty$). $\diamondsuit$

What happens in other metric spaces that contain the standard basis? The reader should show that $\{\delta_n\}_{n \in \mathbb{N}}$ is not a Cauchy sequence in ℓ^∞. Do you think there is *any* normed space of sequences X that contains the standard basis and has a metric such that $\{\delta_n\}_{n \in \mathbb{N}}$ is Cauchy in X? (Compare Problem 2.2.26.)

As we have seen, every convergent sequence is Cauchy, and a sequence that is not Cauchy cannot converge. This does not tell us whether Cauchy sequences converge. Here is an example of a Cauchy sequence that converges, but following this we will give an example of a Cauchy sequence that does not converge.

Example 2.2.6 (A Cauchy Sequence That Converges). For this example we take $X = \mathbb{F}$ (and, as discussed in Example 2.1.5, we implicitly assume that the metric on $\mathbb{F}$ is the *standard metric* $d(x, y) = |x - y|$). Let $x_1, x_2, \ldots$ be the following rational numbers, whose decimal representations are based on the decimal expansion of the irrational number $\pi = 3.14159265\ldots$,

$$x_1 = 3.1, \quad x_2 = 3.14, \quad x_3 = 3.141, \quad x_4 = 3.1415, \quad \ldots.$$

If $m < n$, then $|x_m - x_n| < 10^{-m}$. Choose any $\varepsilon > 0$, and let N be an integer large enough that $10^{-N} < \varepsilon$. Then for all $m, n \geq N$ we have

$$|x_m - x_n| < 10^{-m} \leq 10^{-N} < \varepsilon.$$

Therefore $\{x_n\}_{n \in \mathbb{N}}$ is a Cauchy sequence of scalars. Furthermore, this sequence converges to a point in $X = \mathbb{F}$, because $x_n \to \pi$ as $n \to \infty$. $\diamondsuit$

Our next example shows that in at least some metric spaces, *there can exist Cauchy sequences that do not converge.*

Example 2.2.7 (A Cauchy Sequence That Does Not Converge). If d is a metric on a set X and $Y \subseteq X$, then by restricting d to Y we obtain a metric on Y (see Remark 2.1.2). In particular, since $X = \mathbb{F}$ is a metric space with respect to the standard metric $d(x, y) = |x - y|$ and since the set of rational numbers $Y = \mathbb{Q}$ is a subset of $\mathbb{F}$, the set of rational numbers is also a metric space with respect to $d(x, y) = |x - y|$ (where we now only take $x, y \in \mathbb{Q}$). This metric space $Y = \mathbb{Q}$ will be the setting for this example.

Let $\{x_n\}_{n \in \mathbb{N}}$ be the same sequence that was defined in Example 2.2.6. This is a sequence of rational numbers, so it is a sequence in the metric space $\mathbb{Q}$.

Also, by the same reasoning as before we have $|x_m - x_n| < 10^{-m}$ whenever $m < n$, so it follows that $\{x_n\}_{n\in\mathbb{N}}$ is a Cauchy sequence in $\mathbb{Q}$. However, there is no number x *in the set* $\mathbb{Q}$ that x_n converges to! The sequence does converge to π, but π is irrational, so it does not belong to our metric space $Y = \mathbb{Q}$. Therefore $\{x_n\}_{n\in\mathbb{N}}$ *is not a convergent sequence in the metric space* $\mathbb{Q}$ because there is no rational point $x \in \mathbb{Q}$ such that $x_n \to x$. $\Diamond$

To emphasize, for a sequence to be convergent in a particular space X it must converge *to an element of* X and not just to an element of some larger space. Example 2.2.7 shows that there are sequences in $\mathbb{Q}$ that are Cauchy but do not converge to an element of $\mathbb{Q}$. The following theorem (which is a consequence of Axiom 1.6.2) states that every Cauchy sequence in the scalar field $\mathbb{F}$ does converge to an element of $\mathbb{F}$ with respect to the standard metric. We assign the proof as Problem 2.2.27 (also compare [Rud76, Thm. 3.11]).

Theorem 2.2.8 (All Cauchy Sequences in $\mathbb{F}$ Converge). *If $\{x_n\}_{n\in\mathbb{N}}$ is a Cauchy sequence in $\mathbb{F}$, then there exists some $x \in \mathbb{F}$ such that $x_n \to x$.* $\Diamond$

In summary, *some* metric spaces have the property that every Cauchy sequence in the space converges to an element of the space. Since we can test for Cauchyness without having the limit vector x in hand, this can be very useful. We give such spaces the following name.

Definition 2.2.9 (Complete Metric Space). Let X be a metric space. If every Cauchy sequence in X converges to an element of X, then we say that X is *complete*. $\Diamond$

Remark 2.2.10. The reader should be aware that the term "complete" is heavily overused and has a number of distinct mathematical meanings. In particular, the notion of a *complete space* as given in Definition 2.2.9 is quite different from the notion of a *complete sequence* that will be introduced in Definition 4.4.4. $\Diamond$

We will show that ℓ^1 is a complete metric space. Recall that the metric on ℓ^1 is defined by $d(x,y) = \|x-y\|_1$, so $x_n \to x$ (convergence in ℓ^1) means that $d_1(x, x_n) = \|x - x_n\|_1 \to 0$ as $n \to \infty$. To emphasize that the convergence is taking place with respect to this metric, we often say that $x_n \to x$ *in ℓ^1- norm* if $\|x - x_n\|_1 \to 0$ (the use of the word "norm" instead of "metric" in this context is largely motivated by Chapter 3, where we will focus on norms instead of metrics).

Theorem 2.2.11 (ℓ^1 Is Complete). *If $\{x_n\}_{n\in\mathbb{N}}$ is a Cauchy sequence in ℓ^1, then there exists a vector $x \in \ell^1$ such that $x_n \to x$ in ℓ^1-norm.*

Proof. Assume that $\{x_n\}_{n\in\mathbb{N}}$ is a Cauchy sequence in ℓ^1. Each x_n is a vector in ℓ^1, which means that x_n is an infinite sequence of scalars whose components are summable. For notational clarity, in this proof we will use the alternative notation for writing components that was introduced in equation (1.4). Specifically, we will write x_n as

$$x_n = \big(x_n(1), x_n(2), \dots\big) = \big(x_n(k)\big)_{k \in \mathbb{N}}.$$

That is, $x_n(k)$ will denote the kth component of the vector x_n.

Choose any $\varepsilon > 0$. Then, by the definition of a Cauchy sequence, there is an integer $N > 0$ such that $d_1(x_m, x_n) = \|x_m - x_n\|_1 < \varepsilon$ for all $m, n \geq N$. Therefore, if we fix a particular index $k \in \mathbb{N}$, then for all $m, n \geq N$ we have

$$|x_m(k) - x_n(k)| \leq \sum_{j=1}^{\infty} |x_m(j) - x_n(j)| = \|x_m - x_n\|_1 < \varepsilon.$$

Thus, with k fixed, $\big(x_n(k)\big)_{n \in \mathbb{N}}$ is a Cauchy sequence of *scalars*, and therefore, by Theorem 2.2.8, it must converge to some scalar. Let $x(k)$ be the limit of this sequence, i.e., define

$$x(k) = \lim_{n \to \infty} x_n(k). \tag{2.13}$$

Then let x be the sequence $x = \big(x(k)\big)_{k \in \mathbb{N}} = \big(x(1), x(2), \dots\big)$.

Now, for each fixed integer k, as $n \to \infty$ the scalar $x_n(k)$, which is the kth component of x_n, converges to the scalar $x(k)$, which is the kth component of x. We therefore say that x_n *converges componentwise to* x (see the illustration in Figure 2.3). However, this is not enough. We need to show that $x \in \ell^1$, and that x_n *converges to* x *in* ℓ^1-*norm*.

$$
\begin{array}{llllll}
x_1 = & (x_1(1), & x_1(2), & x_1(3), & x_1(4), & \dots) \quad \text{components of } x_1 \\
x_2 = & (x_2(1), & x_2(2), & x_2(3), & x_2(4), & \dots) \quad \text{components of } x_2 \\
x_3 = & (x_3(1), & x_3(2), & x_3(3), & x_3(4), & \dots) \quad \text{components of } x_3 \\
& \vdots & \vdots & \vdots & \vdots & \vdots \\
& \downarrow & \downarrow & \downarrow & \downarrow & \downarrow \\
x = & (x(1), & x(2), & x(3), & x(4), & \dots) \quad \text{components of } x
\end{array}
$$

Fig. 2.3 Illustration of componentwise convergence. For each k, the kth component of x_n converges to the kth component of x.

To prove that $x_n \to x$ in ℓ^1-norm, choose any $\varepsilon > 0$. Since $\{x_n\}_{n \in \mathbb{N}}$ is Cauchy, there is an $N > 0$ such that $\|x_m - x_n\|_1 < \varepsilon$ for all $m, n \geq N$. Choose any particular $n \geq N$, and fix an integer $M > 0$. Then, since M is finite, we can use Lemma 1.7.2 to compute that

$$\sum_{k=1}^{M} |x(k) - x_n(k)| = \sum_{k=1}^{M} \lim_{m \to \infty} |x_m(k) - x_n(k)|$$

$$= \lim_{m \to \infty} \sum_{k=1}^{M} |x_m(k) - x_n(k)| \qquad \text{(by Lemma 1.7.2)}$$

$$\leq \lim_{m \to \infty} \sum_{k=1}^{\infty} |x_m(k) - x_n(k)| \qquad \text{(all terms nonnegative)}$$

$$= \lim_{m \to \infty} \|x_m - x_n\|_1$$

$$\leq \varepsilon.$$

Since this is true for every M, we conclude that

$$d_1(x_n, x) = \|x - x_n\|_1$$

$$= \sum_{k=1}^{\infty} |x(k) - x_n(k)|$$

$$= \lim_{M \to \infty} \sum_{k=1}^{M} |x(k) - x_n(k)|$$

$$\leq \varepsilon. \qquad (2.14)$$

Even though we do not know yet that $x \in \ell^1$, this computation implies that the vector $y = x - x_n$ has finite ℓ^1-norm (because $\|y\|_1 = \|x - x_n\|_1 \leq \varepsilon$, which is finite). Therefore y belongs to ℓ^1. Since ℓ^1 is closed under addition and since both y and x_n belong to ℓ^1, it follows that their sum $y + x_n$ belongs to ℓ^1. But $x = y + x_n$, so $x \in \ell^1$. Thus our "candidate sequence" x is in ℓ^1. Further, equation (2.14) establishes that $d_1(x_n, x) \leq \varepsilon$ for all $n \geq N$, so we have shown that x_n converges to x in ℓ^1-norm as $n \to \infty$. Therefore ℓ^1 is complete. $\square$

We mentioned componentwise convergence in the proof of Theorem 2.2.11. We make that notion precise in the following definition.

Definition 2.2.12 (Componentwise Convergence). For each $n \in \mathbb{N}$ let $x_n = \big(x_n(k)\big)_{k \in \mathbb{N}}$ be a sequence of scalars, and let $x = \big(x(k)\big)_{k \in \mathbb{N}}$ be another sequence of scalars. We say that x_n *converges componentwise* to x if

$$\forall\, k \in \mathbb{N}, \quad \lim_{n \to \infty} x_n(k) = x(k). \qquad \Diamond$$

The proof of Theorem 2.2.11 shows that if $x_n \to x$ in ℓ^1-norm (i.e., if $\|x - x_n\|_1 \to 0$), then x_n converges componentwise to x. However, the converse statement fails in general. For example, write the components of the nth standard basis vector as $\delta_n = (\delta_n(k))_{k \in \mathbb{N}}$, and let $0 = (0, 0, \dots)$ be the zero sequence. If we fix any particular index k, then $\delta_n(k) = 0$ for all $n > k$. Therefore

$$\lim_{n \to \infty} \delta_n(k) = 0.$$

Thus, for each fixed k the kth component of δ_n converges to the kth component of the zero vector as $n \to 0$ (see Figure 2.4). Consequently δ_n converges componentwise to the zero vector. However, we showed in Example 2.2.5 that δ_n does not converge to the zero vector in ℓ^1-norm (nor does it converge to any other vector in ℓ^1-norm). Thus componentwise convergence does not imply convergence in ℓ^1 in general.

$$
\begin{array}{llllll}
\delta_1 = & (1, & 0, & 0, & 0, & \ldots) \qquad \text{components of } \delta_1 \\
\delta_2 = & (0, & 1, & 0, & 0, & \ldots) \qquad \text{components of } \delta_2 \\
\delta_3 = & (0, & 0, & 1, & 0, & \ldots) \qquad \text{components of } \delta_3 \\
\delta_4 = & (0, & 0, & 0, & 1, & \ldots) \qquad \text{components of } \delta_4 \\
& \vdots & \vdots & \vdots & \vdots & \qquad\quad \vdots \\
& \downarrow & \downarrow & \downarrow & \downarrow & \qquad\quad \downarrow \\
0 = & (0, & 0, & 0, & 0, & \ldots) \qquad \text{components of the zero sequence}
\end{array}
$$

Fig. 2.4 For each k, the kth component of δ_n converges to the kth component of the zero sequence. However, $\|\delta_n - 0\|_1 = 1 \nrightarrow 0$, so δ_n does not converge to the zero sequence in ℓ^1-norm.

Problem 2.2.19 asks the reader to prove that ℓ^∞, like ℓ^1, is a complete metric space. Here is an example of an infinite-dimensional vector space that is not complete.

Example 2.2.13 (c_{00} Is Incomplete). Let c_{00} be the space of all sequences of scalars that have only finitely many nonzero components:

$$c_{00} = \Big\{ x = (x_1, \ldots, x_N, 0, 0, \ldots) \; : \; N > 0, \; x_1, \ldots, x_N \in \mathbb{F} \Big\}. \qquad (2.15)$$

A vector $x \in c_{00}$ is sometimes called a "finite sequence" (note that x does have infinitely many components, but only finitely many of these can be nonzero). Since c_{00} is a subset of ℓ^1, it is a metric space with respect to the ℓ^1-metric defined in equation (2.5). Moreover, c_{00} is closed under both vector addition and scalar multiplication (why?), so c_{00} is a *subspace* of ℓ^1 and not just a *subset*.

For each $n \in \mathbb{N}$, let x_n be the sequence

$$x_n = (2^{-1}, \ldots, 2^{-n}, 0, 0, 0, \ldots),$$

and consider the sequence of vectors $\{x_n\}_{n \in \mathbb{N}}$. This sequence is contained in both c_{00} and ℓ^1. If $m < n$, then

$$x_m - x_n = (0, \ldots, 0, 2^{-m-1}, 2^{-m-2}, \ldots, 2^{-n}, 0, 0, \ldots),$$

so

$$d_1(x_n, x_m) = \|x_n - x_m\|_1 = \sum_{k=m+1}^{n} 2^{-k} < \sum_{k=m+1}^{\infty} 2^{-k} = 2^{-m}.$$

The reader should verify that this implies that $\{x_n\}_{n\in\mathbb{N}}$ is a Cauchy sequence with respect to d_1. If we take our metric space to be $X = \ell^1$, then this sequence does converge. In fact x_n converges in ℓ^1-norm to the sequence

$$x = (2^{-1}, 2^{-2}, \dots) = (2^{-k})_{k\in\mathbb{N}} \in \ell^1,$$

because

$$d_1(x, x_n) = \|x - x_n\|_1 = \sum_{k=n+1}^{\infty} 2^{-k} = 2^{-n} \to 0 \quad \text{as } n \to \infty.$$

However, this vector x does not belong to c_{00}, and there is no other sequence $y \in c_{00}$ such that $x_n \to y$ (why not?). Therefore, if we take our metric space to be $X = c_{00}$, then $\{x_n\}_{n\in\mathbb{N}}$ is not a convergent sequence *in the space* c_{00}. But it is Cauchy, so we conclude that c_{00} is not a complete metric space (with respect to the metric d_1). $\square$

There is an interesting difference between the two incomplete spaces discussed in Examples 2.2.7 and 2.2.13. In Example 2.2.7, the metric space is the set of rationals $\mathbb{Q}$, which is *not* a vector space over the real field or the complex field (because it is not closed under scalar multiplication, since *scalar* means a real or complex number). In contrast, the metric space in Example 2.2.13 is c_{00}, which *is* a vector space with respect to the real field and the complex field (depending on whether we choose $\mathbb{F} = \mathbb{R}$ or $\mathbb{F} = \mathbb{C}$). Both $\mathbb{Q}$ and c_{00} are incomplete, i.e., they contain Cauchy sequences that do not converge in the space with respect to their given metrics, but only one of them is a vector space over $\mathbb{F}$.

Problems

2.2.14. Let $\mathcal{E} = \{\delta_n\}_{n\in\mathbb{N}}$ be the sequence of standard basis vectors. Prove that the finite linear span of $\mathcal{E}$ is precisely the space c_{00} defined in equation (2.15), i.e., $\text{span}(\mathcal{E}) = c_{00}$. Why does the constant sequence $x = (1, 1, 1, \dots)$ not belong to $\text{span}(\mathcal{E})$?

2.2.15. Let X be a metric space, and suppose that $x_n \to x$ and $y_n \to y$ in X. Prove that $d(x_n, y_n) \to d(x, y)$ as $n \to \infty$.

2.2.16. Given points x_n and x in a metric space X, prove that the following four statements are equivalent.

(a) $x_n \to x$, i.e., for each $\varepsilon > 0$ there exists an integer $N > 0$ such that

$$n \geq N \quad \Longrightarrow \quad d(x_n, x) < \varepsilon.$$

(b) For each $\varepsilon > 0$ there exists an integer $N > 0$ such that

$$n \geq N \quad \Longrightarrow \quad d(x_n, x) \leq \varepsilon.$$

(c) For each $\varepsilon > 0$ there exists an integer $N > 0$ such that

$$n > N \quad \Longrightarrow \quad d(x_n, x) < \varepsilon.$$

(d) For each $\varepsilon > 0$ there exists an integer $N > 0$ such that

$$n > N \quad \Longrightarrow \quad d(x_n, x) \leq \varepsilon.$$

Formulate and prove an analogous set of equivalent statements for Cauchy sequences.

2.2.17. Let p_n be the function whose rule is $p_n(t) = t^n$, $t \in [0, 1]$.

(a) Prove directly that $\{p_n\}_{n \in \mathbb{N}}$ is a Cauchy sequence in $C[0, 1]$ with respect to the L^1-metric.

(b) Prove directly that $\{p_n\}_{n \in \mathbb{N}}$ is a not Cauchy sequence in $C[0, 1]$ with respect to the uniform metric.

2.2.18. Let $x = (x_k)_{k \in \mathbb{N}}$ be any sequence in ℓ^1. Give an explicit formula for $\|x - \delta_n\|_1$ in terms of n and the x_k, and show that $\|x - \delta_n\|_1 \not\to 0$ as $n \to \infty$. Conclude that the sequence $\{\delta_n\}_{n \in \mathbb{N}}$ does not converge to x, no matter which $x \in \ell^1$ we choose.

2.2.19. (a) Show that ℓ^∞ is complete with respect to the metric d_∞ defined in equation (2.7).

(b) Show that c_{00} is not complete with respect to the metric d_∞.

(c) Find a proper subspace of ℓ^∞ that is complete with respect to d_∞ (other than the trivial subspace $\{0\}$). Can you find an *infinite-dimensional* subspace that is complete?

2.2.20. Let X be a metric space with metric d_X, and let Y be a metric space with metric d_Y. Given $x_1, x_2 \in X$ and $y_1, y_2 \in Y$, define the distance from the point $(x_1, y_1) \in X \times Y$ to the point $(x_2, y_2) \in X \times Y$ to be

$$d\big((x_1, y_1), (x_2, y_2)\big) = d_X(x_1, x_2) + d_Y(y_1, y_2).$$

Prove that d is a metric on the Cartesian product $X \times Y$, and show that

$$(x_n, y_n) \to (x, y) \text{ in } X \times Y \quad \Longleftrightarrow \quad x_n \to x \text{ in } X \text{ and } y_n \to y \text{ in } Y.$$

2.2.21. Suppose that $\{x_n\}_{n\in\mathbb{N}}$ is a Cauchy sequence in a metric space X, and suppose there exists a subsequence $\{x_{n_k}\}_{k\in\mathbb{N}}$ that converges to a point $x \in X$, i.e., $x_{n_k} \to x$ as $k \to \infty$. Prove that $x_n \to x$ as $n \to \infty$.

2.2.22. Let $X = \mathbb{F}$ and set $x_n = (-1)^n$. Does every subsequence of $(x_n)_{n\in\mathbb{N}}$ converge? Do any subsequences of $(x_n)_{n\in\mathbb{N}}$ converge? Can you determine exactly which subsequences of $(x_n)_{n\in\mathbb{N}}$ converge?

2.2.23. Let $\{x_n\}_{n\in\mathbb{N}}$ be a sequence of points in a metric space X.

(a) Prove that if $d(x_n, x_{n+1}) < 2^{-n}$ for every $n \in \mathbb{N}$, then $\{x_n\}_{n\in\mathbb{N}}$ is a Cauchy sequence.

(b) Prove that if $\{x_n\}_{n\in\mathbb{N}}$ is Cauchy, then there exists a subsequence $\{x_{n_k}\}_{k\in\mathbb{N}}$ such that $d(x_{n_k}, x_{n_{k+1}}) < 2^{-k}$ for each $k \in \mathbb{N}$.

(c) Give an example of a metric space X and a sequence $\{x_n\}_{n\in\mathbb{N}}$ such that $d(x_n, x_{n+1}) < \frac{1}{n}$ for each $n \in \mathbb{N}$, yet $\{x_n\}_{n\in\mathbb{N}}$ is not Cauchy and does not converge.

2.2.24. (a) Let $\{x_n\}_{n\in\mathbb{N}}$ be a sequence in a metric space X, and fix a point $x \in X$. Suppose that every subsequence $\{y_n\}_{n\in\mathbb{N}}$ of $\{x_n\}_{n\in\mathbb{N}}$ has a subsequence $\{z_n\}_{n\in\mathbb{N}}$ of $\{y_n\}_{n\in\mathbb{N}}$ such that $z_n \to x$. Prove that $x_n \to x$.

(b) Give an example of a metric space X and a sequence $\{x_n\}_{n\in\mathbb{N}}$ such that every subsequence $\{y_n\}_{n\in\mathbb{N}}$ has a convergent subsequence $\{z_n\}_{n\in\mathbb{N}}$, yet $\{x_n\}_{n\in\mathbb{N}}$ does not converge. What hypothesis of part (a) does your sequence $\{x_n\}_{n\in\mathbb{N}}$ not satisfy?

2.2.25. Let X be a metric space. Extend the definition of convergence to families indexed by a real parameter by declaring that if $x \in X$ and $x_t \in X$ for t in some interval $(-c, c)$, where $c > 0$, then $x_t \to x$ as $t \to 0$ if for every $\varepsilon > 0$ there exists a $\delta > 0$ such that $d(x_t, x) < \varepsilon$ whenever $|t| < \delta$. Show that $x_t \to x$ as $t \to 0$ if and only if $x_{t_k} \to x$ for every sequence of real numbers $\{t_k\}_{k\in\mathbb{N}}$ such that $t_k \to 0$.

2.2.26. Let $w = (w_k)_{k\in\mathbb{N}}$ be a fixed sequence of positive scalars $w_k > 0$. Given sequences $x = (x_k)_{k\in\mathbb{N}}$ and $y = (y_k)_{k\in\mathbb{N}}$, define

$$d_w(x, y) = \sum_{k=1}^{\infty} |x_k - y_k| w_k.$$

(a) Prove that d_w is a metric on ℓ^1 if and only if $w = (w_k)_{k\in\mathbb{N}}$ is a bounded sequence. In particular, if $w_k = 1$ for every k, then $d_w = d_1$.

(b) Is there any way to choose the scalars w_k so that d_w is a metric on ℓ^1 and the standard basis $\{\delta_n\}_{n\in\mathbb{N}}$ is a convergent sequence *with respect to* d_w?

2.2.27. Challenge: Use Axiom 1.6.2 to prove Theorem 2.2.8 (this is not easy).

2.3 Topology in Metric Spaces

An *open ball* in a metric space is the set of all points that lie within a fixed distance from a central point x. These sets will appear frequently throughout the text, so we introduce a notation to represent them.

Definition 2.3.1 (Open Ball). Let X be a metric space. Given a point $x \in X$ and given a positive number $r > 0$, the *open ball in X with radius r centered at x* is

$$B_r(x) = \{y \in X : d(x, y) < r\}. \qquad \diamond \qquad (2.16)$$

We emphasize that "$r > 0$" means that r is a positive *real number.* In particular, every ball has a *finite* radius. We do not allow a ball to have an infinite radius.

We also emphasize that the definition of a ball in a given space *implicitly depends on the choice of metric!* For example, if $X = \mathbb{R}$ and we are using the standard metric on $\mathbb{R}$, then $B_r(x) = (x - r, x + r)$, the open interval of radius r centered at x. If we use a different metric on $\mathbb{R}$, then the meaning of an open ball can be quite different. What is the open ball $B_r(x)$ when we use the *discrete metric* on $\mathbb{R}$ defined in Problem 2.1.13?

In Figure 2.5 we show the unit ball $B_1(0)$ in $\mathbb{R}^2$ with respect to each of the metrics d_1, d_2, and d_∞ that were introduced in Example 2.1.6, along with the unit ball with respect to the metric $d_{1/2}$ that was defined in Problem 2.1.16. While the colloquial meaning of the word "ball" suggests a sphere or disk, we can see in the figure that only the unit ball that is defined with respect to the Euclidean metric is "spherical" in the ordinary sense. Still, all of the sets depicted in Figure 2.5 are open balls in the sense of Definition 2.3.1, each corresponding to a different choice of metric on $\mathbb{R}^2$. Although the actual shape of a ball depends on the metric that we choose, if X is a generic metric space then for purposes of illustration we often depict a ball in X as if it looked like a disk in $\mathbb{R}^2$ (for example, this is the case in Figure 2.6). Keep in mind that while pictures are suggestive, they are not proofs.

Our first use of open balls is to define the meaning of *boundedness* in a metric space.

Definition 2.3.2 (Bounded Set). Let X be a metric space. We say that a set $E \subseteq X$ is *bounded* if it is contained in some open ball, i.e., if there exists some $x \in X$ and $r > 0$ such that $E \subseteq B_r(x)$. $\diamond$

Example 2.3.3. If we put the discrete metric on $\mathbb{F}^d$, then *every* subset of $\mathbb{F}^d$, including $\mathbb{F}^d$ itself, is bounded (why?). However, if we use the Euclidean metric on $\mathbb{F}^d$, then $\mathbb{F}^d$ is not a bounded set. $\diamond$

The following terminology will be useful in this context.

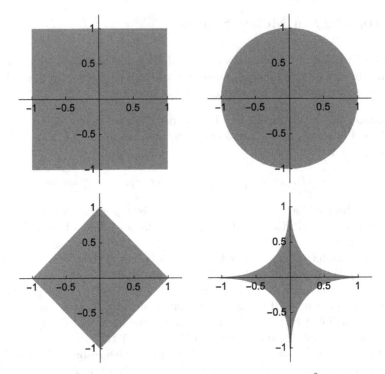

Fig. 2.5 Unit open balls $B_1(0)$ with respect to different metrics on $\mathbb{R}^2$. Top left: d_∞. Top right: d_2 (the Euclidean metric). Bottom left: d_1. Bottom right: $d_{1/2}$.

Definition 2.3.4 (Diameter). The *diameter* of a subset E of X is

$$\operatorname{diam}(E) \;=\; \sup\{d(x,y) : x, y \in E\}. \qquad \diamond$$

The diameter of a set is either a nonnegative real number or ∞. According to Problem 2.3.15, a set is bounded if and only if its diameter is finite.

Next we use open balls to define *open sets* in a metric space. According to the following definition, a set U is open if each point x in U has an open ball centered at x that is entirely contained within U.

Definition 2.3.5 (Open Sets). Let U be a subset of a metric space X. We say that U is an *open subset* of X if for each point $x \in U$ there exists some $r > 0$ such that $B_r(x) \subseteq U$. $\quad \diamond$

Note that the value of r in Definition 2.3.5 depends on the point x, i.e., there can be a different radius r for each point x.

We call the collection $\mathcal{T}$ of all open subsets of X the *topology induced from the metric* d, the *induced topology*, or simply the *topology* of X.

Remark 2.3.6. The letter O could be a natural choice to denote an open set, but we will avoid this because O resembles the symbol for the number zero.

Instead, we will usually use the symbols U, V, W for open sets. The letter G is another traditional choice for an open set, because *Gebiet* means "open set" in mathematical German (the literal meaning is "area," "region," or "neighborhood"). ◇

To illustrate, we will prove that open balls are indeed open sets (note that simply calling a set an "open ball" does not prove that it satisfies the definition of an open set!).

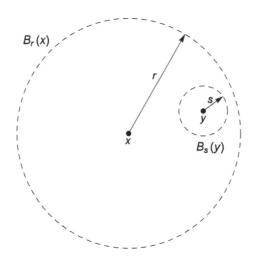

Fig. 2.6 In order for $B_s(y)$ to fit inside $B_r(x)$, we need $d(x,y) + s < r$.

Lemma 2.3.7 (Open Balls Are Open Sets). *If X is a metric space, then every open ball $B_r(x)$ is an open subset of X.*

Proof. Fix $x \in X$ and $r > 0$, and let y be any element of $B_r(x)$. We must show that there is some open ball $B_s(y)$ centered at y that is entirely contained in $B_r(x)$. We will show that the ball $B_s(y)$ with radius

$$s = r - d(x,y)$$

has this property (see Figure 2.6, and observe that $s > 0$ because $d(x,y) < r$). To prove this, choose any point $z \in B_s(y)$. Then $d(y,z) < s$, so by applying the Triangle Inequality, we see that

$$d(x,z) \leq d(x,y) + d(y,z) < d(x,y) + s = r.$$

Hence $z \in B_r(x)$. Thus every point $z \in B_s(y)$ belongs to $B_r(x)$, so we have $B_s(y) \subseteq B_r(x)$. Therefore $B_r(x)$ is an open set. $\square$

We will show that every open set is the union of some collection of open balls. To do this, let U be any open set. Then for each point $x \in U$ there must exist some radius $r_x > 0$ such that $B_{r_x}(x) \subseteq U$. The radius r_x will depend on the point x, but for each x we will have $x \in B_{r_x}(x) \subseteq U$. Consequently

$$U = \bigcup_{x \in U} \{x\} \subseteq \bigcup_{x \in U} B_{r_x}(x) \subseteq U,$$

and therefore

$$U = \bigcup_{x \in U} B_{r_x}(x).$$

Thus U is a union of open balls.

The next lemma states three fundamental properties of open sets.

Lemma 2.3.8. *Let X be a metric space, and let $\mathcal{T}$ be its topology. Then the following statements hold.*

(a) *$\varnothing$ and X belong to $\mathcal{T}$.*

(b) *If I is any index set and $U_i \in \mathcal{T}$ for each $i \in I$, then $\bigcup_{i \in I} U_i \in \mathcal{T}$.*

(c) *If $U, V \in \mathcal{T}$, then $U \cap V \in \mathcal{T}$.*

Proof. (a) If we choose any $x \in X$, then $B_r(x) \subseteq X$ for every $r > 0$, so X is open. On the other hand, the empty set is open because it contains no points, so it is vacuously true that "for each $x \in \varnothing$ there is some $r > 0$ such that $B_r(x) \subseteq \varnothing$."

(b) Suppose that $x \in \bigcup U_i$, where each U_i is open. Then $x \in U_i$ for some particular i. Since U_i is open, there exists some $r > 0$ such that $B_r(x) \subseteq U_i$. Hence $B_r(x) \subseteq \bigcup U_i$, so $\bigcup U_i$ is open.

(c) Suppose that $x \in U \cap V$, where U and V are open. Then $x \in U$, so there is some $r > 0$ such that $B_r(x) \subseteq U$. Likewise, since $x \in V$, there is some $s > 0$ such that $B_s(x) \subseteq V$. Let t be the smaller of r and s. Then $B_t(x) \subseteq B_r(x) \subseteq U$ and $B_t(x) \subseteq B_s(x) \subseteq V$, so $B_t(x) \subseteq U \cap V$. Therefore $U \cap V$ is open. $\square$

By induction, property (c) of Lemma 2.3.8 extends to intersections of *finitely many* open sets. Therefore we often summarize Lemma 2.3.8 by saying that if X is a metric space, then:

- the empty set and the entire space are open,
- an *arbitrary union* of open subsets of X is open, and
- an *intersection of finitely many* open subsets of X is open.

In the field of abstract topology, it is these three properties that are the inspiration for the definition of a topology on an arbitrary set (which need not be a metric space). However, in this text we will always remain within the setting of metric spaces.

The intersection of infinitely many open sets need not be open. For example, the intervals $U_n = (-\frac{1}{n}, 1)$ are open subsets of the real line, but their intersection is $\cap U_n = [0, 1)$, which is not open (why not?).

Next we prove a simple, but important, fact about metric spaces.

Lemma 2.3.9 (Metric Spaces Are Hausdorff). *If X is a metric space and $x \neq y$ are two distinct elements of X, then there exist disjoint open sets U and V such that $x \in U$ and $y \in V$.*

Proof. Suppose that $x \neq y$, and let $r = d(x, y)/2$. If $z \in B_r(x) \cap B_r(y)$, then, by the Triangle Inequality,

$$d(x, y) \leq d(x, z) + d(z, y) < 2r = d(x, y),$$

which is a contradiction. Therefore $B_r(x) \cap B_r(y) = \emptyset$. Since open balls are open sets, the proof is finished by taking $U = B_r(x)$ and $V = B_r(y)$. $\square$

Thus, any two distinct points $x \neq y$ can be "separated" by open sets. Using the terminology of abstract topology, a topological space that has this property is said to be a *Hausdorff space*. Therefore we often summarize Lemma 2.3.9 by saying that *all metric spaces are Hausdorff*.

One additional notion related to open sets is the interior of a set, which is defined as follows.

Definition 2.3.10 (Interior). Let X be a metric space. The *interior* of a set $E \subseteq X$, denoted by E°, is the union of all open sets that are contained in E, i.e.,

$$E^\circ = \bigcup \{U \subseteq X : U \text{ is open and } U \subseteq E\}. \qquad \Diamond$$

According to Problem 2.3.14, E° is an open set, $E^\circ \subseteq E$, and if U is any open subset of E, then $U \subseteq E^\circ$. In this sense, E° is the "largest open set that is contained in E."

Problems

2.3.11. Let d be the discrete metric on a set X. Explicitly describe the open balls $B_r(x)$ for all $x \in X$ and $r > 0$, and prove that every subset of X is open. What does $x_n \to x$ mean with respect to the discrete metric?

2.3.12. Given a point $x \in \mathbb{R}^2$ and a radius $r > 0$, explicitly characterize (with a picture and proof) the open ball $B_r(x)$ in $\mathbb{R}^2$ defined with respect to each of the metrics d_1, d_2, and d_∞ defined in Example 2.1.6.

2.3.13. (a) Let d be the discrete metric on $\mathbb{Z}$, and let d' be the restriction of the standard metric from $\mathbb{R}$ to $\mathbb{Z}$. Show that these two metrics are not identical, yet the topology induced from d is the same as the topology induced from d'.

(b) If we replace $\mathbb{Z}$ by $\mathbb{Q}$ in part (a), are the two topologies still equal?

2.3.14. Let A and B be subsets of a metric space X.

(a) Prove that A° is open, $A^\circ \subseteq A$, and if $U \subseteq A$ is open then $U \subseteq A^\circ$.

(b) Prove that A is open if and only if $A = A^\circ$.

(c) Prove that $(A \cap B)^\circ = A^\circ \cap B^\circ$.

(d) Show by example that $(A \cup B)^\circ$ need not equal $A^\circ \cup B^\circ$.

2.3.15. Let X be a metric space.

(a) Prove that $S \subseteq X$ is bounded if and only if $\text{diam}(S) < \infty$.

(b) If $B_r(x)$ is an open ball in X, must the diameter of $B_r(x)$ be $2r$? Either prove or give a counterexample.

2.3.16. For this problem we take $\mathbb{F} = \mathbb{R}$, so all sequences and functions are real-valued.

(a) Let Q be the "open first quadrant" in $\mathbb{R}^d$, i.e.,

$$Q = \{x = (x_1, \ldots, x_d) \in \mathbb{R}^d : x_1, \ldots, x_d > 0\}.$$

Prove that Q is an open subset of $\mathbb{R}^d$.

(b) Let R be the "open first quadrant" in ℓ^1, i.e.,

$$R = \{x = (x_k)_{k \in \mathbb{N}} \in \ell^1 : x_k > 0 \text{ for every } k\}.$$

Prove that R is *not* an open subset of ℓ^1.

(c) Let S be the "open first quadrant" in $C[0,1]$, i.e.,

$$S = \{f \in C[0,1] : f(t) > 0 \text{ for all } t \in [0,1]\}.$$

Determine, with proof, whether S is an open subset of $C[0,1]$ with respect to the uniform metric. Hint: A continuous real-valued function on a closed finite interval achieves a maximum and a minimum on that interval.

(d) Same as part (c), except this time use the L^1-metric.

2.3.17. Let $\{x_n\}_{n \in \mathbb{N}}$ be a sequence in a metric space X. Show that if $x_n \to x$, then either:

(a) there exists an open set U that contains x and there is some $N > 0$ such that $x_n = x$ for all $n > N$, or

(b) every open set U that contains x must also contain infinitely many *distinct* x_n (that is, the set $\{x_n : n \in \mathbb{N} \text{ and } x_n \in U\}$ contains infinitely many elements).

2.3.18. Let Y be a subset of a metric space X. Recall from Remark 2.1.2 that the restriction of d to Y is the *inherited metric* on Y.

(a) Let $B_r(x)$ denote an open ball in X, and let $B_r^Y(x)$ denote an open ball in Y defined with respect to the inherited metric, i.e.,

$$B_r^Y(x) = \{y \in Y : \mathrm{d}(x, y) < r\}.$$

Prove that $B_r^Y(x) = B_r(x) \cap Y$.

(b) Given $V \subseteq Y$, prove that V is an open subset of Y (with respect to the inherited metric) if and only if there exists an open set $U \subseteq X$ such that $V = U \cap Y$.

(c) Let $X = \mathbb{R}$ with the standard metric, and set $Y = [-1, 1]$. Prove that $(0, 1]$ is an open set in Y. Given $x \in Y$ and $r > 0$, explicitly describe the open ball $B_r^Y(x)$ in Y.

2.4 Closed Sets

The complements of the open sets are very important; we call these the *closed subsets* of X.

Definition 2.4.1 (Closed Sets). Let X be a metric space. We say that a set $F \subseteq X$ is *closed* if its complement $F^C = X \setminus F$ is open. ◇

Following tradition, we often denote a closed set by the letter F, because *fermé* is the French word for *closed*.

By taking complements in Lemma 2.3.8 we see that if X is a metric space then:

- the empty set and the entire space are closed,
- an *arbitrary intersection* of closed subsets of X is closed, and
- a *union of finitely many* closed subsets of X is closed.

Definition 2.4.1 is "indirect" in the sense that it is worded in terms of the complement of E rather than E itself. The next theorem gives a "direct" characterization of closed sets in terms of limits.

Theorem 2.4.2. *If E is a subset of a metric space X, then the following two statements are equivalent.*

(a) *E is closed.*

(b) *If $\{x_n\}_{n \in \mathbb{N}}$ is a sequence of points in E and $x_n \to x \in X$, then $x \in E$.*

Proof. (a) $\Rightarrow$ (b). Assume that E is closed. Suppose that $x \in X$ is a limit of elements of E, i.e., there exist points $x_n \in E$ such that $x_n \to x$. Suppose that x did not belong to E. Then x belongs to E^C, which is an open set, so there must exist some $r > 0$ such that $B_r(x) \subseteq E^C$. Since $x_n \in E$, none of the x_n belong to $B_r(x)$. This implies that $\mathrm{d}(x, x_n) > r$ for every n, which contradicts the assumption that $x_n \to x$. Therefore x must belong to E.

(b) $\Rightarrow$ (a). Suppose that statement (b) holds, but E is not closed. Then E^C is not open, so there must exist some point $x \in E^C$ such that no open ball $B_r(x)$ centered at x is entirely contained in E^C. Considering $r = \frac{1}{n}$ in particular, this tells us that there must exist a point $x_n \in B_{1/n}(x)$ that is not in E^C. But then $x_n \in E$, and we have $\mathrm{d}(x_n, x) < \frac{1}{n}$, so $x_n \to x$. Statement (b) therefore implies that $x \in E$, which is a contradiction since x belongs to E^C. Consequently E must be closed. $\square$

In other words, Theorem 2.4.2 says that

> *a set E is closed if and only if the limit of every*
> *convergent sequence of points of E belongs to E.*

In practice this is often (though not always) the best way to prove that a particular set E is closed.

Problems

2.4.3. If a set $E \subseteq X$ is not open, must it be closed? Either prove that every subset E that is not open must be closed, or exhibit a metric space X and a set E for which this fails.

2.4.4. Let X be a metric space.

(a) Prove that X and $\varnothing$ are simultaneously open and closed subsets of X.

(b) Show that if d is the discrete metric on a set X, then *every* subset of X is both open and closed.

(c) Using the standard metric on $\mathbb{F}$, are there any subsets of $\mathbb{F}$ other than $\mathbb{F}$ and $\varnothing$ that are both open and closed? (The answer is no, but a proof requires more tools than we have developed here.)

2.4.5. Let X be a metric space. Given $x \in X$, prove that $\{x\}$ is a closed subset of X.

2.4.6. Show by example that it is possible for the union of infinitely many closed sets to be open, closed, or neither open nor closed.

2.4.7. Let A be a subset of a metric space X. Given $x \in X$, define the *distance from x to A* to be

$$\mathrm{dist}(x, A) = \inf\{\mathrm{d}(x, y) : y \in A\}.$$

Prove the following statements.

(a) If A is closed, then $x \in A$ if and only if $\operatorname{dist}(x, A) = 0$.

(b) $\operatorname{dist}(x, A) \leq d(x, y) + \operatorname{dist}(y, A)$ for all $x, y \in X$.

(c) $|\operatorname{dist}(x, A) - \operatorname{dist}(y, A)| \leq d(x, y)$ for all $x, y \in X$.

Additionally, show by example that it is possible to have $\operatorname{dist}(x, A) = 0$ even when $x \notin A$.

2.4.8. Let X be a *complete* metric space, and let E be a subset of X. Prove that the following two statements are equivalent.

(a) E is closed.

(b) E is complete, i.e., every Cauchy sequence of points in E converges to a point of E.

2.4.9. Give an example of closed sets $E_n \subseteq \mathbb{R}$ such that $E_1 \supseteq E_2 \supseteq \cdots$ and $\cap E_n = \varnothing$.

2.4.10. Let X be a metric space.

(a) Suppose that X is complete and $F_1 \supseteq F_2 \supseteq \cdots$ is a nested decreasing sequence of closed nonempty subsets of X such that $\operatorname{diam}(F_n) \to 0$. Prove that there exists some $x \in X$ such that $\cap F_n = \{x\}$.

(b) Show by example that the conclusion of part (a) can fail if X is not complete or if the F_n are not closed.

2.4.11. We take $\mathbb{F} = \mathbb{R}$ for this problem.

(a) Let Q be the "closed first quadrant" in $\mathbb{R}^d$, i.e.,

$$Q = \{x = (x_1, \ldots, x_d) \in \mathbb{R}^d : x_1, \ldots, x_d \geq 0\}.$$

Prove that Q is a closed subset of $\mathbb{R}^d$.

(b) Let R be the "closed first quadrant" in ℓ^1, i.e.,

$$R = \{x = (x_k)_{k \in \mathbb{N}} \in \ell^1 : x_k \geq 0 \text{ for every } k\}.$$

Determine, with proof, whether R is a closed subset of ℓ^1.

(c) Let S be the "closed first quadrant" in $C[0, 1]$, i.e.,

$$S = \{f \in C[0, 1] : f(x) \geq 0 \text{ for all } x \in [0, 1]\}.$$

Determine, with proof, whether S is a closed subset of $C[0, 1]$ with respect to the uniform metric.

2.4.12. Let E be the subset of ℓ^1 consisting of sequences whose even components are all zero:

$$E = \{x = (x_k)_{k\in\mathbb{N}} \in \ell^1 : x_{2j} = 0 \text{ for all } j \in \mathbb{N}\}.$$

(a) Prove that E is a proper closed *subspace* of ℓ^1 (not just a subset). Also prove that E is not an open set.

(b) Is c_{00} a closed subspace of ℓ^1? Is it an open subspace?

(c) Let $N > 0$ be a fixed positive integer, and let S_N be the set of all sequences whose entries from the $(N+1)$st component onward are all zero:

$$S_N = \{x = (x_1,\ldots,x_N,0,0,\ldots) : x_1,\ldots,x_N \in \mathbb{F}\}.$$

Prove that S_N is a closed subspace of ℓ^1. Is it open? What is the union of the S_N over $N \in \mathbb{N}$?

(d) Do you think there exist any proper *open* subspaces of ℓ^1? Either prove or give a counterexample.

Hint: Show that if S is open subspace of ℓ^1, then there exists some $r > 0$ such that $B_r(0) \subseteq S$.

2.5 Accumulation Points and Boundary Points

If a point $x \in X$ is such that there exist $x_n \in E$ *with every $x_n \neq x$* such that $x_n \to x$, then in some sense the points of E "accumulate" or "cluster" at x. For example, consider

$$E = [0,1] \cup \{2\}.$$

The elements of E do not "accumulate" at $x = 2$, because there is no way to choose points $x_n \in E$, *with every $x_n \neq 2$*, so that $x_n \to 2$. There are points in E that converge to $x = 2$, e.g., we could take $x_n = \frac{1}{2}$ for $n = 1,\ldots,99$ and $x_n = 2$ for $n \geq 100$ (or just take every $x_n = 2$), but there is no way to find points x_n *all different from* 2 that converge to 2. Therefore $x = 2$ is not an accumulation point of E in the sense of the following definition.

Definition 2.5.1 (Accumulation Point). Let X be a metric space, and let E be a subset of X. A point $x \in X$ is called an *accumulation point* or a *cluster point* of E if there exist points $x_n \in E$, with every $x_n \neq x$, such that $x_n \to x$.   ◇

Some texts also refer to accumulation points as "limit points," but we will avoid that terminology because it does not adequately convey the idea that the points x_n in Definition 2.5.1 must all be distinct from x. To be an accumulation point, it is not enough that x be a limit of points $x_n \in E$, rather it must also be the case that these points x_n are not equal to x.

Example 2.5.2. (a) Consider the set $E = [0, 1] \cup \{2\}$ in the real line. Every point $x \in [0, 1]$ is an accumulation point of E, but $x = 2$ is not an accumulation point. No other point outside of $[0, 1]$ is an accumulation point either (why not?). This set E contains each of its accumulation points, but not every point in E is an accumulation point. What are the accumulation points of the set $(0, 1) \cup \{2\}$?

(b) Now consider $E = \left\{\frac{1}{n}\right\}_{n \in \mathbb{N}} = \left\{1, \frac{1}{2}, \frac{1}{3}, \dots\right\}$ in $\mathbb{R}$. Even though 0 does not belong to E, it is an accumulation point of E, because the points $x_n = \frac{1}{n}$ belong to E, are distinct from 0, and converge to 0. In contrast, 1 is not an accumulation point of E, because we cannot find points $x_n \neq 1$ in E that converge to 1. In fact, 0 is the *only* accumulation point of E (prove this). This set E has an accumulation point that does not belong to E, but no element of E is an accumulation point of E.

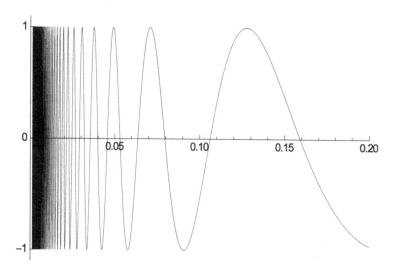

Fig. 2.7 Graph of the function $f(x) = \sin \frac{1}{x}$ for $0.002 < x < 0.2$.

(c) Let $E \subseteq \mathbb{R}^2$ be the graph of $f(x) = \sin \frac{1}{x}$ for $x > 0$, i.e.,

$$E = \left\{\left(x, \sin \frac{1}{x}\right) : x > 0\right\} \subseteq \mathbb{R}^2.$$

Figure 2.7 shows a computer-produced rendering of the graph of f for x between 0.002 and 0.2 (close examination shows inaccuracies due both to approximations made in computing the graph and to the fact that the graph must be pictured as a curve with positive thickness).

We will show that the point $z = (0, 1)$ on the y-axis, which does not belong to E, is an accumulation point of E. Set

$$x_n = \frac{1}{\frac{\pi}{2} + 2\pi n} = \frac{2}{(4n+1)\pi} \quad \text{and} \quad z_n = \left(x_n, \sin\frac{1}{x_n}\right) = (x_n, 1).$$

Then $z_n \in E$ and $z_n \to z = (0, 1)$ as $n \to \infty$. Since no z_n is equal to z, we conclude that z is an accumulation point of E. The reader should check that every point $(0, y)$ with $-1 \leq y \leq 1$ is also an accumulation point of E. Thus this set E has accumulation points that do not belong to E. Are there any accumulation points of E that do belong to E?   ◇

The next theorem (whose proof is Problem 2.5.6) states that a set E is closed if and only if it contains every one of its accumulation points (how is this different from Theorem 2.4.2?). This implies that the set in part (a) of Example 2.5.2 is closed, and the sets in parts (b) and (c) are not closed (can you prove those facts directly?).

Theorem 2.5.3. *If E is a subset of a metric space X, then E is closed if and only if every accumulation point of E belongs to E.*   ◇

Boundary points are somewhat similar to accumulation points, but they are not identical—an accumulation point need not be a boundary point, nor does a boundary point need to be an accumulation point.

Definition 2.5.4 (Boundary Point). Let E be a subset of a metric space X. We say that x is a *boundary point* of E if for each $r > 0$ we have both $B_r(x) \cap E \neq \varnothing$ and $B_r(x) \cap E^C \neq \varnothing$. That is, x is a boundary point if *every* ball centered at x contains both a point of E and a point not in E.

The set of all boundary points of E is called the *boundary* of E, and it is denoted by

$$\partial E = \{x \in X : x \text{ is a boundary point of } E\}. \quad \diamondsuit$$

We characterize closed sets in terms of boundary points in the next theorem (whose proof is assigned as Problem 2.5.6).

Theorem 2.5.5. *If E is a subset of a metric space X, then E is closed if and only if every boundary point of E belongs to E.*   ◇

Problems

2.5.6. Prove Theorems 2.5.3 and 2.5.5.

2.5.7. (a) Prove that every *finite* subset of a metric space is closed and has no accumulation points.

(b) Give an example of a metric space X and an infinite set $Z \subseteq X$ such that Z has no accumulation points. Is your set Z closed? What is ∂Z for your example?

2.5.8. Let E be a subset of a metric space X. Given $x \in X$, prove that the following four statements are equivalent.

(a) x is an accumulation point of E.

(b) If U is an open set that contains x, then $(E \cap U) \setminus \{x\} \neq \varnothing$, i.e., there exists a point $y \in E \cap U$ such that $y \neq x$. (An open set that contains x is sometimes called an *open neighborhood* of x.)

(c) If $r > 0$, then there exists a point $y \in E$ such that $0 < d(x, y) < r$.

(d) Every open set U that contains x also contains *infinitely many* distinct points of E.

2.5.9. Let E be a subset of a metric space X. Given $x \in X$, prove that the following four statements are equivalent.

(a) x is a boundary point of E.

(b) If U is open and $x \in U$, then both $E \cap U$ and $E^C \cap U$ are nonempty.

(c) For each $r > 0$, the open ball $B_r(x)$ contains at least one point of E and at least one point of E^C.

(d) There exist points $x_n \in E$ and $y_n \in E^C$ such that $x_n \to x$ and $y_n \to x$.

2.5.10. Suppose that E is a countably infinite subset of a metric space X, and let $E = \{x_n\}_{n \in \mathbb{N}}$ be a listing of the distinct points in E. Prove that E has an accumulation point x if and only if $\{x_n\}_{n \in \mathbb{N}}$ has some convergent subsequence $\{x_{n_k}\}_{k \in \mathbb{N}}$.

2.5.11. The space c_{00} defined in Example 2.2.13 is a subset of ℓ^∞, and we take the metric on c_{00} to be the ℓ^∞-metric in this problem.

(a) Prove that $x = (1, \frac{1}{2}, \frac{1}{3}, \dots)$ is an accumulation point of c_{00}, but $y = (1, 1, 1, \dots)$ is not.

(b) Exactly which sequences $z = (z_k)_{k \in \mathbb{N}} \in \ell^\infty$ are accumulation points of c_{00}? Prove your claim.

2.5.12. This problem requires some knowledge of Taylor polynomials. Let $\mathcal{P}$ be the set of all polynomial functions, and let f be the function $f(x) = e^x$. Note that $\mathcal{P} \subsetneq C[0, 1]$, and $f \in C[0, 1] \setminus \mathcal{P}$.

(a) Let p_n be the nth Taylor polynomial for e^x about the point $x = 0$, i.e., $p_n(x) = \sum_{k=0}^{n} \frac{f^{(k)}(0)}{k!} x^k$. Show that $p_n \to f$ with respect to the uniform metric, i.e., $d_u(p_n, f) \to 0$ as $n \to \infty$. Use this to show that f is an accumulation point of $\mathcal{P}$ when we place the uniform metric on $C[0, 1]$.

(b) Same as part (a), except this time use the L^1-metric on $C[0, 1]$.

2.5.13. We take $\mathbb{F} = \mathbb{R}$ for this problem. Let R be the "closed first quadrant" in ℓ^1, i.e.,

$$R = \big\{ x = (x_k)_{k \in \mathbb{N}} \in \ell^1 : x_k \geq 0 \text{ for every } k \big\}.$$

(a) Is R a closed subset of ℓ^1?

(b) Suppose that $x = (x_k)_{k \in \mathbb{N}}$ is a point in R that has at least one zero component. Prove that $x \in \partial R$.

(c) More surprisingly, prove that if $x = (x_k)_{k \in \mathbb{N}}$ is *any* point in R (even one whose components are all strictly positive), then x is a boundary point of R. Use this to prove that $R = \partial R$ and $R^\circ = \varnothing$.

2.6 Closure, Density, and Separability

The set of rationals $\mathbb{Q}$ is a subset of the real line, but it is not a *closed* subset, because a sequence of rational points can converge to an irrational point. Can we find a closed set F that "surrounds" $\mathbb{Q}$, i.e., a closed set that satisfies $\mathbb{Q} \subseteq F \subseteq \mathbb{R}$? Of course, we could just take $F = \mathbb{R}$, but is there a smaller closed set that contains $\mathbb{Q}$? By Theorem 2.4.2, any such set F would have to contain every limit of elements of $\mathbb{Q}$, but since *every* $x \in \mathbb{R}$ is a limit of rational points, we conclude that $F = \mathbb{R}$ is the only closed subset of $\mathbb{R}$ that contains $\mathbb{Q}$.

This type of issue arises quite often. Specifically, given a set E, it can be important to find a closed set that not only contains E but is the "smallest closed set" that contains E. The next definition gives a name to this "smallest closed set."

Definition 2.6.1 (Closure). Let E be a subset of a metric space X. The *closure of E*, denoted by $\overline{E}$, is the intersection of all the closed sets that contain E:

$$\overline{E} = \bigcap \{ F \subseteq X : F \text{ is closed and } F \supseteq E \}. \qquad \diamond \qquad (2.17)$$

Observe that $\overline{E}$ is closed, because the intersection of closed sets is closed, and it contains E by definition. Furthermore, if F is any particular closed set that contains E, then F is one of the sets appearing in the intersection on the right-hand side of equation (2.17), so F must contain $\overline{E}$. In summary:

- $\overline{E}$ is a closed set that contains E, and
- if F is any closed set with $E \subseteq F$, then $\overline{E} \subseteq F$.

Thus $\overline{E}$ is closed and contains E, and it is contained within every other closed set that contains E. Hence $\overline{E}$ *is the smallest closed set that contains E.*

The following lemma shows that the closure of E is the set of all possible limits of elements of E (more precisely, $\overline{E}$ is the set of *all limits of convergent sequences of elements of E*). Problem 2.6.16 shows that the closure is also equal to the union of E and the accumulation points of E.

Theorem 2.6.2. *If E is a subset of a metric space X, then*

$$\overline{E} = \{y \in X : \text{there exist } x_n \in E \text{ such that } x_n \to y\}. \tag{2.18}$$

Proof. Let F be the set that appears on the right-hand side of equation (2.18), i.e., F is the set of all possible limits of elements of E. We must show that $F = \overline{E}$.

Choose any point $y \in F$. Then, by definition, there exist points $x_n \in E$ such that $x_n \to y$. Since $E \subseteq \overline{E}$, the points x_n all belong to $\overline{E}$. Hence y is a limit of elements of $\overline{E}$. But $\overline{E}$ is a closed set, so it must contain all of these limits. Therefore y belongs to $\overline{E}$, so we have shown that $F \subseteq \overline{E}$.

In order to prove that $\overline{E}$ is a subset of F, we will first prove that F^C is an open set. To do this, choose any point $y \in F^C$. We must show that there is a ball centered at y that is entirely contained in F^C. That is, we must show that there is some $r > 0$ such that $B_r(y)$ contains no limits of elements of E.

Suppose that for each $k \in \mathbb{N}$, the ball $B_{1/k}(y)$ contained some point from E, say $x_k \in B_{1/k}(y) \cap E$. Then these x_k are points of E that converge to y (why?). Hence y is a limit of points of E, which contradicts the fact that $y \notin F$. Hence there must be at least one k such that $B_{1/k}(y)$ contains no points of E. We will show that $r = 1/k$ is the radius that we seek. That is, we will show that the ball $B_r(y)$, where $r = 1/k$, not only contains no elements of E but furthermore contains no *limits* of elements of E.

Suppose that $B_r(y)$ did contain some point z that was a limit of elements of E, i.e., suppose that there did exist some $x_n \in E$ such that $x_n \to z \in B_r(y)$. Then, since $d(y, z) < r$ and since $d(z, x_n)$ becomes arbitrarily small, by choosing n large enough we will have $d(y, x_n) \le d(y, z) + d(z, x_n) < r$. But then this point x_n belongs to $B_r(y)$, which contradicts the fact that $B_r(y)$ contains no points of E.

Thus, $B_r(y)$ contains no limits of elements of E. Since F is the set of all limits of elements of E, this means that $B_r(y)$ contains no points of F. That is, $B_r(y) \subseteq F^C$.

In summary, we have shown that each point $y \in F^C$ has some ball $B_r(y)$ that is entirely contained in F^C. Therefore F^C is an open set. Hence, by definition, F is a closed set. We also know that $E \subseteq F$ (why?), so F is one of the closed sets that contains E. But $\overline{E}$ is the smallest closed set that contains E, so we conclude that $\overline{E} \subseteq F$. $\square$

We saw earlier that the closure of the set of rationals $\mathbb{Q}$ is the entire real line $\mathbb{R}$. This type of situation can be very important, so we introduce some terminology for it.

Definition 2.6.3 (Dense Subset). Let E be a subset of a metric space X. If $\overline{E} = X$ (i.e., the closure of E is all of X), then we say that E is *dense* in X. $\diamond$

Theorem 2.6.2 tells us that the closure of a set E equals the set of all limits of elements of E. If E is dense, then the closure of E is the entire space X, so *every* point in X must be a limit of points of E. The converse is also true, and so we obtain the following useful equivalent reformulation of the meaning of density (some additional reformulations appear in Problem 2.6.12).

Corollary 2.6.4. *Let E be a subset of a metric space X. Then E is dense in X if and only if for each $x \in X$ there exist points $x_n \in E$ such that $x_n \to x$.*

Proof. $\Rightarrow$. Suppose that E is dense, i.e., $\overline{E} = X$. If we choose any point $x \in X$, then x belongs to the closure of E. Theorem 2.6.2 therefore implies that x is a limit of points of E.

$\Leftarrow$. Suppose that every point $x \in X$ is a limit of points of E. Theorem 2.6.2 tells us that the closure of E is the set of all limits of elements of E, so we conclude that every $x \in X$ belongs to $\overline{E}$, i.e., $\overline{E} = X$. Therefore E is dense in X. $\square$

Here is a short summary of the facts that we proved in Theorem 2.4.2, Theorem 2.6.2, and Corollary 2.6.4:

- *a set E is closed if and only if it contains the limit of every convergent sequence of elements of E,*

- *the closure of a set E is the set of all possible limits of elements chosen from E, and*

- *a set E is dense in X if and only if every point in X is a limit of points of E.*

For example, the set of rationals $\mathbb{Q}$ is not a closed subset of $\mathbb{R}$, because a limit of rational points need not be rational. The closure of $\mathbb{Q}$ is $\mathbb{R}$, because every point in $\mathbb{R}$ can be written as a limit of rational points. Similarly, $\mathbb{Q}$ is a dense subset of $\mathbb{R}$, because every real number can be written as a limit of rational numbers.

Problem 2.6.12 gives another reformulation of density that is useful in some circumstances: E is a dense subset of X if and only if the intersection of E with every nonempty open subset of X is nonempty.

To illustrate, we will show that c_{00}, the space of "finite sequences" introduced in Example 2.2.13, is a dense subspace of ℓ^1.

Theorem 2.6.5. *c_{00} is a dense subspace of ℓ^1.*

Proof. Observe that c_{00} is contained in ℓ^1, and it is a subspace, since it is closed under vector addition and scalar multiplication. So, we just have to show that it is dense. To do this, we will prove that every vector in ℓ^1 is a limit of elements of c_{00}.

Choose any vector $x = (x_k)_{k \in \mathbb{N}} \in \ell^1$. For each $n \in \mathbb{N}$, let y_n be the sequence whose first n components are the same as those of x, but whose remaining components are all zero:

$$y_n = (x_1, \ldots, x_n, 0, 0, \ldots) \in c_{00}.$$

By definition, y_n *converges componentwise* to x. However, this is not enough. We must show that y_n converges *in ℓ^1-norm* to x. Note that

$$x - y_n = (0, \ldots, 0, x_{n+1}, x_{n+2}, \ldots).$$

Applying the definition of the ℓ^1-norm, the fact that $\sum_{k=1}^{\infty} |x_k| < \infty$, and Lemma 1.8.4, we see that

$$\lim_{n \to \infty} d_1(y_n, x) = \lim_{n \to \infty} \|x - y_n\|_1 = \lim_{n \to \infty} \left(\sum_{k=n+1}^{\infty} |x_k| \right) = 0.$$

Thus $y_n \to x$ in ℓ^1-norm. Hence every element of ℓ^1 is a limit of elements of c_{00}, and therefore c_{00} is dense by Corollary 2.6.4. $\square$

Theorem 2.6.5 implies that the closure of c_{00} in the space ℓ^1 is $\overline{c_{00}} = \ell^1$. What is the closure of c_{00} when we consider it to be a subset of ℓ^∞ and use the ℓ^∞-metric instead? (See Problem 2.6.18.)

We introduce one more notion related to density. To motivate this, recall that the set of rationals $\mathbb{Q}$ is a dense subset of the real line. Since $\mathbb{Q}$ is countable, it follows that $\mathbb{R}$ contains a subset that is "small" in terms of cardinality, but is "large" in the sense that it is dense. In higher dimensions, the set $\mathbb{Q}^d$ consisting of vectors with rational components is a countable yet dense subset of $\mathbb{R}^d$. It may seem unlikely that an infinite-dimensional space could contain such a subset, but we show next that this is true of the infinite-dimensional space ℓ^1.

Example 2.6.6. Recall that c_{00}, the set of "finite sequences," is dense in ℓ^1. However, c_{00} is not a countable set (why not?). To construct a countable dense subset of ℓ^1, first consider the sets

$$A_N = \{(r_1, \ldots, r_N, 0, 0, \ldots) : r_1, \ldots, r_N \text{ rational}\}, \quad \text{where } N \in \mathbb{N}.$$

Each A_N is countable (why?). Since the union of countably many countable sets is countable, it follows that

$$A = \{(r_1, \ldots, r_N, 0, 0, \ldots) : N > 0, \ r_1, \ldots, r_N \text{ rational}\} \quad (2.19)$$

is also countable. Furthermore, the reader will prove in Problem 2.6.20 that A is dense in ℓ^1. Thus ℓ^1 contains a countable dense subset. $\diamond$

Thus, some metric spaces (even including some infinite-dimensional vector spaces) contain countable dense subsets, but it is also true that there exist metric spaces that do not contain any countable dense subsets. This gives us one way to distinguish between "small" and "large" spaces. We introduce a name for the "small" spaces that contain countable dense subsets.

Definition 2.6.7 (Separable Space). A metric space that contains a countable dense subset is said to be *separable*.   ◇

Using this terminology, Example 2.6.6 demonstrates that ℓ^1 is a separable space. In contrast, we will show that ℓ^∞ is nonseparable.

Theorem 2.6.8. ℓ^∞ *does not contain a countable dense subset and therefore is not separable.*

Proof. Recall from equation (2.7) that the metric on ℓ^∞ is

$$d_\infty(x, y) = \|x - y\|_\infty = \sup_{k \in \mathbb{N}} |x_k - y_k|.$$

Let S be the set of all sequences whose components are either 0 or 1, i.e.,

$$S = \big\{ x = (x_k)_{k \in \mathbb{N}} : x_k \in \{0, 1\} \text{ for every } k \big\}.$$

This is an uncountable set (see Problem 1.5.1), and if $x \neq y$ are any two distinct elements of S, then $d_\infty(x, y) = 1$ (why?).

Suppose that ℓ^∞ were separable. Then there would exist countably many vectors $y_1, y_2, \ldots \in \ell^\infty$ such that the set $A = \{y_n\}_{n \in \mathbb{N}}$ was dense in ℓ^∞. Choose any point $x \in S$. Since A is dense in ℓ^∞, Corollary 2.6.4 implies that there must exist points of A that lie as close to x as we like. In particular, there must exist some integer n_x such that $d_\infty(x, y_{n_x}) < \frac{1}{2}$. If we define $f(x) = n_x$, then this gives us a mapping $f \colon S \to \mathbb{N}$. Since S is uncountable and $\mathbb{N}$ is countable, f cannot be injective. Hence there must exist two distinct points $x, z \in S$ such that $n_x = n_z$. Since x and z are two different points from S, we have $d_\infty(x, z) = 1$. Applying the Triangle Inequality and the fact that $y_{n_x} = y_{n_z}$, we therefore obtain

$$1 = d_\infty(x, z) \leq d_\infty(x, y_{n_x}) + d_\infty(y_{n_z}, z) < \frac{1}{2} + \frac{1}{2} < 1.$$

This is a contradiction, so no countable subset of ℓ^∞ can be dense.   □

Problems

2.6.9. The empty set $\varnothing$ is a subset of every metric space X. What are the interior, closure, and boundary of the empty set? Is the empty set open or closed? Answer the same questions with the empty set replaced by X.

2.6.10. Take $X = \mathbb{R}$, and find E°, ∂E, and $\overline{E}$ for each of the following sets.

(a) $E = \left\{ 1, \frac{1}{2}, \frac{1}{3}, \dots \right\}$.

(b) $E = [0, 1)$.

(c) $E = \mathbb{Z}$, the set of integers.

(d) $E = \mathbb{Q}$, the set of rationals.

(e) $E = \mathbb{R} \backslash \mathbb{Q}$, the set of irrationals.

2.6.11. Let E be a subset of a metric space X. Show that if E is not dense in X, then $X \backslash E$ contains an open ball.

2.6.12. Let E be a subset of a metric space X. Prove that the following four statements are equivalent.

(a) E is dense in X.

(b) Given $x \in X$ and $\varepsilon > 0$, there exists some $y \in E$ such that $\mathrm{d}(x, y) < \varepsilon$.

(c) $E \cap U \neq \varnothing$ for every nonempty open set $U \subseteq X$.

(d) Every point $x \in E^C$ is an accumulation point of E.

2.6.13. Let E be a subset of a metric space X. Prove that E has empty interior ($E^\circ = \varnothing$) if and only if $X \backslash E$ is dense in X.

2.6.14. Let X be a metric space.

(a) Given sets $A, B \subseteq X$, prove that $\overline{A \cup B} = \overline{A} \cup \overline{B}$. Use induction to extend this to the union of finitely many subsets of X.

(b) Let I be an arbitrary index set. Given sets $E_i \subseteq X$ for $i \in I$, prove that

$$\overline{\bigcap_{i \in I} E_i} \subseteq \bigcap_{i \in I} \overline{E_i}.$$

Show by example that equality can fail if the index set I is infinite.

2.6.15. Let X be a metric space.

(a) Show that the closure of the open ball $B_r(x)$ satisfies

$$\overline{B_r(x)} \subseteq \{ y \in X : \mathrm{d}(x, y) \leq r \}.$$

Must equality hold?

(b) Show that if $x \in X$ and $y \in B_r(x)$, then there exists some $s > 0$ such that $\overline{B_s(y)} \subseteq B_r(x)$. Is it possible for equality to hold?

2.6.16. Given a subset E of a metric space X, prove the following statements.

(a) E is closed if and only if $E = \overline{E}$.

(b) $\overline{E}$ is the union of E and all of the accumulation points of E.

(c) $\overline{E} = E \cup \partial E$.

2.6.17. This is an extension of Problem 2.4.10. Given a metric space X, prove that the following two statements are equivalent.

(a) X is complete.

(b) If $F_1 \supseteq F_2 \supseteq \cdots$ is any decreasing sequence of nested, closed, nonempty subsets of X such that $\mathrm{diam}(F_n) \to 0$, then $\cap F_n \neq \varnothing$.

Hint: For one direction, let F_n be the closure of $\{x_n, x_{n+1}, \dots\}$.

2.6.18. In this problem we take $X = \ell^\infty$, with respect to its standard metric $d_\infty(x, y) = \|x - y\|_\infty$.

(a) Is c_{00} a dense subspace of ℓ^∞? Is ℓ^1 a dense subspace of ℓ^∞?

(b) Let c_0 be the space of all sequences whose components converge to zero:

$$c_0 = \left\{ x = (x_k)_{k \in \mathbb{N}} : \lim_{k \to \infty} x_k = 0 \right\}. \tag{2.20}$$

Prove that $c_{00} \subsetneq c_0 \subsetneq \ell^\infty$, and show that the closure of c_{00} with respect to the ℓ^∞-metric is c_0, i.e., $\overline{c_{00}} = c_0$.

(c) Why does part (b) not contradict the statement made after Theorem 2.6.5 that $\overline{c_{00}} = \ell^1$?

2.6.19. Let S be the subset of $\mathbb{F}^d$ consisting of all sequences with rational components:

$$S = \left\{ (r_1, \dots, r_d) : r_1, \dots, r_d \text{ are rational} \right\}$$

(recall from Section 1.1 that if $\mathbb{F} = \mathbb{C}$, then we say that a complex number is rational if both its real and imaginary parts are rational). Prove that S is a countable dense subset of $\mathbb{F}^d$. Conclude that $\mathbb{F}^d$ is separable.

2.6.20. Let S be the set of all "finite sequences" with rational components defined in equation (2.19).

(a) Explain why S is contained in ℓ^1, but is not a subspace of ℓ^1.

(b) Prove that S is a countable dense subset of ℓ^1, and conclude that ℓ^1 is separable.

2.6.21. Let X be a metric space. A collection of open subsets $\{V_i\}_{i \in I}$ is called a *base* for the topology of X if for each $x \in X$ and each open set U that contains x there is some base element V_i such that $x \in V_i \subseteq U$. Prove that if X is separable, then X has a countable base.

Remark: Using the language of abstract topology, a topological space that has a countable base is said to be *second countable*. Therefore this problem shows that every separable metric space is second countable.

2.6.22. Let (X, d) be a separable metric space, and let Y be a subset of X. Show that Y is separable (using the metric d on X restricted to the set Y).

2.6.23. Let E be a nonempty subset of a metric space X. We define the distance from a point $x \in X$ to E to be $\operatorname{dist}(x, E) = \inf\{d(x, y) : y \in E\}$. For each $r > 0$, let $G_r(E)$ be the set of all points that are within a distance of r from E:
$$G_r(E) = \{x \in E : \operatorname{dist}(x, E) < r\}.$$
Prove that $G_r(E)$ is an open set, and $\overline{E} = \cap\{G_r(E) : r > 0\}$.

2.7 The Cantor Set

We will construct an interesting example of a closed subset of $\mathbb{R}$.

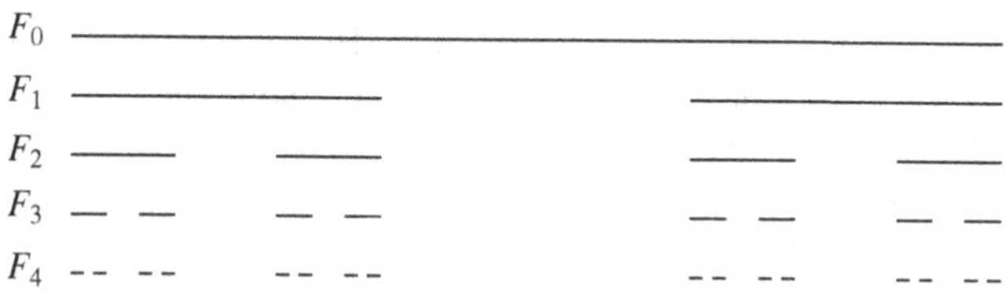

Fig. 2.8 The Cantor set C is the intersection of the sets F_n over all $n \geq 0$.

Example 2.7.1. Define
$$F_0 = [0, 1],$$
$$F_1 = [0, \tfrac{1}{3}] \cup [\tfrac{2}{3}, 1],$$
$$F_2 = [0, \tfrac{1}{9}] \cup [\tfrac{2}{9}, \tfrac{1}{3}] \cup [\tfrac{2}{3}, \tfrac{7}{9}] \cup [\tfrac{8}{9}, 1],$$

and so forth (see Figure 2.8). That is, we start with the interval $F_0 = [0, 1]$, and then remove the middle third of this interval to obtain $F_1 = [0, \tfrac{1}{3}] \cup [\tfrac{2}{3}, 1]$. We then remove the middle third of each of $[0, \tfrac{1}{3}]$ and $[\tfrac{2}{3}, 1]$ to obtain F_2, remove each middle third of the four intervals left in F_2 to obtain F_3, and so on. At stage n, the set F_n is a union of 2^n closed intervals, and we remove the middle third of each of these intervals to create the next set F_{n+1}. In this way we construct a nested decreasing sequence of closed sets
$$F_0 \supseteq F_1 \supseteq F_2 \supseteq \cdots.$$

The *Cantor middle-thirds set* (or simply the *Cantor set*) is the intersection of all of these F_n:
$$C = \bigcap_{n=0}^{\infty} F_n.$$

The Cantor set is closed because it is an intersection of closed sets. $\diamond$

While it may seem that nothing is left in C except a few points like 0, $\frac{1}{3}$, $\frac{2}{3}$, 1, in fact the Cantor set contains *uncountably many* points! One proof of the uncountability of C is given in Problem 2.7.2. Problem 2.7.3 shows that the Cantor set has many other remarkable properties:

- it contains no intervals,
- its interior is the empty set,
- it equals its own boundary, and
- every point in C is an accumulation point of C.

Although the Cantor set is very large in terms of cardinality, in other senses the Cantor set is quite small. For example, if we try to assign a "length" or "measure" to C, we are forced to conclude that its measure is zero. To see why, observe that for any given integer n, the set F_n is the union of 2^n disjoint closed intervals, each of which has length 3^{-n}. Hence the total length of the set F_n is $(2/3)^n$. The Cantor set is a subset of F_n, so if the Cantor set has a "length" or "measure," then this measure can be no more than $(2/3)^n$. This is true for every n, so the measure of C must be zero.

It is not a trivial matter to make the idea of the "measure" of subsets of $\mathbb{R}^d$ precise. This is part of the subject of *measure theory*, and there are surprising difficulties and interesting complications in the construction of what is known as *Lebesgue measure*. Measure theory is usually developed in courses on real analysis; for details we refer to texts such as [WZ77], [Fol99], [SS05], and [Heil19]. It turns out the Cantor set C has a well-defined Lebesgue measure, and this measure is zero, even though C is uncountable.

Problems

2.7.2. The *ternary expansion* of $x \in [0, 1]$ is

$$x = \sum_{n=1}^{\infty} \frac{c_n}{3^n},$$

where each "digit" c_n is either 0, 1, or 2. Every point $x \in [0, 1]$ has a unique ternary expansion, except for points of the form $x = m/3^n$ with m, n integer, which have two ternary expansions (one ending with infinitely many 0's, and one ending with infinitely many 2's). Show that x belongs to the Cantor set C if and only if x has at least one ternary expansion for which every digit c_n is either 0 or 2, and use this to show that C is uncountable. Does the point $x = \frac{1}{4}$ belong to C? Is $x = \frac{1}{4}$ a boundary point of any of the sets F_n?

2.7.3. Show that the Cantor set C is closed, C contains no open intervals, $C^\circ = \varnothing$ (the interior of C is empty), $C = \partial C$ (every point in C is a boundary point), every point in C is an accumulation point of C, and $[0,1] \setminus C$ is dense in $[0,1]$.

2.7.4. Construct a two-dimensional analogue of the Cantor set C. Subdivide the unit square $[0,1]^2$ into nine equal subsquares, and keep only the four closed corner squares. Repeat this process forever, and let S be the intersection of all of these sets. Prove that S has empty interior, equals its own boundary, and equals $C \times C$.

2.7.5. The Cantor set construction removes 2^{n-1} intervals from F_n, each of length 3^{-n}, to obtain F_{n+1}. Modify this construction by removing 2^{n-1} intervals from F_n that each have length a_n instead of 3^{-n}, and set $P = \cap F_n$ (we call P the *Smith–Volterra–Cantor set* or the *fat Cantor set*).

(a) Show that P is closed, P contains no open intervals, $P^\circ = \varnothing$, $P = \partial P$, and $[0,1] \setminus P$ is dense in $[0,1]$.

(b) Although we have not rigorously defined the measure of arbitrary sets, can you give a heuristic (informal) argument that shows that if $a_n \to 0$ quickly enough (e.g., consider $a_n = 2^{-3n}$), then P must have strictly positive measure? Hence this P is a closed set *with positive measure* that *contains no intervals*!

2.8 Compact Sets in Metric Spaces

A *compact set* is a special type of closed set. The abstract definition of a compact set is phrased in terms of coverings by open sets, but we will also derive several equivalent reformulations of the definition for subsets of metric spaces. One of these will show that if K is a compact set, then every infinite sequence $\{x_n\}_{n \in \mathbb{N}}$ of points of K has a subsequence that converges to an element of K. We will also prove that in the Euclidean space $\mathbb{F}^d$, a subset is compact if and only if it is both closed and bounded. However, in other metric spaces the characterization of compact sets is a far more subtle issue. For example, we will see that the closed unit disk in ℓ^1 is a *closed and bounded set that is not compact!*

By a *cover* of a set S, we mean a collection of sets $\{E_i\}_{i \in I}$ whose union contains S. If each set E_i is open, then we call $\{E_i\}_{i \in I}$ an *open cover* of S. The index set I may be finite or infinite (even uncountable). If I is finite, then we call $\{E_i\}_{i \in I}$ a *finite cover* of S. Thus a *finite open cover* of S is a collection of finitely many open sets whose union contains S. Also, if $\{E_i\}_{i \in I}$ is a cover of S, then a *finite subcover* is a collection of finitely many of the E_i, say $E_{i_1}, \ldots, E_{i_N}$, that still covers S.

Definition 2.8.1 (Compact Set). A subset K of a metric space X is *compact* if every open cover of K contains a finite subcover. Stated precisely, K is compact if it is the case that whenever

$$K \subseteq \bigcup_{i \in I} U_i,$$

where $\{U_i\}_{i \in I}$ is *any* collection of open subsets of X, then there exist *finitely* many indices $i_1, \ldots, i_N \in I$ such that

$$K \subseteq \bigcup_{k=1}^{N} U_{i_k}.$$

If X itself is a compact set, then we say that X is a *compact metric space*. ◇

In order for a set K to be called *compact*, it must be the case that *every* open cover of K has a finite subcover. If there is even one particular open covering of K that does not have a finite subcover, then K is not compact.

Example 2.8.2. (a) Consider the interval $(0, 1]$ in the real line $\mathbb{R}$. For each integer $k \in \mathbb{N}$, let U_k be the open interval

$$U_k = \left(\tfrac{1}{k}, 2\right).$$

Then $\{U_k\}_{k \in \mathbb{N}}$ is an open cover of $(0, 1]$, because $\cup U_k = (0, 2) \supseteq (0, 1]$, but we cannot cover the interval $(0, 1]$ using only finitely many of the sets U_k (why not?). Consequently, this open cover $\{U_k\}_{k \in \mathbb{N}}$ contains no finite subcover, so $(0, 1]$ is not compact.

Even though $(0, 1]$ is not compact, there do exist *some* open coverings that have finite subcoverings. For example, if we set $U_1 = (0, 1)$ and $U_2 = (0, 2)$, then $\{U_1, U_2\}$ is an open cover of $(0, 1]$, and $\{U_2\}$ is a finite subcover. However, as we showed above, it is not true that *every* open cover of $(0, 1]$ contains a finite subcover. This is why $(0, 1]$ is not compact.

(b) Now consider the interval $[1, \infty)$. Even though this interval is a closed subset of $\mathbb{R}$, we will prove that it is not compact. To see this, set $U_k = (k - 1, k + 1)$ for each $k \in \mathbb{N}$. Then $\{U_k\}_{k \in \mathbb{N}}$ is an open cover of $[1, \infty)$, but we cannot cover $[0, \infty)$ using only finitely many of the intervals U_k. Hence there is at least one open cover of $[0, \infty)$ that has no finite subcover, so $[0, \infty)$ is not compact. ◇

We prove next that all compact subsets of a metric space are both closed and bounded.

Lemma 2.8.3. *If K is a compact subset of a metric space X, then K is closed and bounded.*

Proof. Suppose that K is a compact set, and fix any particular point $x \in X$. Then the union of the open balls $B_n(x)$ over $n \in \mathbb{N}$ is all of X (why?), so $\{B_n(x)\}_{n \in \mathbb{N}}$ an open cover of K. This cover must have a finite subcover, say

$$\{B_{n_1}(x), B_{n_2}(x), \ldots, B_{n_M}(x)\}.$$

By reordering these sets if necessary we can assume that $n_1 < \cdots < n_M$. With this ordering the balls are *nested*:

$$B_{n_1}(x) \subseteq B_{n_2}(x) \subseteq \cdots \subseteq B_{n_M}(x).$$

Since K is contained in the union of these balls it is therefore contained in the ball of largest radius, i.e., $K \subseteq B_{n_M}(x)$. According to Definition 2.3.2, this says that K is bounded.

It remains to show that K is closed. If $K = X$ then we are done, so assume that $K \neq X$. Fix any point y in $K^C = X \backslash K$. If x is a point in K then $x \neq y$, so by the *Hausdorff property* stated in Lemma 2.3.9 there must exist disjoint open sets U_x, V_x such that $x \in U_x$ and $y \in V_x$. The collection $\{U_x\}_{x \in K}$ is an open cover of K, so it must contain some finite subcover. That is, there must exist finitely many points $x_1, \ldots, x_N \in K$ such that

$$K \subseteq U_{x_1} \cup \cdots \cup U_{x_N}. \tag{2.21}$$

Each V_{x_j} is disjoint from U_{x_j}, so it follows from equation (2.21) that the set

$$V = V_{x_1} \cap \cdots \cap V_{x_N}$$

is entirely contained in the complement of K. Hence V is an open set that satisfies

$$y \in V \subseteq K^C.$$

This shows that K^C is open (why?), so we conclude that K is closed. □

While every compact subset of a metric space is closed and bounded, there can exist sets that are closed and bounded but not compact. For example, we will see in Example 2.8.6 that the "closed unit disk" in ℓ^1 is closed and bounded but *not compact*. On the other hand, we prove next that, *in the Euclidean space* $\mathbb{F}^d$, a set is compact *if and only if* it closed and bounded.

Theorem 2.8.4 (Heine–Borel Theorem). *If* $K \subseteq \mathbb{F}^d$, *then* K *is compact if and only if* K *is closed and bounded.*

Proof. For simplicity of presentation, we will assume in this proof that $\mathbb{F} = \mathbb{R}$, but the same ideas can be used to prove the result when $\mathbb{F} = \mathbb{C}$ (this is Problem 2.8.19).

⇒. According to Lemma 2.8.3, every compact set is closed and bounded.

⇐. First we will show that the closed cube $Q = [-R, R]^d$ is a compact subset of $\mathbb{R}^d$. If Q were not compact, then there would exist an open cover $\{U_i\}_{i \in I}$ of Q that has no finite subcover. By bisecting each side of Q, we can divide Q into 2^d closed subcubes whose union is Q. At least one of these subcubes cannot be covered by finitely many of the sets U_i (why not?). Call that subcube Q_1. Then subdivide Q_1 into 2^d closed subcubes. At least one of these, which we call Q_2, cannot be covered by finitely many U_i. Continuing

in this way we obtain nested decreasing closed cubes $Q \supseteq Q_1 \supseteq Q_2 \supseteq \cdots$.
Appealing to Problem 2.8.17, $\cap Q_k$ is nonempty. Hence there is a point $x \in Q$
that belongs to every Q_k. This point must belong to some set U_i (since these
sets cover Q), and since U_i is open there is some $r > 0$ such that $B_r(x) \subseteq U_i$.
However, the sidelengths of the cubes Q_k decrease to zero as k increases,
so if we choose k large enough then we will have $x \in Q_k \subseteq B_r(x) \subseteq U_i$
(check this!). Thus Q_k is covered by a single set U_i, which is a contradiction.
Therefore Q must be compact.

Now let K be an arbitrary closed and bounded subset of $\mathbb{R}^d$. Then K is
contained in $Q = [-R, R]^d$ for some $R > 0$. Hence K is a closed subset of the
compact set Q. Therefore K is compact by Problem 2.8.11. $\square$

We will give several equivalent reformulations of compactness in terms of
the following concepts.

Definition 2.8.5. Let E be a subset of a metric space X.

(a) E is *sequentially compact* if every sequence $\{x_n\}_{n \in \mathbb{N}}$ of points from E
    contains a convergent subsequence $\{x_{n_k}\}_{k \in \mathbb{N}}$ whose limit belongs to E.

(b) E is *totally bounded* if given any $r > 0$ we can cover E by finitely many
    open balls of radius r. That is, for each $r > 0$ there must exist finitely
    many points $x_1, \ldots, x_N \in X$ such that

$$E \subseteq \bigcup_{k=1}^{N} B_r(x_k). \qquad \Diamond$$

Here are some examples illustrating sequential compactness.

Example 2.8.6. (a) Every finite subset of a metric space is sequentially compact (why?).

(b) The interval $(0, 1]$ is not sequentially compact. To see why, consider
the sequence $\{\frac{1}{n}\}_{n \in \mathbb{N}}$. This is a convergent sequence contained in $(0, 1]$, but
it does not converge to an element of $(0, 1]$, and no subsequence converges to
an element of $(0, 1]$.

(c) Suppose that X is a metric space, and E is a subset of X that is not
closed. Then E does not contain all of its accumulation points. Hence there
must be some point $x \in X$ that is an accumulation point of E but does
not belong to E. Consequently, there exist points $x_n \in E$ such that $x_n \to x$.
Hence $\{x_n\}_{n \in \mathbb{N}}$ is an infinite sequence in E that converges to a point outside of
E. Therefore (why?) no subsequence can converge to an element of E, so E is
not sequentially compact. Consequently we have proved, by a contrapositive
argument, that all sequentially compact sets are closed.

(d) The interval $[1, \infty)$ is closed, but it is not sequentially compact because
the sequence $\mathbb{N} = \{1, 2, 3, \ldots\}$ has no convergent subsequences. Thus there
can exist closed sets that are not sequentially compact.

(e) Let D be the closed unit disk in ℓ^1, i.e.,

$$D = \{x \in \ell^1 : \|x\|_1 \le 1\}.$$

This set is both closed and bounded, but it is not sequentially compact because the standard basis $\{\delta_n\}_{n\in\mathbb{N}}$ is contained in D yet it contains *no convergent subsequences* (prove this). $\diamondsuit$

Here are some examples illustrating total boundedness.

Example 2.8.7. (a) We will show that the interval $[0,1)$ is totally bounded. Given $r > 0$, let n be large enough that $\frac{1}{n} < r$. Open balls in one dimension are just open intervals, and the finitely many balls with radius $\frac{1}{n}$ centered at the points $0, \frac{1}{n}, \frac{2}{n}, \dots, 1$ cover $[0,1)$. Consequently the balls with radius r centered at these points cover $[0,1)$. Therefore $[0,1)$ is totally bounded (even though it is not closed).

(b) The interval $[1, \infty)$ is not totally bounded, because we cannot cover it with finitely many balls that each have the same fixed radius.

(c) Based on our intuition from finite dimensions, we might guess that every bounded set is totally bounded. However, according to Problem 2.8.21, the closed unit disk in ℓ^1 is not totally bounded, even though it is both closed and bounded. $\diamondsuit$

We will need the following lemma.

Lemma 2.8.8. *Let E be a sequentially compact subset of a metric space X. If $\{U_i\}_{i\in I}$ is an open cover of E, then there exists a number $\delta > 0$ such that if B is an open ball of radius δ that intersects E, then there is an $i \in I$ such that $B \subseteq U_i$.*

Proof. Let $\{U_i\}_{i\in I}$ be an open cover of E. We want to prove that there is a $\delta > 0$ such that

$$B_\delta(x) \cap E \neq \varnothing \quad \Longrightarrow \quad B_\delta(x) \subseteq U_i \text{ for some } i \in I.$$

Suppose that no $\delta > 0$ has this property. Then for each positive integer n, there must exist some open ball with radius $\frac{1}{n}$ that intersects E but is not contained in any U_i. Call this open ball G_n. For each n, choose a point $x_n \in G_n \cap E$. Then since $\{x_n\}_{n\in\mathbb{N}}$ is a sequence in E and E is sequentially compact, there must be a subsequence $\{x_{n_k}\}_{k\in\mathbb{N}}$ that converges to a point $x \in E$. Since $\{U_i\}_{i\in I}$ is a cover of E, we must have $x \in U_i$ for some $i \in I$, and since U_i is open, there must exist some $r > 0$ such that $B_r(x) \subseteq U_i$. Now choose k large enough that we have both

$$\frac{1}{n_k} < \frac{r}{3} \quad \text{and} \quad d(x, x_{n_k}) < \frac{r}{3}.$$

Keeping in mind that G_{n_k} contains x_{n_k}, G_{n_k} is an open ball with radius $1/n_k$, the distance from x to x_{n_k} is less than $r/3$, and $B_r(x)$ has radius r, it follows (why?) that $G_{n_k} \subseteq B_r(x) \subseteq U_i$. But this is a contradiction. $\square$

Now we prove some reformulations of compactness that hold for subsets of metric spaces.

Theorem 2.8.9. *If K is a subset of a metric space X, then the following three statements are equivalent.*

(a) *K is compact.*

(b) *K is sequentially compact.*

(c) *K is totally bounded and complete (where complete means that every Cauchy sequence of points from K converges to a point in K).*

Proof. (a) $\Rightarrow$ (b). We will prove the contrapositive statement. Suppose that K is not sequentially compact. Then there exists a sequence $\{x_n\}_{n \in \mathbb{N}}$ in K that has no subsequence that converges to an element of K. Choose any point $x \in K$. If every open ball centered at x contains infinitely many of the points x_n, then by considering radii $r = \frac{1}{k}$ we can construct a subsequence $\{x_{n_k}\}_{k \in \mathbb{N}}$ that converges to x. This is a contradiction, so there must exist some open ball centered at x that contains only finitely many x_n. If we call this ball B_x, then $\{B_x\}_{x \in K}$ is an open cover of K that contains no finite subcover (because any finite union $B_{x_1} \cup \cdots \cup B_{x_M}$ can contain only finitely many of the x_n). Consequently, K is not compact.

(b) $\Rightarrow$ (c). Suppose that K is sequentially compact, and suppose that $\{x_n\}_{n \in \mathbb{N}}$ is a Cauchy sequence in K. Then since K is sequentially compact, there is a subsequence $\{x_{n_k}\}_{k \in \mathbb{N}}$ that converges to some point $x \in K$. Hence $\{x_n\}_{n \in \mathbb{N}}$ is Cauchy and has a convergent subsequence. Appealing to Problem 2.2.21, this implies that $x_n \to x$. Therefore K is complete.

Suppose that K were not totally bounded. Then there would be a radius $r > 0$ such that K cannot be covered by finitely many open balls of radius r centered at points of X. Choose any point $x_1 \in K$. Since K cannot be covered by a single r-ball, K cannot be a subset of $B_r(x_1)$. Hence there exists a point $x_2 \in K \backslash B_r(x_1)$. In particular, $d(x_2, x_1) \geq r$. But K cannot be covered by two r-balls, so there must exist a point x_3 that belongs to $K \backslash (B_r(x_1) \cup B_r(x_2))$. In particular, we have both $d(x_3, x_1) \geq r$ and $d(x_3, x_2) \geq r$. Continuing in this way, we obtain a sequence of points $\{x_n\}_{n \in \mathbb{N}}$ in K that has no convergent subsequence, which is a contradiction.

(c) $\Rightarrow$ (b). Assume that K is complete and totally bounded, and let $\{x_n\}_{n \in \mathbb{N}}$ be any sequence of points in K. Since K is totally bounded, it can be covered by finitely many open balls of radius $\frac{1}{2}$. Each x_n belongs to K and hence must be contained in one or more of these balls. But there are only finitely many of the balls, so at least one ball must contain x_n for infinitely many different indices n. That is, there is some infinite subsequence

$\{x_n^{(1)}\}_{n\in\mathbb{N}}$ of $\{x_n\}_{n\in\mathbb{N}}$ that is contained in an open ball of radius $\frac{1}{2}$. The Triangle Inequality therefore implies that

$$\forall\, m, n \in \mathbb{N}, \quad d(x_m^{(1)}, x_n^{(1)}) < 1.$$

Similarly, since K can be covered by finitely many open balls of radius $\frac{1}{4}$, there is some subsequence $\{x_n^{(2)}\}_{n\in\mathbb{N}}$ of $\{x_n^{(1)}\}_{n\in\mathbb{N}}$ such that

$$\forall\, m, n \in \mathbb{N}, \quad d(x_m^{(2)}, x_n^{(2)}) < \frac{1}{2}.$$

Continuing by induction, for each $k > 1$ we find a subsequence $\{x_n^{(k)}\}_{n\in\mathbb{N}}$ of $\{x_n^{(k-1)}\}_{n\in\mathbb{N}}$ such that $d(x_m^{(k)}, x_n^{(k)}) < \frac{1}{k}$ for all $m, n \in \mathbb{N}$.

Now consider the diagonal subsequence $\{x_k^{(k)}\}_{k\in\mathbb{N}}$. Given $\varepsilon > 0$, let N be large enough that $\frac{1}{N} < \varepsilon$. If $j \geq k > N$, then $x_j^{(j)}$ is one element of the sequence $\{x_n^{(k)}\}_{n\in\mathbb{N}}$ (why?), say $x_j^{(j)} = x_n^{(k)}$. Hence

$$d(x_j^{(j)}, x_k^{(k)}) = d(x_n^{(k)}, x_k^{(k)}) < \frac{1}{k} < \frac{1}{N} < \varepsilon.$$

Thus $\{x_k^{(k)}\}_{k\in\mathbb{N}}$ is a Cauchy subsequence of the original sequence $\{x_n\}_{n\in\mathbb{N}}$. Since K is complete, this subsequence must converge to some element of E. Hence K is sequentially compact.

(b) $\Rightarrow$ (a). Assume that K is sequentially compact. Since we have already proved that statement (b) implies statement (c), we know that K is complete and totally bounded.

Suppose that $\{U_i\}_{i\in I}$ is any open cover of K. By Lemma 2.8.8, there exists a $\delta > 0$ such that if B is an open ball of radius δ that intersects E, then there is an $i \in I$ such that $B \subseteq U_i$. However, K is totally bounded, so we can cover K by finitely many open balls of radius δ. Each of these balls is contained in some U_i, so K is covered by finitely many U_i. $\square$

The type of reasoning that was used to prove the implication (c) $\Rightarrow$ (b) of Theorem 2.8.9 is often called a *Cantor diagonalization argument*. We will see another argument of this type in the proof of Theorem 7.4.6.

We saw in Example 2.8.6 that the closed unit disk D in ℓ^1 is not sequentially compact. Applying Theorem 2.8.9, it follows that D is neither compact nor totally bounded, even though it is both closed and bounded.

Problems

2.8.10. (a) Prove directly from the definition that every finite subset of a metric space, including the empty set, is compact.

(b) In $X = \mathbb{R}$, exhibit an open cover of $E = \{\frac{1}{n}\}_{n \in \mathbb{N}}$ that has no finite subcover.

2.8.11. Given a metric space X, prove the following statements.

(a) If K is a compact subset of X and $E \subseteq K$, then E is closed if and only if E is compact.

(b) If E is a subset of X, then E is contained in some compact set K if and only if $\overline{E}$ is compact.

2.8.12. Let X be a metric space. Prove that an arbitrary intersection of compact subsets of X is compact, and a finite union of compact subsets of X is compact.

2.8.13. (a) Suppose that F and K are nonempty disjoint subsets of a metric space X. Prove that if F is closed and K is compact, then $\mathrm{dist}(F, K) > 0$, where $\mathrm{dist}(F, K) = \inf\{d(x, y) : x \in F, y \in K\}$.

(b) Show by example that the distance between two nonempty disjoint *closed* sets E and F can be zero.

2.8.14. Assume that K is a compact subset of $\mathbb{R}$. Prove directly that $K' = \{(x, 0) : x \in K\}$ is a compact subset of $\mathbb{R}^2$.

Note: *Directly* means that you should not use the Heine–Borel Theorem in your proof; instead, prove that K' is compact by using the definition of a compact set.

2.8.15. Prove that a compact metric space is both complete and separable.

2.8.16. Suppose that $f: \mathbb{R} \to \mathbb{R}$ is differentiable at every point, but there is no x such that we have both $f(x) = 0$ and $f'(x) = 0$. Prove that there are at most finitely many points $x \in [0, 1]$ such that $f(x) = 0$.

2.8.17. (a) Exhibit closed intervals $I_n \subseteq \mathbb{R}$ that satisfy $I_1 \supseteq I_2 \supseteq \cdots$ and $\cap I_n = \varnothing$.

(b) Assume that $I_1 \supseteq I_2 \supseteq \cdots$ is a nested decreasing sequence of closed finite intervals in $\mathbb{R}$. Prove that $\cap I_n \neq \varnothing$.

(c) Let $Q_1 \supseteq Q_2 \supseteq \cdots$ be a nested decreasing sequence of closed cubes in $\mathbb{R}^d$. Prove that $\cap Q_n \neq \varnothing$.

Remark: This problem is a special case of Problem 2.8.18, but it is instructive to prove it directly.

2.8.18. (a) Prove the *Cantor Intersection Theorem* for metric spaces: If $K_1 \supseteq K_2 \supseteq \cdots$ is a nested decreasing sequence of nonempty compact subsets of a metric space X, then $\cap K_n \neq \varnothing$.

(b) Show by example that it is possible for the intersection of a nested decreasing sequence of nonempty *closed* sets to be empty.

2.8.19. We take $\mathbb{F} = \mathbb{C}$ in this problem.

(a) Given $R > 0$, prove that

$$Q = \{(a_1 + ib_1, \ldots, a_d + ib_d) : -R \le a_i, b_i \le R\}$$

is a compact subset of $\mathbb{C}^d$.

(b) Prove the Heine–Borel Theorem (Theorem 2.8.4) for the case $\mathbb{F} = \mathbb{C}$.

2.8.20. Let E be a totally bounded subset of a metric space X.

(a) Prove directly that $\overline{E}$ is totally bounded.

(b) Show that if X is complete, then $\overline{E}$ is compact.

2.8.21. The closed unit disk in ℓ^1 is $D = \{x \in \ell^1 : \|x\|_1 \le 1\}$. Give direct proofs of the following statements.

(a) D is a closed and bounded subset of ℓ^1.

(b) D is not sequentially compact.

(c) D is not totally bounded.

(d) The collection of open balls $\{B_2(\pm \delta_n)\}_{n \in \mathbb{N}}$ covers D but does not have a finite subcover, so D is not compact.

2.8.22. Let K be the "rectangular box" in ℓ^1 that has sides of length $\frac{1}{2}, \frac{1}{4}, \ldots$, i.e.,

$$K = \{y = (y_k)_{k \in \mathbb{N}} : 0 \le y_k \le 2^{-k}\}.$$

This problem will show that K, which is known as a *Hilbert cube*, is a compact subset of ℓ^1.

(a) Suppose that $\{z_k\}_{k \in \mathbb{N}}$ is a sequence of elements of K and z_k converges *componentwise* to $y \in \ell^1$. Prove that $z_k \to y$ *in ℓ^1-norm*.

(b) Suppose that $\{x_n\}_{n \in \mathbb{N}}$ is an infinite sequence of elements of K. Use a Cantor diagonalization argument to prove that there is a subsequence $\{x_k^{(k)}\}_{k \in \mathbb{N}}$ that converges componentwise to some sequence y.

(c) Prove that K is sequentially compact in ℓ^1 (and therefore is also compact and totally bounded).

(d) Is there anything special about the numbers 2^{-k}, or can we replace them with numbers $a_k > 0$ that satisfy $\sum a_k < \infty$?

2.9 Continuity for Functions on Metric Spaces

Our focus up to now has mostly been on particular metric spaces and on particular points in those spaces. Now we will look at *functions* that transform points in one metric space into points in another space. We will be especially interested in *continuous functions*, which are defined as follows.

Definition 2.9.1 (Continuous Function). Let X and Y be metric spaces. We say that a function $f \colon X \to Y$ is *continuous* if given any open set $V \subseteq Y$, its inverse image $f^{-1}(V)$ is an open subset of X. $\quad \diamond$

That is, f is continuous if the *inverse image* of every open set is open. However, the *direct image* of an open set under a continuous function need not be open. For example, $f(x) = \sin x$ is continuous on $\mathbb{R}$, but the direct image of the open interval $(0, 2\pi)$ is $f((0, 2\pi)) = [-1, 1]$. Thus, a continuous function need not send open sets *to* open sets.

By taking complements, we can infer that a continuous function also need not map closed sets to closed sets. For a concrete example, take $f(x) = \arctan x$. This function is continuous on $\mathbb{R}$, which is a closed set, but the the direct image of $\mathbb{R}$ under f is $f(\mathbb{R}) = (-\frac{\pi}{2}, \frac{\pi}{2})$, which is not closed. Consequently, the following lemma, which states that *a continuous function maps compact sets to compact sets*, may seem a bit surprising at first glance.

Lemma 2.9.2. *Let X and Y be metric spaces. If $f \colon X \to Y$ is continuous and K is a compact subset of X, then $f(K)$ is a compact subset of Y.*

Proof. Let $\{V_i\}_{i \in I}$ be any open cover of $f(K)$. Then each set $U_i = f^{-1}(V_i)$ is open, and $\{U_i\}_{i \in I}$ is an open cover of K (why?). Since K is compact, this cover must have a finite subcover, say $\{U_{i_1}, \dots, U_{i_N}\}$. The reader should check that this implies that $\{V_{i_1}, \dots, V_{i_N}\}$ is a finite subcover of $f(K)$. Therefore $f(K)$ is compact. $\quad \square$

As a corollary, we prove that a continuous function on a compact set must be bounded.

Corollary 2.9.3. *Let X and Y be metric spaces, and assume that K is a compact subset of X. If $f \colon K \to Y$ is continuous, then f is bounded on K, i.e., $\mathrm{range}(f)$ is a bounded subset of Y.*

Proof. Lemma 2.9.2 implies that $\mathrm{range}(f) = f(K)$ is a compact subset of Y. By Lemma 2.8.3, all compact sets are bounded. $\quad \square$

In particular, if $f \colon K \to \mathbb{F}$ is a continuous function on a compact domain K, then the range of f is a bounded set of scalars, and therefore

$$\sup_{x \in K} |f(x)| < \infty.$$

The next lemma gives a useful reformulation of continuity in terms of preservation of limits. In particular, part (c) of this lemma says that a function on a metric space is continuous if and only if f *maps convergent sequences to convergent sequences*.

Lemma 2.9.4. *Let X be a metric space with metric d_X, and let Y be a metric space with metric d_Y. Given a function $f \colon X \to Y$, the following three statements are equivalent.*

(a) f *is continuous.*

(b) *If x is any point in X, then for every $\varepsilon > 0$ there exists a $\delta > 0$ such that*

$$d_X(x, y) < \delta \quad \Longrightarrow \quad d_Y\big(f(x), f(y)\big) < \varepsilon.$$

(c) *Given any point $x \in X$ and any sequence $\{x_n\}_{n \in \mathbb{N}}$ in X,*

$$x_n \to x \text{ in } X \quad \Longrightarrow \quad f(x_n) \to f(x) \text{ in } Y.$$

Proof. For this proof we let $B_r^X(x)$ and $B_s^Y(y)$ denote open balls in X and Y, respectively.

(a) $\Rightarrow$ (b). Suppose that f is continuous, and choose any point $x \in X$ and any $\varepsilon > 0$. Then the open ball $V = B_\varepsilon^Y(f(x))$ is an open subset of Y, so $U = f^{-1}(V)$ must be an open subset of X. Since $x \in U$, there exists some $\delta > 0$ such that $B_\delta^X(x) \subseteq U$. If $y \in X$ is any point that satisfies $d_X(x, y) < \delta$, then $y \in B_\delta^X(x) \subseteq U$, and therefore

$$f(y) \in f(U) \subseteq V = B_\varepsilon^Y(f(x)).$$

Consequently, $d_Y\big(f(x), f(y)\big) < \varepsilon$.

(b) $\Rightarrow$ (c). Assume that statement (b) holds, choose $x \in X$, and let $x_n \in X$ be any points such that $x_n \to x$. Fix $\varepsilon > 0$, and let $\delta > 0$ be the number whose existence is given by statement (b). Since $x_n \to x$, there must exist some $N > 0$ such that $d_X(x, x_n) < \delta$ for all $n > N$. Statement (b) therefore implies that $d_Y\big(f(x), f(x_n)\big) < \varepsilon$ for all $n > N$, so we conclude that $f(x_n) \to f(x)$ in Y.

(c) $\Rightarrow$ (a). Suppose that statement (c) holds, and let V be any open subset of Y. Suppose that $f^{-1}(V)$ were not open. Then $f^{-1}(V)$ is not empty, so there is a point $x \in f^{-1}(V)$. Therefore, by definition, $y = f(x) \in V$. Since V is open, there exists some $\varepsilon > 0$ such that $B_\varepsilon^Y(y) \subseteq V$. But $f^{-1}(V)$ is not open, so the ball $B_{1/n}^X(x)$ cannot be entirely contained in $f^{-1}(V)$ for any integer n. Hence for every n there must be some $x_n \in B_{1/n}^X(x)$ such that $x_n \notin f^{-1}(V)$. Since $d_X(x, x_n) < 1/n$, we have $x_n \to x$. Applying statement (c), it follows that $f(x_n) \to f(x)$. Therefore there must exist some $N > 0$ such that $d_Y\big(f(x), f(x_n)\big) < \varepsilon$ for all $n > N$. Hence for every $n > N$ we have $f(x_n) \in B_\varepsilon^Y(y) \subseteq V$, and therefore $x_n \in f^{-1}(V)$. This contradicts the definition of x_n, so we conclude that $f^{-1}(V)$ must be open. Therefore f is continuous. $\square$

The number δ that appears in statement (b) of Lemma 2.9.4 depends both on the point x and the number ε. We say that a function f is *uniformly continuous* if δ can be chosen independently of x. Here is the precise definition.

Definition 2.9.5 (Uniform Continuity). Let X be a metric space with metric d_X, and let Y be a metric space with metric d_Y. If $E \subseteq X$, then we say that a function $f : X \to Y$ is *uniformly continuous* on E if for every $\varepsilon > 0$ there exists a $\delta > 0$ such that for all $x, y \in E$ we have

$$ d_X(x, y) < \delta \implies d_Y\big(f(x), f(y)\big) < \varepsilon. \qquad \diamondsuit $$

The reader should be aware that "uniform" is another overused term in mathematical English. In particular, *uniform continuity* as introduced in Definition 2.9.5 is distinct from *uniform convergence*, which is convergence with respect to the uniform norm (see Definition 3.5.5).

Next we prove that a function that is continuous on a compact domain is uniformly continuous on that set.

Lemma 2.9.6. *Let X and Y be metric spaces. If K is a compact subset of X and $f : K \to Y$ is continuous, then f is bounded and uniformly continuous on K.*

Proof. For this proof, let d_X and d_Y denote the metrics on X and Y, and let $B_r^X(x)$ and $B_s^Y(y)$ denote open balls in X and Y, respectively.

Suppose that $f : K \to Y$ is continuous. Corollary 2.9.3 showed that f is bounded, so we just have to prove that f is uniformly continuous. Fix $\varepsilon > 0$. For each point $z \in Y$, let $U_z = f^{-1}(B_\varepsilon^Y(z))$. Then $\{U_z\}_{z \in Y}$ is an open cover of K. Since K is compact, it is sequentially compact. Therefore Lemma 2.8.8 implies that there exists a number $\delta > 0$ such that if B is any open ball of radius δ in X that intersects K, then $B \subseteq U_z$ for some $z \in Y$.

Now choose any points $x, y \in K$ with $d_X(x, y) < \delta$. Then $x, y \in B_\delta^X(x)$, and $B_\delta^X(x)$ must be contained in some set U_z, so

$$ f(x), f(y) \in f(B_\delta^X(x)) \subseteq f(U_z) \subseteq B_\varepsilon^Y(z). $$

Thus $f(x)$ and $f(y)$ are both contained in the same ball. Since that ball has radius ε, it follows that $d_Y(f(x), f(y)) < 2\varepsilon$. This shows that f is uniformly continuous. $\square$

Can you give an example of a function $f : \mathbb{R} \to \mathbb{R}$ that is bounded and continuous but not uniformly continuous?

Sometimes we need to refer to the set where a function is nonzero, but we also often encounter the closure of that set. We introduce the following terminology.

Definition 2.9.7 (Support). Let X be a metric space. The *support* of a function $f : X \to \mathbb{F}$ is the closure in X of the set of points where f is nonzero:

$$ \mathrm{supp}(f) = \overline{\{x \in X : f(x) \neq 0\}}. $$

We say that f has *compact support* if $\mathrm{supp}(f)$ is a compact subset of X. $\quad \diamondsuit$

For example, if we define $f \colon \mathbb{R} \to \mathbb{R}$ by

$$f(x) = \begin{cases} \sin x, & 0 \le x \le 2\pi, \\ 0, & \text{otherwise,} \end{cases}$$

then $f(x)$ is nonzero for all $x \in (0, \pi) \cup (\pi, 2\pi)$, and the support of f is the closed interval $\operatorname{supp}(f) = [0, 2\pi]$.

By definition, the support of a continuous function is a closed set. By the Heine–Borel Theorem, a subset of the Euclidean space $\mathbb{F}^d$ is compact if and only if it is both closed and bounded. Therefore, if the domain of f is $\mathbb{F}^d$, then f has compact support if and only if $\operatorname{supp}(f)$ is a bounded set. This happens if and only if f is identically zero outside of some ball. Consequently:

> A continuous function $f \colon \mathbb{F}^d \to \mathbb{F}$ has compact support if and only if f is identically zero outside of some ball $B_r(x)$.

Problems

2.9.8. Let $X = [0, \infty)$, and prove the following statements.

(a) $f(x) = x^2$ is continuous but not uniformly continuous on $[0, \infty)$.

(b) $g(x) = x^{1/2}$ is continuous and uniformly continuous on $[0, \infty)$.

(c) $h(x) = \sin x^2$ is continuous but not uniformly continuous on $[0, \infty)$, even though it is bounded.

(d) If $f \colon [0, \infty) \to \mathbb{F}$ is continuous and there is some $R > 0$ such that $f(x) = 0$ for all $x > R$, then f is uniformly continuous on $[0, \infty)$.

(e) If $f \colon [0, \infty) \to \mathbb{F}$ is continuous and f "vanishes at infinity," i.e., if

$$\lim_{x \to \infty} f(x) = 0,$$

then f is uniformly continuous on $[0, \infty)$.

2.9.9. Let X and Y be metric spaces. Show that a function $f \colon X \to Y$ is continuous if and only if $f^{-1}(E)$ is a closed subset of X for each closed set $E \subseteq Y$.

2.9.10. Let X and Y be metric spaces. Assume that E is a dense subset of X, and $f, g \colon X \to Y$ are continuous functions such that $f(x) = g(x)$ for each $x \in E$. Prove that $f = g$, i.e., $f(x) = g(x)$ for every $x \in X$.

2.9.11. Prove that every function $f \colon \mathbb{Z} \to \mathbb{F}$ is continuous, where the metric on $\mathbb{Z}$ is the restriction of the standard metric from $\mathbb{R}$ to $\mathbb{Z}$.

2.9.12. Let X, Y, and Z be metric spaces, and suppose that $f \colon X \to Y$ and $g \colon Y \to Z$ are continuous functions. Prove that the composition $g \circ f \colon X \to Z$ is continuous.

2.9.13. Given continuous functions f, $g\colon X \to \mathbb{F}$ on a metric space X, prove the following statements.

(a) If a and b are scalars, then $af + bg$ is continuous.

(b) fg is continuous.

(c) If $g(x) \neq 0$ for every x, then $1/g$ and f/g are continuous.

(d) $h(x) = |f(x)|$ is continuous.

(e) If $\mathbb{F} = \mathbb{R}$, then $m(x) = \min\{f(x), g(x)\}$ and $M(x) = \max\{f(x), g(x)\}$ are continuous.
Hint: Show that $m(x) = \frac{1}{2}(f + g - |f - g|)$ and $M(x) = \frac{1}{2}(f + g + |f - g|)$.

(f) $Z_f = \{x \in X : f(x) = 0\}$ is a closed subset of X.

2.9.14. Let K be a compact subset of a metric space X, and assume that $f\colon K \to \mathbb{R}$ is continuous. Prove that f achieves a maximum and a minimum on K, i.e., there exist points $x, y \in K$ such that

$$f(x) = \inf\{f(t) : t \in K\} \quad \text{and} \quad f(y) = \sup\{f(t) : t \in K\}.$$

2.9.15. Assume $f\colon X \to Y$ is continuous, where X and Y are metric spaces. Given $E \subseteq X$, prove that $f(\overline{E}) \subseteq \overline{f(E)}$, but show by example that equality need not hold.

2.9.16. Let X and Y be metric spaces. The *graph* of a function $f\colon X \to Y$ is $\mathrm{graph}(f) = \{(x, f(x)) \in X \times Y : x \in X\}$. Prove that if f is continuous and X is compact, then $\mathrm{graph}(f)$ is a compact subset of $X \times Y$ (with respect to the metric defined in Problem 2.2.20).

2.9.17. Suppose that X is a vector space. We say that a metric d on X is *translation-invariant* if $\mathrm{d}(x + z, y + z) = \mathrm{d}(x, y)$ for all x, y, $z \in X$. Show that if d is a translation-invariant metric, then vector addition is continuous. That is, show that if we define $f\colon X \times X \to X$ by $f(x, y) = x + y$, then f is continuous (see Problem 2.2.20 for the definition of the metric on $X \times X$).

2.9.18. (a) Let X and Y be metric spaces, and suppose that $f\colon X \to Y$ is uniformly continuous. Prove that f maps Cauchy sequences to Cauchy sequences, i.e.,

$$\{x_n\}_{n \in \mathbb{N}} \text{ is Cauchy in } X \quad \Longrightarrow \quad \{f(x_n)\}_{n \in \mathbb{N}} \text{ is Cauchy in } Y.$$

(b) Show by example that part (a) can fail if we only assume that f is continuous instead of uniformly continuous. (Contrast this with Lemma 2.9.4, which shows that every continuous function must map convergent sequences to convergent sequences.)

2.9.19. Let (a, b) be a finite open interval. Given a function $f\colon (a, b) \to \mathbb{F}$, prove that the following three statements are equivalent.

(a) f is uniformly continuous on (a, b).

(b) If $\{x_n\}_{n \in \mathbb{N}}$ is a sequence in (a, b) that converges to some point $x \in [a, b]$, then $\{f(x_n)\}_{n \in \mathbb{N}}$ is a convergent sequence of scalars. (Note that x might not belong to (a, b).)

(c) There is a continuous function $g\colon [a, b] \to \mathbb{F}$ that satisfies $g(x) = f(x)$ for every $x \in (a, b)$.

2.9.20. Let (X, d_X) and (Y, d_Y) be metric spaces. We say that a function $f\colon X \to Y$ is *Lipschitz continuous* on X with *Lipschitz constant* $C \geq 0$ if

$$\forall\, x, y \in X, \quad d_Y\big(f(x), f(y)\big) \leq C\, d_X(x, y).$$

Prove that a Lipschitz continuous function is uniformly continuous on X.

2.9.21. This problem will establish the *Banach Fixed Point Theorem* (also known as the *Contractive Mapping Theorem*). Let X be a *complete* metric space. We say that $f\colon X \to X$ is a *contraction* if f is Lipschitz continuous in the sense of Problem 2.9.20 and its Lipschitz constant C satisfies $0 \leq C < 1$. Let f^0 be the identity function on X, and let $f^1 = f$, $f^2 = f \circ f$, $f^3 = f \circ f \circ f$, and so forth.

(a) Given $x \in X$, prove that

$$d(f^n(x), x) \leq \frac{1 - C^n}{1 - C}\, d(f(x), x), \qquad \text{for all } n \geq 0.$$

(b) Show that the sequence $\{f^n(x)\}_{n \in \mathbb{N}}$ is Cauchy, and therefore converges to some point $x_0 \in X$.

(c) Show that f has a unique *fixed point*, i.e., there is a unique point $x_0 \in X$ that satisfies $f(x_0) = x_0$.

(d) Prove that $d(x_0, x) \leq \dfrac{1}{1 - C}\, d(f(x), x)$ for all $x \in X$.

(e) Fix $0 < \alpha < 1$ and define $T\colon \mathbb{R}^2 \to \mathbb{R}^2$ by $T(x, y) = (x, \alpha y)$. Prove that T is not contractive on $\mathbb{R}^2$, but if we fix an x and let $L_x = \{(x, y) : y \in \mathbb{R}\}$, then T is contractive on the line L_x. Identify the fixed point of T on L_x.

2.9.22. Let X be a *compact* metric space, and assume $f\colon X \to X$ satisfies

$$d\big(f(x), f(y)\big) < d(x, y), \qquad \text{for all } x \neq y \in X.$$

Note that f need not be a contraction in the sense of Problem 2.9.21. Define a function $g\colon X \to \mathbb{R}$ by $g(x) = d\big(x, f(x)\big)$ for $x \in X$. Prove the following statements.

(a) $|g(x) - g(y)| < 2\,\mathrm{d}(x, y)$ for all $x, y \in X$.

(b) g is continuous and $g(x) = 0$ for some $x \in X$.

(c) f has a unique fixed point, i.e., there is a unique point $x_0 \in X$ that satisfies $f(x_0) = x_0$.

2.9.23. We say that a function $f: \mathbb{R}^d \to \mathbb{R}$ is *upper semicontinuous* (abbreviated usc) at a point $x \in \mathbb{R}^d$ if for every $\varepsilon > 0$ there exists some $\delta > 0$ such that

$$|x - y| < \delta \implies f(y) \leq f(x) + \varepsilon.$$

An analogous definition is made for *lower semicontinuity* (lsc). Prove the following statements.

(a) If $g: \mathbb{R}^d \to \mathbb{R}$ and $r > 0$ are fixed, then $h(x) = \inf\{g(y) : y \in B_r(x)\}$ is usc at every point where $h(x) \neq -\infty$.

(b) If $f: \mathbb{R}^d \to \mathbb{R}$ is given, then f is continuous at x if and only if f is both usc and lsc at x.

(c) If $\{f_i\}_{i \in J}$ is a family of functions on $\mathbb{R}^d$ that are each usc at a point x, then $g(t) = \inf_{i \in J} f_i(t)$ is usc at x.

(d) $f: \mathbb{R}^d \to \mathbb{R}$ is usc at every point $x \in \mathbb{R}^d$ if and only if

$$f^{-1}[a, \infty) = \{x \in \mathbb{R}^d : f(x) \geq a\}$$

is closed for each $a \in \mathbb{R}$. Likewise, f is lsc at every point x if and only if

$$f^{-1}(a, \infty) = \{x \in \mathbb{R}^d : f(x) > a\}$$

is open for each $a \in \mathbb{R}$.

(e) If K is a compact subset of $\mathbb{R}^d$ and $f: \mathbb{R}^d \to \mathbb{R}$ is usc at every point of K, then f is bounded above on K.

2.10 Urysohn's Lemma

Urysohn's Lemma is a basic result about the existence of continuous functions. To prove it we will need the following lemma, which states that the distance from a point x to a subset E of a metric space is a uniformly continuous function of x.

Lemma 2.10.1. *Let X be a metric space. If E is a nonempty subset of X, then the function $f: X \to \mathbb{R}$ defined by*

$$f(x) = \mathrm{dist}(x, E) = \inf\{\mathrm{d}(x, z) : z \in E\}, \qquad x \in X,$$

is uniformly continuous on X.

Proof. Fix $\varepsilon > 0$, and set $\delta = \varepsilon/2$. Choose any points $x, y \in X$ such that $d(x, y) < \delta$. By definition of an infimum, there exist points $a, b \in E$ such that

$$d(x, a) < \text{dist}(x, E) + \delta \quad \text{and} \quad d(y, b) < \text{dist}(y, E) + \delta.$$

Consequently,

$$\begin{aligned}
f(y) = \text{dist}(y, E) &\leq d(y, a) \\
&\leq d(y, x) + d(x, a) \\
&< \delta + \big(\text{dist}(x, E) + \delta\big) \\
&= f(x) + \varepsilon.
\end{aligned}$$

Interchanging the roles of x and y, we similarly obtain $f(x) < f(y) + \varepsilon$. Therefore $|f(x) - f(y)| < \varepsilon$ whenever $d(x, y) < \delta$, so f is uniformly continuous on X. □

Now we prove Urysohn's Lemma, which states that if E and F are disjoint closed sets, then we can find a continuous function that is identically 0 on E and identically 1 on F. In this sense, E and F can be "separated" by a continuous function.

Theorem 2.10.2 (Urysohn's Lemma). *If E, F are disjoint closed subsets of a metric space X, then there exists a continuous function $\theta\colon X \to \mathbb{R}$ such that:*

(a) $0 \leq \theta \leq 1$ *on X,*

(b) $\theta = 0$ *on E, and*

(c) $\theta = 1$ *on F.*

Proof. If $E = \varnothing$ then we just take $\theta = 1$, and likewise if $F = \varnothing$ then we can take $\theta = 0$. Therefore we assume that E and F are both nonempty. Since E is closed, if $x \notin E$ then $\text{dist}(x, E) > 0$ (see Problem 2.4.7). Since Lemma 2.10.1 tells us that $\text{dist}(x, E)$ and $\text{dist}(x, F)$ are each continuous functions of x, it follows (verify this!) that the function

$$\theta(x) = \frac{\text{dist}(x, E)}{\text{dist}(x, E) + \text{dist}(x, F)}$$

has the required properties. □

We can sometimes improve on Urysohn's Lemma when dealing with particular concrete domains. For example, the following result says that if we take $X = \mathbb{R}$ and fix a compact set $K \subseteq \mathbb{R}$, then we can find an *infinitely differentiable* function that is identically 1 on K and identically 0 outside of a larger set.

Theorem 2.10.3. *If $K \subseteq \mathbb{R}$ is compact and $U \supseteq K$ is open, then there exists an infinitely differentiable function $f\colon \mathbb{R} \to \mathbb{R}$ such that*

(a) $0 \le f \le 1$ *everywhere,*

(b) $f = 1$ *on K, and*

(c) $f = 0$ *on U^C.* ◇

One way to prove Theorem 2.10.3 is by combining Urysohn's Lemma with the operation of *convolution*. For details on convolution we refer to texts such as [DM72], [Ben97], [Kat04], and [Heil19]. For comparison, Problem 2.10.6 presents a slightly less ambitious result. Specifically, that problem shows that there exists an infinitely differentiable function g that is identically zero outside of the finite interval $[-1, 1]$ and nonzero on the interior of this interval. Problems 2.10.5 and 2.10.7 give some related constructions.

Problems

2.10.4. Let E, F be disjoint closed subsets of a metric space X. Set $f(x) = \mathrm{dist}(x, E)$ and $g(x) = \mathrm{dist}(x, F)$ for $x \in X$, and define

$$U = \{x \in X : f(x) < g(x)\} \qquad \text{and} \qquad V = \{x \in X : g(x) < f(x)\}.$$

Prove that U and V are disjoint open sets that satisfy $E \subseteq U$ and $F \subseteq V$.

Remark: In the language of abstract topology, this shows that every metric space is a *normal topological space*.

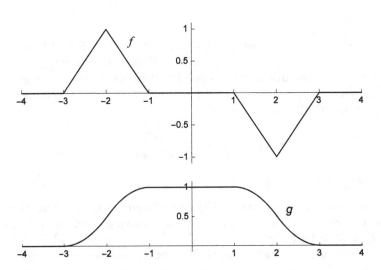

Fig. 2.9 The functions f (top) and g (bottom) from Problem 2.10.5.

2.10.5. Let f be the piecewise linear function pictured in Figure 2.9. Show that

$$g(x) \;=\; \int_{-3}^{x} f(t)\,dt, \qquad x \in \mathbb{R},$$

has the following properties.

(a) $0 \leq g(x) \leq 1$ for every $x \in \mathbb{R}$.

(b) $g(x) = 1$ for $|x| \leq 1$.

(c) $g(x) = 0$ for $|x| \geq 3$.

(d) g is differentiable on $\mathbb{R}$, but g is not twice differentiable on $\mathbb{R}$.

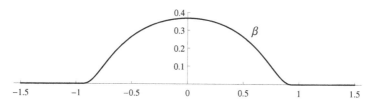

Fig. 2.10 The infinitely differentiable function β from Problem 2.10.6.

2.10.6. Define

$$\gamma(x) \;=\; \begin{cases} e^{-1/x}, & x > 0, \\ 0, & x \leq 0, \end{cases}$$

and set $\beta(x) = \gamma(1 - x^2)$. The graph of β is shown in Figure 2.10. Prove the following statements.

(a) $\gamma(x) = 0$ for all $x \leq 0$, and $\gamma(x) > 0$ for all $x > 0$.

(b) For each $n \in \mathbb{N}$, there exists a polynomial p_n of degree $n - 1$ such that the nth derivative of γ is

$$\gamma^{(n)}(x) \;=\; \frac{p_n(x)}{x^{2n}}\, \gamma(x).$$

(c) γ is infinitely differentiable, and $\gamma^{(n)}(0) = 0$ for every $n \geq 0$.

(d) β is infinitely differentiable, $\beta(x) > 0$ for $|x| < 1$, and $\beta(x) = 0$ for $|x| \geq 1$.

2.10.7. Prove that there exists an infinitely differentiable function θ that satisfies:

(a) $0 \leq \theta(x) \leq 1$ for every $x \in \mathbb{R}$,

(b) $\theta(x) = 1$ for all $|x| \leq 1$,

(c) $\theta(x) = 0$ for all $|x| \geq 3$.

2.11 The Baire Category Theorem

It is not possible to write the Euclidean plane $\mathbb{R}^2$ as the union of *countably many* straight lines. This statement is a special case of the Baire Category Theorem, which we will prove in this section. This theorem says that a complete metric space cannot be written as a countable union of sets that are "nowhere dense" in the following sense.

Definition 2.11.1 (Nowhere Dense and Meager Sets). Let X be a metric space, and fix $E \subseteq X$.

(a) E is *nowhere dense* or *rare* if $X \backslash \overline{E}$ is dense in X.

(b) E is *meager* or *first category* if it can be written as a countable union of nowhere dense sets.

(c) E is *nonmeager* or *second category* if it is not meager. $\diamond$

Although the terms "first category" and "second category" are in common use, they are easily confusable, so we will prefer the more descriptive terms "meager" and "nonmeager."

Here is a useful reformulation of the meaning of "nowhere dense" (the proof is assigned as Problem 2.11.7).

Lemma 2.11.2. *If E is a nonempty subset of a metric space X, then the following two statements are equivalent.*

(a) *E is nowhere dense.*

(b) *$\overline{E}$ has empty interior, i.e., $\overline{E}$ contains no nonempty open subsets.* $\diamond$

In particular, a nonempty *closed* set is nowhere dense if and only if it contains no open balls. For example, the Cantor set (see Example 2.7.1) is a nowhere dense subset of $[0, 1]$ because it is closed and contains no intervals. The set of rationals $\mathbb{Q}$ also contains no intervals, but it is not a nowhere dense subset of $\mathbb{R}$ because its closure is $\mathbb{R}$, which has a nonempty interior. On the other hand, $\mathbb{Q}$ is meager in $\mathbb{R}$ since it is the countable union of the singletons $\{x\}$ with x ranging through $\mathbb{Q}$, and each singleton $\{x\}$ is nowhere dense.

Now we prove the Baire Category Theorem.

Theorem 2.11.3 (Baire Category Theorem). *If X is a nonempty complete metric space, then X is a nonmeager subset of X. That is, if*

$$X = \bigcup_{n=1}^{\infty} E_n, \tag{2.22}$$

then at least one of the sets E_n is not nowhere dense.

Proof. Suppose that $X = \cup E_n$, where each set E_n is nowhere dense. Then each of the sets

$$U_n = X \setminus \overline{E_n}, \qquad n \in \mathbb{N},$$

is a dense, open subset of X.

Since U_1 is dense, it is nonempty. Hence there exists some point $x_1 \in U_1$. Since U_1 is open, there is some radius $r_1 > 0$ such that

$$B_1 = B_{r_1}(x_1) \subseteq U_1.$$

Since U_2 is dense, there is a point $x_2 \in U_2 \cap B_1$. Since U_2 and B_1 are both open, there exists a radius $r_2 > 0$ such that

$$B_2 = B_{r_2}(x_2) \subseteq U_2 \cap B_1.$$

By Problem 2.6.15, by making r_2 smaller we can ensure that we also have $\overline{B_2} \subseteq B_1$. And if needed, by taking r_2 to be even smaller yet, we can further assume that $r_2 < r_1/2$.

Continuing in this way, we obtain points $x_n \in U_n$ and balls B_n such that

$$B_n = B_{r_n}(x_n) \subseteq U_n, \qquad \overline{B_n} \subseteq B_{n-1}, \qquad \text{and} \qquad r_n < \frac{r_{n-1}}{2}.$$

In particular, $r_n \to 0$, and the balls B_n form a nested decreasing sequence.

Fix $\varepsilon > 0$, and let N be large enough that $r_N < \varepsilon$. If $m, n > N$, then we have $x_m, x_n \in B_N$, so $d(x_m, x_n) < 2r_N < 2\varepsilon$. Thus the sequence $\{x_n\}_{n \in \mathbb{N}}$ is Cauchy. Since X is complete, this implies that there is some point $x \in X$ such that $x_n \to x$ as $n \to \infty$.

Now let $m > 0$ be any fixed integer. Since the balls B_n form a nested decreasing sequence, we have $x_n \in B_n \subseteq B_{m+1}$ for all $n > m$. Since $x_n \to x$, this implies that $x \in \overline{B_{m+1}} \subseteq B_m$. This is true for every m, so

$$x \in \bigcap_{m=1}^{\infty} B_m \subseteq \bigcap_{n=1}^{\infty} U_n = \bigcap_{n=1}^{\infty} (X \setminus \overline{E_n}) = X \setminus \left(\bigcup_{n=1}^{\infty} \overline{E_n} \right).$$

But this implies that $x \notin \cup \overline{E_n} = X$, which is a contradiction. $\square$

Remark 2.11.4. The hypothesis of completeness in the Baire Category Theorem is necessary, i.e., the theorem can fail if X is not complete. For example, the set of rationals $\mathbb{Q}$ is a metric space with respect to $d(x,y) = |x - y|$. However, $\mathbb{Q}$ is not complete, yet $\mathbb{Q}$ is a meager set in $\mathbb{Q}$. $\diamond$

Specializing to the case that all of the sets E_n in Theorem 2.11.3 are closed gives us the following corollary. This is the form of the Baire Category Theorem that we typically apply in practice.

Corollary 2.11.5. *Assume that X is a nonempty complete metric space. If*

$$X = \bigcup_{n=1}^{\infty} E_n, \tag{2.23}$$

*where each E_n is a **closed** subset of X, then at least one set E_n must contain a nonempty open subset.*

Proof. If none of the E_n contain a nonempty open subset, then they are all nowhere dense (because they are closed and have empty interiors). Consequently, if equation (2.23) holds then X is a meager set, which contradicts the Baire Category Theorem. $\square$

We will give one application of the Baire Category Theorem (more can be found in Theorem 4.5.3 and Problems 2.11.11, 4.2.16, and 4.4.11). Before we state the theorem, note that if U is an open set that is dense in X, then its complement $E = X \setminus U$ is closed and contains no balls (why?), so $E = X \setminus U$ is a nowhere dense subset of X.

Corollary 2.11.6. *Let X be a complete metric space. If U_n is an open, dense subset of X for each $n \in \mathbb{N}$, then $\cap U_n \neq \varnothing$. In fact, $\cap U_n$ is dense in X.*

Proof. Each set $E_n = U_n^C = X \setminus U_n$ is closed and nowhere dense in X.

Suppose that $\cap U_n$ is not dense in X. Then by Problem 2.6.11, there is some open ball $B_r(x)$ entirely contained in the complement of $\cap U_n$. That is,

$$B_r(x) \subseteq \left(\bigcap_{n=1}^{\infty} U_n\right)^C = \bigcup_{n=1}^{\infty} E_n.$$

The closed ball

$$Y = \overline{B_{r/2}(x)}$$

is a closed subset of X and therefore, by Problem 2.8.11, is itself a complete metric space. We have $Y = \bigcup_{n=1}^{\infty} (E_n \cap Y)$. Each set $E_n \cap Y$ is closed in Y. The complement of $E_n \cap Y$ in Y is

$$Y \setminus (E_n \cap Y) = Y \cap U_n,$$

which is dense in Y (because U_n is dense in X). Therefore $E_n \cap Y$ is nowhere dense in Y. But then Y is a countable union of the nowhere dense sets $E_n \cap Y$, which contradicts the Baire Category Theorem. $\square$

For an exercise that uses Corollary 2.11.6, see Problem 2.11.12.

Problems

2.11.7. Prove Lemma 2.11.2.

2.11.8. We take $\mathbb{F} = \mathbb{R}$ for this problem. Determine whether the following subsets of ℓ^1 are nowhere dense.

(a) The set S consisting of sequences whose components past the 99th entry are all zero: $S = \{x = (x_1, \ldots, x_{99}, 0, 0, \ldots) : x_1, \ldots, x_{99} \in \mathbb{F}\}$.

(b) c_{00} (the space of all "finite sequences").

(c) $R = \{x = (x_k)_{k \in \mathbb{N}} \in \ell^1 : x_k \geq 0 \text{ for every } k\}$ (the "closed first quadrant" in ℓ^1; compare Problem 2.5.13).

2.11.9. Which of the sets listed in Problem 2.11.8 are meager subsets of ℓ^1?

2.11.10. Prove the following statements.

(a) If $x \in \mathbb{R}$, then $\mathbb{R} \setminus \{x\}$ is an open, dense subset of $\mathbb{R}$.

(b) The set of irrationals, $\mathbb{R} \setminus \mathbb{Q}$, is a dense subset of $\mathbb{R}$, but it is neither open nor closed.

2.11.11. Suppose that f is an infinitely differentiable function on $\mathbb{R}$ such that for each $x \in \mathbb{R}$ there exists some integer $n_x \geq 0$ such that $f^{(n_x)}(x) = 0$ (where $f^{(k)}$ denotes the kth derivative of f).

(a) Prove that $E_n = \{x \in \mathbb{R} : f^{(n)}(x) = 0\}$ is a closed subset of $\mathbb{R}$ for each integer $n \geq 0$. What is the union of all of these sets E_n?

(b) Use the Baire Category Theorem to show that there exist some open interval (a, b) and some polynomial p such that $f(x) = p(x)$ for all $x \in (a, b)$.

2.11.12. Suppose that U is an unbounded open subset of $\mathbb{R}$, and let

$$A = \{x \in \mathbb{R} : kx \in U \text{ for infinitely many } k \in \mathbb{Z}\}.$$

(a) Given $n \in \mathbb{N}$, set $A_n = \bigcup_{|k| > n} U/k$, where $U/k = \{x/k : x \in U\}$. Prove that A_n is dense in $\mathbb{R}$.

Hint: Show that $A_n \cap (a, b) \neq \emptyset$ for every open interval (a, b) and apply Problem 2.6.12.

(b) Show that $A = \bigcap_{n=1}^{\infty} A_n$, and prove that A is dense in $\mathbb{R}$.

Chapter 3
Norms and Banach Spaces

We studied metric spaces in detail in Chapter 2. A metric on a set X provides us with a notion of the *distance between points* in X. In this chapter we will study norms, which are special types of metrics. However, in order for us to be able to to define a norm, our set X must be a *vector space*. Hence it has extra "structure" that a generic metric space need not possess. In a vector space we can add vectors and multiply vectors by scalars. A *norm* assigns to each vector x in X a *length* $\|x\|$ in a way that respects the structure of X. Specifically, a norm must be *homogeneous* in the sense that $\|cx\| = |c| \, \|x\|$ for all scalars c and all vectors x, and a norm must satisfy the Triangle Inequality, which in this setting takes the form $\|x + y\| \le \|x\| + \|y\|$. When we have a norm we also have a metric, which is defined by letting the *distance between points* be the *length of their difference*, i.e., $d(x, y) = \|x - y\|$. Since a normed space is a metric space, all of the definitions and theorems from Chapter 2 are valid for normed spaces. For example, convergence in a normed space means that the distance between vectors x_n and the limit x shrinks to zero as n increases.

3.1 The Definition of a Norm

Just as a metric generalizes the idea of "distance between points" from Euclidean space to arbitrary sets, a norm generalizes the idea of the "length of a vector" from Euclidean space to vector spaces. Here is the precise definition of a norm, and the slightly weaker notion of a seminorm. Recall our convention (see Section 1.10) that all vector spaces in this volume are over either the real field $\mathbb{R}$ or the complex plane $\mathbb{C}$, and we let the symbol $\mathbb{F}$ stand for a generic choice of $\mathbb{R}$ or $\mathbb{C}$.

Definition 3.1.1 (Seminorms and Norms). Let X be a vector space. A *seminorm* on X is a function $\| \cdot \| \colon X \to \mathbb{R}$ such that for all vectors $x, y \in X$ and all scalars $c \in \mathbb{F}$ we have:

© Springer International Publishing AG, part of Springer Nature 2018
C. Heil, *Metrics, Norms, Inner Products, and Operator Theory*, Applied and
Numerical Harmonic Analysis, https://doi.org/10.1007/978-3-319-65322-8_3

(a) Nonnegativity: $0 \le \|x\| < \infty$,

(b) Homogeneity: $\|cx\| = |c| \, \|x\|$, and

(c) The Triangle Inequality: $\|x + y\| \le \|x\| + \|y\|$.

A seminorm is a *norm* if we also have:

(d) Uniqueness: $\|x\| = 0$ if and only if $x = 0$.

A vector space X together with a norm $\| \cdot \|$ is called a *normed vector space*, a *normed linear space*, or simply a *normed space*. ◇

We will deal mostly with norms, but seminorms do arise naturally in certain situations.

We refer to the number $\|x\|$ as the *length* of a vector x, and we say that $\|x - y\|$ is the *distance* between the vectors x and y. A vector x that has length 1 is called a *unit vector*, or is said to be *normalized*. If x is a nonzero vector, then

$$y = \frac{1}{\|x\|} x = \frac{x}{\|x\|}$$

is a unit vector.

We usually use the symbols $\| \cdot \|$ to denote a norm or seminorm, sometimes with subscripts to distinguish among different norms (for example, $\| \cdot \|_a$ and $\| \cdot \|_b$). Other common symbols for norms and seminorms are $| \cdot |$, $\| | \cdot | \|$, and $\rho(\cdot)$.

The *trivial vector space* $\{0\}$ is a normed space (just set $\|0\| = 0$). Since $\{0\}$ is a subspace of every vector space, we do have to deal with the trivial space on occasion.

The absolute value function $|x|$ is a norm on the vector space $\mathbb{F}$ (recall that $\mathbb{F}$ is either $\mathbb{R}$ or $\mathbb{C}$). In fact, up to multiplication by a positive scalar this is the only norm on $\mathbb{F}$ (see Problem 3.1.8). We will give some examples of norms on other spaces, beginning with the finite-dimensional space $\mathbb{F}^d$.

Example 3.1.2. Direct calculations (assigned as Problem 3.1.6) show that each of the following is a norm on $\mathbb{F}^d$.

(a) The ℓ^1-*norm*:

$$\|x\|_1 = |x_1| + \cdots + |x_d|, \qquad x = (x_1, \ldots, x_d) \in \mathbb{F}^d.$$

(b) The *Euclidean norm* or ℓ^2-*norm*:

$$\|x\|_2 = \left(|x_1|^2 + \cdots + |x_d|^2\right)^{1/2}, \qquad x = (x_1, \ldots, x_d) \in \mathbb{F}^d.$$

(c) The ℓ^∞-*norm*:

$$\|x\|_\infty = \max\{|x_1|, \ldots, |x_d|\}, \qquad x = (x_1, \ldots, x_d) \in \mathbb{F}^d. \qquad ◇$$

The Euclidean norm of a vector in $\mathbb{R}^d$ is the ordinary physical length of the vector (as measured by a ruler in d-dimensional space). The Triangle Inequality for the Euclidean norm on $\mathbb{R}^2$ is illustrated in Figure 3.1. When dealing with $\mathbb{F}^d$, we always assume that we are using the Euclidean norm unless we explicitly state otherwise.

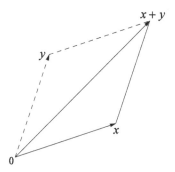

Fig. 3.1 The Triangle Inequality for the Euclidean norm on $\mathbb{R}^2$. Two vectors x, y and their sum $x + y$ are pictured. The lengths of the three edges of the triangle drawn with solid lines are $\|x\|_2$, $\|y\|_2$, and $\|x + y\|_2$, and we can see that $\|x\|_2 + \|y\|_2 \geq \|x + y\|_2$.

Note the similarity between the norms $\|\cdot\|_1$, $\|\cdot\|_2$, and $\|\cdot\|_\infty$ introduced in Example 3.1.2 and the metrics d_1, d_2, and d_∞ presented in equations (2.2), (2.1), and (2.3), respectively. In fact, given any points $x, y \in \mathbb{F}^d$ we have

$$d_1(x,y) = \|x - y\|_1, \qquad d_2(x,y) = \|x - y\|_2, \qquad d_\infty(x,y) = \|x - y\|_\infty.$$

We will see in Section 3.3 that every norm $\|\cdot\|$ "induces" a corresponding metric d by the rule $d(x, y) = \|x - y\|$. However, not every metric has a norm that it is induced from.

Two infinite-dimensional vector spaces that we studied in Section 2.1 are ℓ^1 and ℓ^∞. We will define a norm on each of these spaces, beginning with ℓ^1.

Example 3.1.3. Recall from Example 2.1.7 that ℓ^1 is the set of all infinite sequences $x = (x_k)_{k \in \mathbb{N}}$ that are *summable*, i.e., the quantity

$$\|x\|_1 = \sum_{k=1}^{\infty} |x_k|,$$

which we called the "ℓ^1-norm of x," is finite. However, simply calling something a norm does not prove that it is a norm, so we check now that $\|\cdot\|_1$ actually is a norm.

The nonnegativity and finiteness condition $0 \leq \|x\|_1 < \infty$ is certainly satisfied for each $x \in \ell^1$, and the homogeneity condition is straightforward. The Triangle Inequality follows from the calculation

$$\|x+y\|_1 = \sum_{k=1}^{\infty} |x_k + y_k|$$

$$\leq \sum_{k=1}^{\infty} (|x_k| + |y_k|)$$

$$= \sum_{k=1}^{\infty} |x_k| + \sum_{k=1}^{\infty} |y_k| = \|x\|_1 + \|y\|_1.$$

Finally, if $\|x\|_1 = 0$ then we must have $x_k = 0$ for every k, so $x = (0, 0, \dots)$, which is the zero vector in ℓ^1. This shows that $\|\cdot\|_1$ is a norm on ℓ^1. There are many other norms on ℓ^1, but unless we explicitly state otherwise, when we deal with ℓ^1 we always assume that $\|\cdot\|_1$ is the norm we are using. $\diamond$

Now we turn to ℓ^{∞}.

Example 3.1.4. In Example 2.1.9 we defined ℓ^{∞} to be the space of all bounded sequences, and we declared the "ℓ^{∞}-norm" or "sup-norm" of a sequence $x = (x_k)_{k \in \mathbb{N}}$ to be

$$\|x\|_{\infty} = \sup_{k \in \mathbb{N}} |x_k|.$$

The reader should verify that this is a norm on the space ℓ^{∞}. Unless we explicitly state otherwise, we always assume that this is the norm on ℓ^{∞}. $\diamond$

As we mentioned earlier, each norm on a space determines, or induces, a metric on that space (we will formulate this precisely in Theorem 3.3.1). Indeed, the norms $\|\cdot\|_1$ and $\|\cdot\|_{\infty}$ in Examples 3.1.3 and 3.1.4 are related to the metrics d_1 and d_{∞} defined in equations (2.5) and (2.7) by the rules

$$d_1(x, y) = \|x - y\|_1 \quad \text{and} \quad d_{\infty}(x, y) = \|x - y\|_{\infty}.$$

Next we define two different norms on the space of continuous functions on an interval $[a, b]$.

Example 3.1.5. Let $[a, b]$ be a bounded closed interval, and let $C[a, b]$ denote the space of all continuous, scalar-valued functions on $[a, b]$:

$$C[a, b] = \{f : [a, b] \to \mathbb{F} : f \text{ is continuous}\}.$$

We can define many norms on this space. Two important examples are the *L^1-norm*, defined by

$$\|f\|_1 = \int_a^b |f(t)| \, dt, \qquad f \in C[a, b], \tag{3.1}$$

and the *uniform norm*, defined by

$$\|f\|_u = \sup_{t \in [a,b]} |f(t)|, \qquad f \in C[a, b]. \tag{3.2}$$

The proof that these are norms on $C[a, b]$ is assigned as Problem 3.1.9. We will see later that $C[a, b]$ is complete with respect to the uniform norm, but not with respect to the L^1-norm (this is Lemma 3.4.7). Consequently, we often choose to work with the uniform norm on $C[a, b]$, but the L^1-norm and other norms are still very important. Unless we explicitly state otherwise, we will always assume that the norm on $C[a, b]$ is the uniform norm.   ◇

Once again, the norms in Example 3.1.5 are related to metrics that we introduced in Chapter 2. In particular, the L^1-norm is related to the L^1-metric d_1 defined in equation (2.11) by $d_1(f, g) = \|f - g\|_1$, and the uniform norm is related to the uniform metric d_u defined in equation (2.10) by $d_u(f, g) = \|f - g\|_u$.

Problems

3.1.6. Prove directly that the three norms listed in Example 3.1.2 are indeed norms on $\mathbb{F}^d$.

3.1.7. (a) Prove that $\rho_A(x) = |x_2| + \cdots + |x_d|$, $x = (x_1, x_2, \ldots, x_d) \in \mathbb{F}^d$, defines a seminorm but not a norm on $\mathbb{F}^d$.

(b) Let $B = \begin{bmatrix} 1 & 3 \\ 3 & 9 \end{bmatrix}$. Prove that $\rho_B(x) = ((Bx) \cdot x)^{1/2}$ is a seminorm but not a norm on $\mathbb{R}^2$, where $x \cdot y = x_1 y_1 + x_2 y_2$ is the dot product of vectors in $\mathbb{R}^2$.

(c) Let $C = \begin{bmatrix} 1 & 3 \\ 3 & 10 \end{bmatrix}$. Prove that $\rho_C(x) = ((Cx) \cdot x)^{1/2}$ is a norm on $\mathbb{R}^2$ (compare Problem 5.2.13).

3.1.8. Show that if $\| \cdot \|$ is a norm on the scalar field $\mathbb{F}$, then there exists a positive number $\lambda > 0$ such that $\|x\| = \lambda |x|$ for all $x \in \mathbb{F}$.

3.1.9. (a) Prove that the L^1-norm $\| \cdot \|_1$ defined in equation (3.1) and the uniform norm $\| \cdot \|_u$ defined in equation (3.2) are each norms on $C[a, b]$.

(b) Challenge: Prove that $\|f\|_2 = \left(\int_a^b |f(t)|^2 \, dt \right)^{1/2}$, which we call the L^2-*norm*, defines another norm on $C[a, b]$.

Hint: Recall the arithmetic–geometric mean inequality, $\sqrt{ab} \le \frac{a+b}{2}$ for a, $b \ge 0$.

3.2 The ℓ^p Spaces

We introduced the space ℓ^1 in Example 2.1.7. Now we will define an entire family of related spaces.

Given a finite number $1 \le p < \infty$, we say that a sequence of scalars $x = (x_k)_{k \in \mathbb{N}}$ is *p-summable* if

$$\sum_{k=1}^{\infty} |x_k|^p < \infty.$$

In this case, we set

$$\|x\|_p = \|(x_k)_{k \in \mathbb{N}}\|_p = \left(\sum_{k=1}^{\infty} |x_k|^p \right)^{1/p}. \tag{3.3}$$

If the sequence x is not p-summable then we take $\|x\|_p = \infty$. If $p = 1$ then we usually write "absolutely summable" or just "summable" instead of "1-summable," and for $p = 2$ we usually write "square summable" instead of "2-summable."

We also allow $p = \infty$, although in this case the definition is different. As we declared in Example 2.1.9, the ℓ^∞-*norm* (or the *sup-norm*) of a sequence $x = (x_k)_{k \in \mathbb{N}}$ is

$$\|x\|_\infty = \sup_{k \in \mathbb{N}} |x_k|.$$

The sup-norm of x is finite if and only if x is a bounded sequence. Given a sequence $x = (x_k)_{k \in \mathbb{N}}$, two simple but useful observations are that

$$|x_k| \le \|x\|_\infty \qquad \text{for each } k \in \mathbb{N},$$

and (why?)

$$\|x\|_\infty \le \|x\|_p, \qquad 1 \le p \le \infty.$$

We call $\|x\|_p$ the ℓ^p-*norm* of x (though we should note that we have not yet established that it is a norm on any space). We collect the sequences that have finite ℓ^p-norm to form a space that we call ℓ^p:

$$\ell^p = \{ x = (x_k)_{k \in \mathbb{N}} : \|x\|_p < \infty \}. \tag{3.4}$$

Thus, if $1 \le p < \infty$ then ℓ^p is the set of all p-summable sequences, while ℓ^∞ is the set of all bounded sequences. The space ℓ^1 defined by this method is the same space that we introduced in Example 2.1.7, and likewise ℓ^∞ is precisely the space defined in Example 2.1.9.

Each element of ℓ^p is a sequence of scalars, but because we will often need to consider *sequences of elements* of ℓ^p we will usually refer to $x \in \ell^p$ as a *point* or *vector* in ℓ^p. For example, the standard basis $\{\delta_n\}_{n \in \mathbb{N}}$ is a sequence of vectors in ℓ^p, and each particular vector δ_n is a sequence of scalars.

The ℓ^p spaces do not all contain the same vectors. For example, the sequence

$$x = \left(\tfrac{1}{k} \right)_{k \in \mathbb{N}} = \left(1, \tfrac{1}{2}, \tfrac{1}{3}, \dots \right)$$

belongs to ℓ^p for each $1 < p \le \infty$, but it does not belong to ℓ^1. Problem 3.2.9 shows that the ℓ^p spaces are nested in the sense that

$$1 \le p < q \le \infty \quad \Longrightarrow \quad \ell^p \subsetneq \ell^q. \tag{3.5}$$

Thus ℓ^∞ is the largest space in this family, while ℓ^1 is the smallest. According to Problem 3.2.10, if $x \in \ell^p$ for some finite p, then

$$\lim_{p \to \infty} \|x\|_p = \|x\|_\infty.$$

This is one reason why we defined $\| \cdot \|_\infty$ the way that we did.

It is clear that $\| \cdot \|_p$ satisfies the nonnegativity, homogeneity, and uniqueness properties of a norm, but it is not obvious whether the Triangle Inequality is satisfied. We will prove that $\| \cdot \|_p$ is a norm on ℓ^p, but before doing so we need a fundamental result known as *Hölder's Inequality*. This inequality gives a relation between ℓ^p and $\ell^{p'}$, where p' is the *dual index* to p. If $1 < p < \infty$, then p' is the unique number that satisfies

$$\frac{1}{p} + \frac{1}{p'} = 1. \tag{3.6}$$

Explicitly,

$$p' = \frac{p}{p-1}, \quad 1 < p < \infty.$$

If we adopt the standard real analysis convention that

$$\frac{1}{\infty} = 0, \tag{3.7}$$

then equation (3.6) still has a unique solution if $p = 1$ ($p' = \infty$ in that case) or if $p = \infty$ ($p' = 1$). We take these to be the definition of p' for these endpoint cases. Hence some particular examples of dual indices are

$$1' = \infty, \quad \left(\frac{4}{3}\right)' = 4, \quad \left(\frac{3}{2}\right)' = 3, \quad 2' = 2, \quad 3' = \frac{3}{2}, \quad 4' = \frac{4}{3}, \quad \infty' = 1.$$

We have $(p')' = p$ for each $1 \le p \le \infty$. If $1 \le p \le 2$ then $2 \le p' \le \infty$, while if $2 \le p \le \infty$ then $1 \le p' \le 2$ (hence analysts often consider $p = 2$ to be the "midpoint" of the extended interval $[1, \infty] = [1, \infty) \cup \{\infty\}$).

To motivate the next lemma, recall the arithmetic–geometric mean inequality,

$$\sqrt{ab} \le \frac{a+b}{2}, \quad a, b \ge 0.$$

Replacing a with a^2 and b with b^2 gives

$$ab \le \frac{a^2}{2} + \frac{b^2}{2}, \quad a, b \ge 0.$$

We generalize this inequality to values of p between 1 and ∞ as follows (the proof of this lemma is assigned as Problem 3.2.11).

Lemma 3.2.1. *If* $1 < p < \infty$, *then*

$$ab \leq \frac{a^p}{p} + \frac{b^{p'}}{p'}, \qquad a, b \geq 0. \qquad \diamond \tag{3.8}$$

Now we prove Hölder's Inequality. In this result, given two sequences $x = (x_k)_{k \in \mathbb{N}}$ and $y = (y_k)_{k \in \mathbb{N}}$, we let xy be the sequence obtained by multiplying corresponding components of x and y, i.e., $xy = (x_k y_k)_{k \in \mathbb{N}} = (x_1 y_1, x_2 y_2, x_3 y_3, \dots)$.

Theorem 3.2.2 (Hölder's Inequality). *Fix* $1 \leq p \leq \infty$ *and let* p' *be the dual index to* p. *If* $x = (x_k)_{k \in \mathbb{N}} \in \ell^p$ *and* $y = (y_k)_{k \in \mathbb{N}} \in \ell^{p'}$, *then the sequence* $xy = (x_k y_k)_{k \in \mathbb{N}}$ *belongs to* ℓ^1, *and*

$$\|xy\|_1 \leq \|x\|_p \|y\|_{p'}. \tag{3.9}$$

If $1 < p < \infty$, *then equation* (3.9) *is*

$$\sum_{k=1}^{\infty} |x_k y_k| \leq \left(\sum_{k=1}^{\infty} |x_k|^p \right)^{1/p} \left(\sum_{k=1}^{\infty} |y_k|^{p'} \right)^{1/p'}. \tag{3.10}$$

If $p = 1$, *then equation* (3.9) *is*

$$\sum_{k=1}^{\infty} |x_k y_k| \leq \left(\sum_{k=1}^{\infty} |x_k| \right) \left(\sup_{k \in \mathbb{N}} |y_k| \right). \tag{3.11}$$

If $p = \infty$, *then equation* (3.9) *is*

$$\sum_{k=1}^{\infty} |x_k y_k| \leq \left(\sup_{k \in \mathbb{N}} |x_k| \right) \left(\sum_{k=1}^{\infty} |y_k| \right). \tag{3.12}$$

Proof. *Case* $p = 1$. In this case we have $p' = \infty$. Hence $y \in \ell^{\infty}$, which means that y is bounded. In fact, by the definition of $\|y\|_{\infty}$ we have $|y_k| \leq \|y\|_{\infty}$ for every k. Therefore

$$\sum_{k=1}^{\infty} |x_k y_k| \leq \sum_{k=1}^{\infty} |x_k| \, \|y\|_{\infty} = \|y\|_{\infty} \sum_{k=1}^{\infty} |x_k|,$$

which is equation (3.11). The case $p = \infty$ is entirely symmetric, because $p' = 1$ when $p = \infty$.

Case $1 < p < \infty$. If either x or y is the zero sequence, then equation (3.10) holds trivially, so we may assume that $x \neq 0$ and $y \neq 0$ (i.e., neither x nor y is the sequence of all zeros, and hence $\|x\|_p > 0$ and $\|y\|_{p'} > 0$).

Suppose first that $x \in \ell^p$ and $y \in \ell^{p'}$ are *unit vectors* in their respective spaces, i.e., $\|x\|_p = 1$ and $\|y\|_{p'} = 1$. Then by applying equation (3.8) we see that

$$\|xy\|_1 = \sum_{k=1}^{\infty} |x_k y_k| \le \sum_{k=1}^{\infty} \left(\frac{|x_k|^p}{p} + \frac{|y_k|^{p'}}{p'} \right)$$

$$= \frac{\|x\|_p^p}{p} + \frac{\|y\|_{p'}^{p'}}{p'} = \frac{1}{p} + \frac{1}{p'} = 1. \qquad (3.13)$$

Now let x be any nonzero sequence in ℓ^p, and let y be any nonzero sequence in $\ell^{p'}$. Let

$$u = \frac{x}{\|x\|_p} \quad \text{and} \quad v = \frac{y}{\|y\|_{p'}}.$$

Then u is a unit vector in ℓ^p, and v is a unit vector in $\ell^{p'}$, so equation (3.13) implies that $\|uv\|_1 \le 1$. However,

$$uv = \frac{xy}{\|x\|_p \|y\|_{p'}},$$

so by homogeneity we obtain

$$\frac{\|xy\|_1}{\|x\|_p \|y\|_{p'}} = \left\| \frac{x}{\|x\|_p} \frac{y}{\|y\|_{p'}} \right\|_1 = \|uv\|_1 \le 1.$$

Rearranging yields $\|xy\|_1 \le \|x\|_p \|y\|_{p'}$. $\square$

We will use Hölder's Inequality to prove that $\| \cdot \|_p$ satisfies the Triangle Inequality (which for $\| \cdot \|_p$ is often called *Minkowski's Inequality*).

Theorem 3.2.3 (Minkowski's Inequality). *If $1 \le p \le \infty$, then for all $x, y \in \ell^p$ we have*

$$\|x + y\|_p \le \|x\|_p + \|y\|_p. \qquad (3.14)$$

If $1 \le p < \infty$, then equation (3.14) is

$$\left(\sum_{k=1}^{\infty} |x_k + y_k|^p \right)^{1/p} \le \left(\sum_{k=1}^{\infty} |x_k|^p \right)^{1/p} + \left(\sum_{k=1}^{\infty} |y_k|^p \right)^{1/p},$$

while if $p = \infty$, then equation (3.14) is

$$\sup_{k \in \mathbb{N}} |x_k + y_k| \le \left(\sup_{k \in \mathbb{N}} |x_k| \right) + \left(\sup_{k \in \mathbb{N}} |y_k| \right).$$

Proof. The cases $p = 1$ and $p = \infty$ are straightforward, so we assign the proof for those indices as Problem 3.2.13.

Assume that $1 < p < \infty$, and $x = (x_k)_{k \in \mathbb{N}}$ and $y = (y_k)_{k \in \mathbb{N}}$ are given. Since $p > 1$, we have $p - 1 > 0$, so we can write

$$\|x+y\|_p^p = \sum_{k=1}^{\infty} |x_k + y_k|^p = \sum_{k=1}^{\infty} |x_k + y_k| \, |x_k + y_k|^{p-1}$$

$$\leq \sum_{k=1}^{\infty} |x_k| \, |x_k + y_k|^{p-1} + \sum_{k=1}^{\infty} |y_k| \, |x_k + y_k|^{p-1}$$

$$= S_1 + S_2.$$

To simplify the series S_1, let $z_k = |x_k + y_k|^{p-1}$, so that

$$S_1 = \sum_{k=1}^{\infty} |x_k| \, |x_k + y_k|^{p-1} = \sum_{k=1}^{\infty} |x_k| \, |z_k|.$$

We apply Hölder's Inequality, and then substitute $p' = p/(p-1)$, to compute as follows:

$$S_1 = \sum_{k=1}^{\infty} |x_k| \, |z_k|$$

$$\leq \left(\sum_{k=1}^{\infty} |x_k|^p \right)^{1/p} \left(\sum_{k=1}^{\infty} |z_k|^{p'} \right)^{1/p'} \qquad \text{(Hölder)}$$

$$= \left(\sum_{k=1}^{\infty} |x_k|^p \right)^{1/p} \left(\sum_{k=1}^{\infty} |x_k + y_k|^p \right)^{(p-1)/p} \qquad \text{(substitute)}$$

$$= \|x\|_p \, \|x+y\|_p^{p-1}.$$

A similar calculation shows that

$$S_2 \leq \|y\|_p \, \|x+y\|_p^{p-1}.$$

Combining these inequalities, we see that

$$\|x+y\|_p^p \leq S_1 + S_2 \leq \|x+y\|_p^{p-1} \left(\|x\|_p + \|y\|_p \right).$$

If $x + y$ is not the zero vector, then we can divide both sides by $\|x+y\|_p^{p-1}$ to obtain $\|x+y\|_p \leq \|x\|_p + \|y\|_p$. On the other hand, this inequality holds trivially if $x + y = 0$, so we are done. $\square$

Since $\| \cdot \|_p$ also satisfies the nonnegativity, homogeneity, and uniqueness requirements of a norm, we have proved the following theorem.

Theorem 3.2.4. *If* $1 \leq p \leq \infty$, *then* $\| \cdot \|_p$ *is a norm on* ℓ^p. $\diamond$

By making minor changes to the arguments above, we can create a corresponding family of norms on $\mathbb{F}^d$ (and Problem 3.6.5 extends this idea to arbitrary finite-dimensional vector spaces).

Theorem 3.2.5. *Given $1 \le p \le \infty$, for each $x = (x_1, \ldots, x_d) \in \mathbb{F}^d$ define*

$$\|x\|_p = \begin{cases} \left(|x_1|^p + \cdots + |x_d|^p\right)^{1/p}, & \text{if } 1 \le p < \infty, \\ \max\{|x_1|, \ldots, |x_d|\}, & \text{if } p = \infty. \end{cases}$$

Then $\|\cdot\|_p$ is a norm on $\mathbb{F}^d$. ◇

By substituting integrals for sums, we can create a related family of norms on the space of continuous functions $C[a, b]$. The proof that these are norms is assigned as Problem 3.2.7.

Theorem 3.2.6. *Given $1 \le p \le \infty$, for each $f \in C[a, b]$ define*

$$\|f\|_p = \begin{cases} \left(\displaystyle\int_a^b |f(t)|^p \, dt\right)^{1/p}, & \text{if } 1 \le p < \infty, \\ \displaystyle\sup_{t \in [a,b]} |f(t)|, & \text{if } p = \infty. \end{cases}$$

Then $\|\cdot\|_p$ is a norm on $C[a, b]$. ◇

We call $\|\cdot\|_p$ the L^p-*norm* on $C[a, b]$. In particular, the "L^∞-norm" $\|\cdot\|_\infty$ on $C[a, b]$ is the same as the uniform norm $\|\cdot\|_u$ on $C[a, b]$. However, the fact that the L^∞-norm and the uniform norm coincide on $C[a, b]$ is somewhat misleading, because when we study larger function spaces that include discontinuous functions, these two norms need not be the same. The "correct" definition of the L^∞-norm for general functions requires the use of the theory of *Lebesgue measure*, which is a topic developed in courses on real analysis (see texts such as [WZ77], [Fol99], [SS05], and [Heil19]).

Problems

3.2.7. Prove Theorem 3.2.6.

3.2.8. (a) Find the ℓ^p-norm of the sequence $x = (2^{-k})_{k \in \mathbb{N}}$ for $1 \le p \le \infty$.

(b) Given $1 \le p \le \infty$, for what values of $\alpha \in \mathbb{R}$ does the sequence $x = (k^\alpha)_{k \in \mathbb{N}}$ belong to ℓ^p?

3.2.9. Given fixed indices $1 \le p < q \le \infty$, prove the following statements.

(a) $\ell^p \subseteq \ell^q$.

(b) There exists a sequence x that belongs to ℓ^q but not ℓ^p.

(c) $\|x\|_q \le \|x\|_p$ for each $x \in \ell^p$. Hint: First consider sequences x that satisfy $\|x\|_\infty = 1$.

3.2.10. Prove that if $x \in \ell^q$ for some finite index q, then

$$\lim_{p \to \infty} \|x\|_p = \|x\|_\infty. \tag{3.15}$$

Give an example of a sequence $x \in \ell^\infty$ such that equation (3.15) fails.

3.2.11. (a) Show that if $0 < \theta < 1$, then $t^\theta \le \theta t + (1 - \theta)$ for all $t \ge 0$, and equality holds if and only if $t = 1$.

Hint: Consider the derivatives of t^θ and $\theta t + (1 - \theta)$.

(b) Suppose that $1 < p < \infty$ and $a, b \ge 0$. Apply part (a) with $t = a^p b^{-p'}$ and $\theta = 1/p$ to show that

$$ab \le \frac{a^p}{p} + \frac{b^{p'}}{p'}.$$

(c) Prove that equality holds in part (b) if and only if $b = a^{p-1}$.

3.2.12. Given $1 < p < \infty$, show that equality holds in Hölder's Inequality if and only if there exist scalars α, β, not both zero, such that $\alpha |x_k|^p = \beta |y_k|^{p'}$ for each $k \in \mathbb{N}$. What about the cases $p = 1$ and $p = \infty$?

3.2.13. Prove that equation (3.14) holds if $p = 1$ or $p = \infty$.

3.2.14. Fix $1 < p < \infty$. Given $x, y \in \ell^p$, prove that $\|x + y\|_p = \|x\|_p + \|y\|_p$ if and only if $x = 0$ or y is a positive scalar multiple of x. Does this hold if $p = 1$ or $p = \infty$?

3.2.15. Fix $1 \le p < \infty$.

(a) Let $x = (x_k)_{k \in \mathbb{N}}$ be a sequence of scalars that decay on the order of $k^{-\alpha}$ where $\alpha > \frac{1}{p}$. That is, assume that $\alpha > \frac{1}{p}$ and there exists a constant $C > 0$ such that

$$|x_k| \le C k^{-\alpha} \quad \text{for all } k \in \mathbb{N}. \tag{3.16}$$

Show that $x \in \ell^p$.

(b) Set $\alpha = \frac{1}{p}$. Exhibit a sequence $x \notin \ell^p$ that satisfies equation (3.16) for some $C > 0$, and another sequence $x \in \ell^p$ that satisfies equation (3.16) for some $C > 0$.

(c) Given $\alpha > 0$, show that there exists a sequence $x = (x_k)_{k \in \mathbb{N}} \in \ell^p$ such that there is *no* constant $C > 0$ that satisfies equation (3.16). Conclude that no matter how small we choose α, there exist sequences in ℓ^p whose decay rate is slower than $k^{-\alpha}$.

(d) Suppose that the components of $x = (x_k)_{k \in \mathbb{N}} \in \ell^p$ are nonnegative and *monotonically decreasing*. Show that there exist some $\alpha \ge \frac{1}{p}$ and some $C > 0$ such that equation (3.16) holds.

Hint: Consider $\sum_{k=n+1}^{2n} |x_k|^p$.

3.2.16. Prove the following generalization of Hölder's Inequality. Assume that $1 \le p, q, r \le \infty$ satisfy

$$\frac{1}{p} + \frac{1}{q} = \frac{1}{r}.$$

Given $x = (x_k)_{k \in \mathbb{N}} \in \ell^p$ and $y = (y_k)_{k \in \mathbb{N}} \in \ell^q$, prove that $xy = (x_k y_k)_{k \in \mathbb{N}}$ belongs to ℓ^r, and

$$\|xy\|_r \le \|x\|_p \|y\|_q.$$

Hint: Consider $w = (|x_k|^r)_{k \in \mathbb{N}}$ and $z = (|y_k|^r)_{k \in \mathbb{N}}$.

3.3 The Induced Metric

The following theorem shows that every normed space is also a metric space.

Theorem 3.3.1. *If X is a normed space, then*

$$d(x, y) = \|x - y\|, \qquad x, y \in X, \tag{3.17}$$

defines a metric on X.

Proof. Because $\|\cdot\|$ is a norm, we have $0 \le \|x - y\| < \infty$ for all $x, y \in X$. Thus d is nonnegative and finite on X. The symmetry requirement of a metric is satisfied because

$$d(x, y) = \|x - y\| = \|(-1)(y - x)\| = |-1| \|y - x\| = d(y, x),$$

and the uniqueness requirement holds since

$$d(x, y) = 0 \quad \Longleftrightarrow \quad \|x - y\| = 0 \quad \Longleftrightarrow \quad x - y = 0.$$

Finally, the Triangle Inequality for d follows from the Triangle Inequality for $\|\cdot\|$, because

$$d(x, z) = \|(x - y) + (y - z)\| \le \|x - y\| + \|y - z\| = d(x, y) + d(y, z). \quad \diamond$$

The metric d defined by equation (3.17) is called the *metric on X induced from* $\|\cdot\|$, or simply the *induced metric* on X.

Thus, whenever we are given a normed space X we automatically have a metric on X as well as a norm. Therefore all of the definitions and results that we obtained for metric spaces in Chapter 2 apply to normed spaces, using the induced metric $d(x, y) = \|x - y\|$. For example, convergence in a normed space is defined by

$$x_n \to x \quad \Longleftrightarrow \quad \lim_{n \to \infty} \|x - x_n\| = 0.$$

It may be possible to place a metric on X other than the induced metric, but unless we explicitly state otherwise, all metric-related statements on a normed space are taken with respect to the induced metric.

Example 3.3.2. If $1 \leq p < \infty$, then the metric induced from the ℓ^p-norm defined in equation (3.3) is

$$d_p(x, y) = \|x - y\|_p = \left(\sum_{k=1}^{\infty} |x_k - y_k|^p \right)^{1/p}, \qquad x, y \in \ell^p.$$

Unless we specify otherwise, we always assume that the norm on ℓ^p is $\|\cdot\|_p$, and the corresponding metric is the induced metric. Thus, for example, when we speak of an open ball $B_r(x)$ in ℓ^p we implicitly assume that this ball is defined using the ℓ^p-norm, i.e.,

$$B_r(x) = \{y \in \ell^p : \|x - y\|_p < r\},$$

the set of all sequences y that lie within a distance of r from x, as measured by the ℓ^p-norm. What is the distance between two standard basis vectors δ_m and δ_n in ℓ^p?   ◇

Here are some useful properties of norms and convergence with respect to a norm.

Lemma 3.3.3. *If X is a normed space and x_n, x, y_n, y are vectors in X, then the following statements hold.*

(a) Uniqueness of limits: *If $x_n \to x$ and $x_n \to y$, then $x = y$.*

(b) Reverse Triangle Inequality: $\big| \|x\| - \|y\| \big| \leq \|x - y\|$.

(c) Convergent implies Cauchy: *If $x_n \to x$, then $\{x_n\}_{n \in \mathbb{N}}$ is Cauchy.*

(d) Boundedness of Cauchy sequences: *If $\{x_n\}_{n \in \mathbb{N}}$ is a Cauchy sequence, then* $\sup \|x_n\| < \infty$.

(e) Continuity of the norm: *If $x_n \to x$, then $\|x_n\| \to \|x\|$.*

(f) Continuity of vector addition: *If $x_n \to x$ and $y_n \to y$, then $x_n + y_n \to x + y$.*

(g) Continuity of scalar multiplication: *If $x_n \to x$ and $c_n \to c$ (where c_n, c are scalars), then $c_n x_n \to cx$.*

Proof. We will prove one part, and assign the proof of the remaining statements as Problem 3.3.8.

(d) Suppose that $\{x_n\}_{n \in \mathbb{N}}$ is Cauchy. Then, by applying the definition of a Cauchy sequence with the specific choice $\varepsilon = 1$, we see that there exists some $N > 0$ such that $\|x_m - x_n\| < 1$ for all $m, n \geq N$. Therefore, for all $n \geq N$ we have

$$\|x_n\| = \|x_n - x_N + x_N\| \leq \|x_n - x_N\| + \|x_N\| \leq 1 + \|x_N\|.$$

Hence, for an arbitrary n we have

$$\|x_n\| \leq \max\{\|x_1\|, \ldots, \|x_{N-1}\|, \|x_N\| + 1\}.$$

Thus $\{x_n\}_{n \in \mathbb{N}}$ is a bounded sequence. $\square$

Since a normed space is a vector space, we can define lines, planes, and other related notions. In particular, if x and y are vectors in X, then the *line segment* joining x to y is the set of all points of the form $tx + (1-t)y$, where $0 \leq t \leq 1$.

Definition 3.3.4 (Convex Set). We say that a subset K of a vector space X is *convex* if given any two points $x, y \in K$, the line segment joining x to y is entirely contained within K. That is, K is convex if

$$x, y \in K, \ 0 \leq t \leq 1 \quad \Longrightarrow \quad tx + (1-t)y \in K. \qquad \diamondsuit$$

All subspaces are convex by definition, and we prove now that every open ball $B_r(x)$ in a normed space X is convex.

Lemma 3.3.5. *If X is a normed space, $x \in X$, and $r > 0$, then the open ball $B_r(x) = \{y \in X : \|x - y\| < r\}$ is convex.*

Proof. Choose any two points $y, z \in B_r(x)$ and fix $0 \leq t \leq 1$. Then

$$
\begin{aligned}
\|x - ((1-t)y + tz)\| &= \|(1-t)(x-y) + t(x-z)\| \\
&\leq (1-t)\|(x-y)\| + t\|x-z\| \\
&< (1-t)r + tr = r,
\end{aligned}
$$

so $(1-t)y + tz \in B_r(x)$. $\square$

Theorem 3.2.5 established that $\|\cdot\|_p$ is a norm on $\mathbb{R}^2$ for each p in the range $1 \leq p \leq \infty$. Consequently, by Lemma 3.3.5, every open ball defined with respect to these norms must be convex. The unit open balls $B_1(0)$ defined with respect to $\|\cdot\|_p$ for $p = 3/2$ and $p = 3$ are shown in Figure 3.2, and we can see that those two balls are indeed convex. The unit balls $B_1(0)$ with respect to $\|\cdot\|_p$ for $p = 1$, 2, and ∞ were depicted earlier in Figure 2.5, and we can see that those balls are convex as well.

Figure 2.5 also shows the unit ball $B_1(0)$ that corresponds to the metric $d_{1/2}$ on $\mathbb{R}^2$ that was introduced in Problem 2.1.16. We can see by inspection that that unit ball is not convex. As a consequence, the metric $d_{1/2}$ cannot be induced from any norm. That is, there is no way to define a norm $\|\cdot\|$ on $\mathbb{R}^2$ that satisfies $\|x - y\| = d_{1/2}(x, y)$, because if there were, then the unit ball $B_1(0)$ corresponding to this norm would be convex. Thus, while every norm induces a metric, it is not true that every metric is induced from some norm. In this sense there are "more metrics than norms." Problem 3.3.14 shows how to define a metric on ℓ^p for each index in the range $0 < p < 1$, but

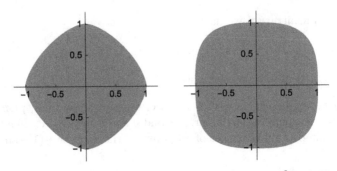

Fig. 3.2 Unit open balls $B_1(0)$ with respect to different norms on $\mathbb{R}^2$. Left: Norm $\|\cdot\|_{3/2}$. Right: Norm $\|\cdot\|_3$.

again that metric is not induced from any norm, because balls are *not convex* when $p < 1$. Even so, the ℓ^p spaces with indices $0 < p < 1$ play an important role in certain applications, such as those requiring "sparse representations."

Problems

3.3.6. Fix $1 \leq p \leq \infty$, and assume that the norm on $\mathbb{F}^d$ is the ℓ^p-norm defined in Theorem 3.2.5. Let $\{x_n\}_{n \in \mathbb{N}}$ be a sequence of vectors in $\mathbb{F}^d$. Prove that $x_n \to x$ in ℓ^p-norm if and only if x_n converges componentwise to x.

3.3.7. Given $1 \leq p \leq \infty$, prove the following statements.

(a) If $x_n \to x$ in ℓ^p, then x_n converges componentwise to x.

(b) There exist $x_n, x \in \ell^p$ such that x_n converges componentwise to x but x_n does not converge to x in ℓ^p-norm.

3.3.8. Prove the remaining statements in Lemma 3.3.3.

3.3.9. Given a normed space X, prove that the function $\rho \colon X \to \mathbb{R}$ defined by $\rho(x) = \|x\|$ is continuous on X.

3.3.10. Let x be a nonzero vector in a normed space X. Prove directly that the one-dimensional subspace spanned by x, span$\{x\} = \{cx : c \in \mathbb{F}\}$, is a closed subspace of X.

3.3.11. Show that if S is an open *subspace* of a normed space X, then $S = X$.

3.3.12. Let X be a normed space. Given a subspace M of X, prove the following statements.

(a) The closure $\overline{M}$ of M is also a subspace of X.

(b) If $M \neq X$, then the interior of M is empty, i.e., $M^\circ = \varnothing$.

3.3.13. Let X be a vector space.

(a) Prove that the intersection of any family of convex subsets of X is convex, i.e., if K_i is a convex subset of X for each i in some index set I, then $\bigcap_{i \in I} K_i$ is convex.

(b) Given nested convex sets $K_1 \subseteq K_2 \subseteq \cdots$, show that $\bigcup K_n$ is convex.

(c) Prove that if $X \neq \{0\}$ and $\|\cdot\|$ is a norm on X, then every open ball $B_r(x)$ contains infinitely many distinct points of X.

(d) Let d be the discrete metric on X. Are open balls $B_r(x)$ defined with respect to the discrete metric convex? Is the discrete metric induced from any norm on X?

3.3.14. This problem extends Problem 2.1.16(c). Given $0 < p < 1$, let $\|x\|_p$ and ℓ^p be defined by equations (3.3) and (3.4), respectively. Prove the following statements.

(a) $(1+t)^p \leq 1 + t^p$ for all $t > 0$.

(b) If $a, b > 0$, then $(a+b)^p \leq a^p + b^p$.

(c) $\|x+y\|_p^p \leq \|x\|_p^p + \|y\|_p^p$ for all $x, y \in \ell^p$.

(d) $d_p(x, y) = \|x - y\|_p^p = \sum_{k=1}^{\infty} |x_k - y_k|^p$ defines a metric on ℓ^p.

(e) The unit open ball $B_1(0) = \{y \in \ell^p : d_p(0, y) < 1\}$ is not convex.

(f) d_p is not induced from any norm on ℓ^p.

Also prove that an analogous metric can be defined on $\mathbb{F}^d$ by the formula $d_p(x, y) = \sum_{k=1}^{d} |x_k - y_k|^p$, and plot the corresponding unit ball $B_1(0)$ in $\mathbb{R}^2$ for various values of $p < 1$. Is the ball convex?

3.4 Banach Spaces

Every convergent sequence in a normed vector space must be Cauchy, but the converse need not hold. In *some* normed spaces it is true that all Cauchy sequences are convergent. We give spaces that have this property the following name.

Definition 3.4.1 (Banach Space). A normed space X is a *Banach space* if it is complete, i.e., if every Cauchy sequence in X converges to an element of X. ◇

Thus a Banach space is precisely a normed space that is complete. The terms "Banach space" and "complete normed space" are interchangeable, and we will use whichever is more convenient in a given context.

Remark 3.4.2. A rather uninteresting example of a Banach space is the vector space $X = \{0\}$ with norm defined by $\|0\| = 0$. We call this the *trivial Banach space*. A *nontrivial normed space* is any normed space other than $\{0\}$, and a *nontrivial Banach space* is any complete normed space other than $\{0\}$. ◇

We proved in Theorem 2.2.11 that the space ℓ^1 is complete. Therefore ℓ^1 is an example of an infinite-dimensional Banach space. A similar argument shows that ℓ^p is a Banach space for $1 \leq p \leq \infty$. We state this as the following theorem, and assign the proof as Problem 3.4.11. We will see some other examples of Banach spaces in Section 3.5.

Theorem 3.4.3. *ℓ^p is a Banach space for each index $1 \leq p \leq \infty$.* $\Diamond$

With only minor modifications, the same argument shows that $\mathbb{F}^d$ is complete with respect to each of the norms defined in Theorem 3.2.5.

We will prove in Section 3.7 that every *finite-dimensional* vector space X is complete with respect to any norm that we place on X. However, not every infinite-dimensional normed space is complete. One example of an incomplete infinite-dimensional normed space is the space of "finite sequences" c_{00} that we studied in Example 2.2.13 (see Theorem 3.4.6 for details).

If X is a Banach space and Y is a subspace of X, then by restricting the norm on X to vectors in Y we obtain a norm on Y. When is Y complete with respect to this inherited norm? The following result (whose proof is Problem 3.3.8) answers this.

Lemma 3.4.4. *Let Y be a subspace of a Banach space X, and let the norm on Y be the restriction of the norm on X to the set Y. Then Y is a Banach space with respect to this norm if and only if Y is a closed subset of X.* $\Diamond$

To give an application of Lemma 3.4.4, let c_0 be the set of all sequences of scalars that "vanish at infinity," i.e.,

$$c_0 = \left\{ x = (x_k)_{k \in \mathbb{N}} : \lim_{k \to \infty} x_k = 0 \right\}. \tag{3.18}$$

The reader should check that we have the inclusions

$$c_{00} \subsetneq \ell^1 \subsetneq \ell^p \subsetneq c_0 \subsetneq \ell^\infty, \qquad 1 < p < \infty.$$

We will use Lemma 3.4.4 to prove that c_0 is a Banach space with respect to the norm of ℓ^∞.

Lemma 3.4.5. *c_0 is a closed subspace of ℓ^∞ with respect to $\|\cdot\|_\infty$, and hence is a Banach space with respect to that norm.*

Proof. We will show that c_0 is closed by proving that the ℓ^∞-norm limit of every sequence of elements of c_0 belongs to c_0. To do this, suppose that $\{x_n\}_{n \in \mathbb{N}}$ is a sequence of vectors from c_0 and there exists some vector $x \in \ell^\infty$ such that $\|x - x_n\|_\infty \to 0$. For convenience, denote the components of x_n and x by

$$x_n = \big(x_n(k)\big)_{k \in \mathbb{N}} \quad \text{and} \quad x = \big(x(k)\big)_{k \in \mathbb{N}}.$$

We must show that $x \in c_0$, which means that we must prove that $x(k) \to 0$ as $k \to \infty$.

Fix any $\varepsilon > 0$. Since $x_n \to x$ in ℓ^∞-norm, there exists some n such that $\|x - x_n\|_\infty < \frac{\varepsilon}{2}$ (in fact, this will be true for all large enough n, but we need only one n for this proof). Since $x_n \in c_0$, we know that $x_n(k) \to 0$ as $k \to \infty$. Therefore there exists some $K > 0$ such that $|x_n(k)| < \frac{\varepsilon}{2}$ for all $k > K$. Consequently, for every $k > K$ we have

$$|x(k)| \leq |x(k) - x_n(k)| + |x_n(k)| < \|x - x_n\|_\infty + \frac{\varepsilon}{2} < \frac{\varepsilon}{2} + \frac{\varepsilon}{2} = \varepsilon.$$

This shows that $x(k) \to 0$ as $k \to \infty$, so x belongs to c_0. Therefore Theorem 2.6.2 implies that c_0 is a closed subset of ℓ^∞, and hence Lemma 3.4.4 implies that c_0 is a Banach space with respect to the sup-norm. $\quad\square$

In contrast, we will show that c_{00} is not a closed subspace of ℓ^∞ with respect to the sup-norm, and therefore it is incomplete with respect to that norm.

Theorem 3.4.6. *The sup-norm is a norm on c_{00}, but c_{00} is not complete with respect to $\|\cdot\|_\infty$.*

Proof. Since c_{00} is a subspace of ℓ^∞, the restriction of the sup-norm $\|\cdot\|_\infty$ to c_{00} is a norm on c_{00}. For each $n \in \mathbb{N}$, let x_n be the sequence

$$x_n = \left(1, \tfrac{1}{2}, \tfrac{1}{3}, \dots, \tfrac{1}{n}, 0, 0, \dots\right),$$

and let

$$x = \left(1, \tfrac{1}{2}, \tfrac{1}{3}, \dots\right).$$

Each vector x_n belongs to c_{00}. The vector x belongs to ℓ^∞, but it does not belong to c_{00}. Given any $n \in \mathbb{N}$, we have

$$x - x_n = \left(0, \dots, 0, \tfrac{1}{n+1}, \tfrac{1}{n+2}, \dots\right),$$

so

$$\|x - x_n\|_\infty = \frac{1}{n+1} \to 0 \qquad \text{as } n \to \infty.$$

Thus $x_n \to x$ in ℓ^∞-norm. Thinking of c_{00} as a subset of ℓ^∞, this tells us that x is a limit of elements of c_{00}, even though x does not belong to c_{00}. Therefore Theorem 2.4.2 implies that c_{00} is not a closed subset of ℓ^∞. Applying Lemma 3.4.4, we conclude that c_{00} is not complete with respect to $\|\cdot\|_\infty$. $\quad\square$

We introduced the L^p-norms on $C[a, b]$ in Theorem 3.2.6. These are similar to the ℓ^p-norms, but (for finite p) they use a Riemann integral in their definition instead of an infinite sum. We saw in Theorem 3.4.3 that ℓ^p is a Banach space with respect to the norm $\|\cdot\|_p$. We prove next that $C[a, b]$ is *not* a Banach space with respect to $\|\cdot\|_p$ when p is finite. In contrast, we will see in Section 3.5 that $C[a, b]$ is a Banach space with respect to the uniform norm.

Lemma 3.4.7. *If $1 \leq p < \infty$, then $C[a, b]$ is not complete with respect to the L^p-norm $\| \cdot \|_p$.*

Proof. For simplicity of presentation we will take $a = -1$ and $b = 1$. For each $n \in \mathbb{N}$, define a continuous function $f_n \colon [-1, 1] \to \mathbb{R}$ by

$$
f_n(t) = \begin{cases} -1, & -1 \leq t \leq -\frac{1}{n}, \\ \text{linear}, & -\frac{1}{n} < t < \frac{1}{n}, \\ 1, & \frac{1}{n} \leq t \leq 1. \end{cases}
$$

Figure 3.3 shows the graph of the function f_5.

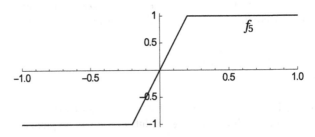

Fig. 3.3 Graph of the function f_5.

Choose any positive integers m and n. One of m and n must be larger, so let us say $m \leq n$. Since $f_m(t)$ and $f_n(t)$ are always between -1 and 1, we have $|f_m(t) - f_n(t)| \leq |f_m(t)| + |f_n(t)| \leq 2$ for every t (in fact, $|f_m(t) - f_n(t)| \leq 1$ for every t; can you see why?). Since we also have $f_m(t) = f_n(t) = 0$ for $t \notin [-\frac{1}{m}, \frac{1}{m}]$, we compute that

$$
\begin{aligned}
\|f_m - f_n\|_p^p &= \int_{-1}^{1} |f_m(t) - f_n(t)|^p \, dt \\
&= \int_{-1/m}^{1/m} |f_m(t) - f_n(t)|^p \, dt \\
&\leq \int_{-1/m}^{1/m} 2^p \, dt = \frac{2^{p+1}}{m},
\end{aligned}
$$

and therefore

$$
\|f_m - f_n\|_p \leq \left(\frac{2^{p+1}}{m} \right)^{1/p} = \frac{2^{(p+1)/p}}{m^{1/p}} \leq \frac{4}{m^{1/p}}, \qquad \text{for all } n \geq m.
$$

Consequently, if we fix $\varepsilon > 0$, then for all large enough m and n we will have $\|f_m - f_n\|_p < \varepsilon$. This shows that $\{f_n\}_{n \in \mathbb{N}}$ is a Cauchy sequence in $C[-1, 1]$

with respect to the norm $\| \cdot \|_p$. However, we will prove that there is *no function* $g \in C[-1,1]$ such that $\|g - f_n\|_p \to 0$ as $n \to \infty$.

Suppose that there were some $g \in C[-1,1]$ such that $\|g - f_n\|_p \to 0$. Fix $0 < c < 1$, and suppose that g is not identically 1 on $[c,1]$. Then $|g - 1|^p$ is continuous but not identically zero on $[c,1]$, so its integral on this interval must be nonzero:

$$C = \int_c^1 |g(t) - 1|^p \, dt > 0.$$

On the other hand, f_n is identically 1 on $[c,1]$ for all $n > 1/c$, so for all large enough n we have

$$\|g - f_n\|_p^p = \int_{-1}^1 |g(t) - f_n(t)|^p \, dt$$

$$\geq \int_c^1 |g(t) - f_n(t)|^p \, dt$$

$$= \int_c^1 |g(t) - 1|^p \, dt = C.$$

Since C is a fixed positive constant, it follows that $\|g - f_n\|_p \geq C^{1/p} \not\to 0$, which is a contradiction. Therefore g must be identically 1 on $[c,1]$. This is true for every $c > 1$, so $g(t) = 1$ for all $0 < t \leq 1$. A similar argument shows that $g(t) = -1$ for all $-1 \leq t < 0$. But there is no continuous function that takes these values, so we have obtained a contradiction.

Thus, although $\{f_n\}_{n \in \mathbb{N}}$ is Cauchy in $C[-1,1]$, there is no function in $C[-1,1]$ that this sequence can converge to with respect to $\| \cdot \|_p$. Therefore $C[-1,1]$ is not complete with respect to this norm. $\square$

Each incomplete normed space that we have seen so far has been contained in some larger complete space. For example, c_{00} is incomplete with respect to the norm $\| \cdot \|_1$, yet it is a dense subset of ℓ^1, which is complete with respect to $\| \cdot \|_1$ (see Example 2.2.13). Problem 6.6.19 shows that if Y is an incomplete normed space, then there is a unique complete normed space X such that Y is a dense subset of X. While we know that such a space X exists, in practice it can be difficult to concretely characterize or identify that complete space X. However, for $Y = C[a,b]$, which is incomplete with respect to $\| \cdot \|_p$ when p is finite, that larger complete space is the *Lebesgue space* $X = L^p[a,b]$. Rigorously defining $L^p[a,b]$ is beyond the scope of this text, since it requires the theory of the *Lebesgue integral*, but we give the following formal definition (formal because the integrals in this definition are undefined *Lebesgue integrals*, not Riemann integrals).

Definition 3.4.8 (The Lebesgue Space $L^p[a,b]$). Given $1 \leq p < \infty$, the *Lebesgue space* $L^p[a,b]$ consists of all *Lebesgue measurable functions* $f \colon [a,b] \to \mathbb{F}$ such that the *Lebesgue integral* of $|f|^p$ is finite, i.e.,

$$\int_a^b |f(t)|^p \, dt \, < \, \infty.$$

The L^p-*norm* of $f \in L^p[a,b]$ is

$$\|f\|_p \; = \; \left(\int_a^b |f(t)|^p \, dt \right)^{1/p}. \qquad \Diamond$$

The Lebesgue space $L^p[a,b]$ is a Banach space (although there is a technical requirement that we must identify functions that are *equal almost everywhere* in order for the uniqueness requirement of a norm to be satisfied). If a function f is Riemann integrable on $[a,b]$, then it is also Lebesgue integrable, and the Riemann integral of f equals the Lebesgue integral of f. However, the Lebesgue integral is more general than the Riemann integral, and also can be defined on domains other than intervals.

For $p = \infty$, the Lebesgue space $L^\infty[a,b]$ consists of all Lebesgue measurable functions that are *essentially bounded* on $[a,b]$. The L^∞-norm of a *continuous* bounded function is the same as its uniform norm, but in general $\|f\|_u$ and $\|f\|_\infty$ can be different. For example, recall from our discussion in Section 2.7 that the Cantor set C is an uncountable subset of $[0,1]$ that has measure zero. It follows from this that if $g(t) = 0$ for all $t \notin C$, then g is essentially bounded *no matter how we define it on* C, and its L^∞-norm is $\|g\|_\infty = 0$. Yet by choosing points $t_n \in C$ and setting $g(t_n) = n$ we can make g unbounded, with uniform norm $\|g\|_u = \infty$.

We refer to texts such as [Fol99], [SS05], [WZ77], and [Heil19] for precise definitions and discussion of the Lebesgue integral, the space $L^p[a,b]$, and more general Lebesgue spaces $L^p(E)$.

Problems

3.4.9. Prove Lemma 3.4.4.

3.4.10. (a) Given $x \in c_0$, prove that there exist vectors $x_n \in c_{00}$ such that $\|x - x_n\|_\infty \to 0$ as $n \to \infty$.

(b) Prove that c_{00} is dense in c_0 (with respect to the norm $\| \cdot \|_\infty$). Hence $\overline{c_{00}} = c_0$ when we use the sup-norm.

(c) Given $1 \le p < \infty$, prove that c_{00} is dense in ℓ^p (with respect to the norm $\| \cdot \|_p$). Hence $\overline{c_{00}} = \ell^p$ when we use the ℓ^p-norm.

3.4.11. Given $1 \le p \le \infty$, prove the following statements.

(a) ℓ^p is complete.

(b) If p is finite, then ℓ^p is separable (compare Problem 2.6.20).

3.4.12. Fix $1 \leq p < \infty$ and $x = (x_k)_{k \in \mathbb{N}} \in \ell^p$. Prove that $\sum_{k=1}^{\infty} \frac{|x_k|}{k} < \infty$, but show by example that this can fail if $p = \infty$.

3.4.13. For each $n \in \mathbb{N}$, let $y_n = (1, \frac{1}{2}, \dots, \frac{1}{n}, 0, 0, \dots)$. Note that $y_n \in \ell^1$ for every n.

(a) Assume that the norm on ℓ^1 is its usual norm $\|\cdot\|_1$. Prove that $\{y_n\}_{n \in \mathbb{N}}$ is not a Cauchy sequence in ℓ^1 with respect to this norm.

(b) Now assume that the norm on ℓ^1 is $\|\cdot\|_2$. Prove that $\{y_n\}_{n \in \mathbb{N}}$ is a Cauchy sequence in ℓ^1 with respect to $\|\cdot\|_2$, yet there is no vector $y \in \ell^1$ such that $\|y - y_n\|_2 \to 0$. Conclude that ℓ^1 is not complete with respect to the norm $\|\cdot\|_2$.

3.4.14. (a) Let $w = (w_k)_{k \in \mathbb{N}}$ be a fixed sequence of *positive* scalars $w_k > 0$. Given a sequence of scalars $x = (x_k)_{k \in \mathbb{N}}$, set

$$
\|x\|_{p,w} = \begin{cases} \left(\displaystyle\sum_{k=1}^{\infty} |x_k|^p w_k^p \right)^{1/p}, & \text{if } 1 \leq p < \infty, \\[2mm] \displaystyle\sup_{k \in \mathbb{N}} |x_k| w_k, & \text{if } p = \infty. \end{cases}
$$

Define $\ell^p_w = \{ x : \|x\|_{p,w} < \infty \}$. Prove that this *weighted ℓ^p-space* is a Banach space for each index $1 \leq p \leq \infty$.

(b) Given a fixed index $1 \leq p \leq \infty$, prove that $\ell^p = \ell^p_w$ if and only if there exist constants $A, B > 0$ such that $A \leq w_k \leq B$ for every $k \in \mathbb{N}$.

3.5 Uniform Convergence of Functions

Suppose that X is a set, $f_1, f_2, \dots$ are functions that map X into $\mathbb{F}$, and $f \colon X \to \mathbb{F}$ is another function. There are many ways in which we can quantify what it means for the functions f_n to converge to f. Perhaps the simplest is the following notion of *pointwise convergence*.

Definition 3.5.1 (Pointwise Convergence). Let X be a set, and let functions $f_n, f \colon X \to \mathbb{F}$ be given. We say that f_n *converges pointwise to f on X* if

$$
\lim_{n \to \infty} f_n(x) = f(x), \quad \text{for every } x \in X.
$$

In this case we write $f_n \to f$ *pointwise.*   ◇

Here is an example.

Example 3.5.2 (Shrinking Triangles). Set $X = [0, 1]$, and for each $n \in \mathbb{N}$ let $f_n \colon [0, 1] \to \mathbb{F}$ be the continuous function given by

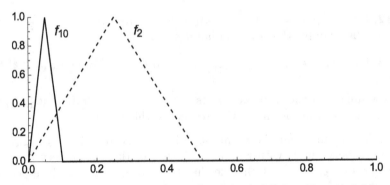

Fig. 3.4 Graphs of the functions f_2 (dashed) and f_{10} (solid) from Example 3.5.2.

$$
f_n(t) = \begin{cases}
0, & t = 0, \\
\text{linear}, & 0 < t < \frac{1}{2n}, \\
1, & t = \frac{1}{2n}, \\
\text{linear}, & \frac{1}{2n} < t < \frac{1}{n}, \\
0, & \frac{1}{n} \le t \le 1.
\end{cases}
$$

Then $f_n(t) \to 0$ for each $t \in [0, 1]$ (see the illustration in Figure 3.4). Therefore f_n converges pointwise to the zero function on $[0, 1]$. ◇

However, we often need other, more stringent, notions of convergence of functions. We defined the *uniform norm* of functions on the domain $[0, 1]$ in equation (2.9). Now we extend that definition to functions on other sets, and then in Definition 3.5.5 we will use this notion of the uniform norm to define *uniform convergence* of functions.

Definition 3.5.3 (Uniform Norm). Let X be a set. The *uniform norm* of a function $f \colon X \to \mathbb{F}$ is

$$
\|f\|_u = \sup_{x \in X} |f(x)|. \qquad ◇
$$

The uniform norm of an arbitrary function f can be infinite; indeed, $\|f\|_u < \infty$ if and only if f is bounded on X. We collect the bounded functions to form the space

$$
\mathcal{F}_b(X) = \{f \colon X \to \mathbb{F} : f \text{ is bounded}\}.
$$

By definition, the uniform norm of each element of $\mathcal{F}_b(X)$ is finite.

Lemma 3.5.4. *The uniform norm* $\| \cdot \|_u$ *is a norm on* $\mathcal{F}_b(X)$.

Proof. The nonnegativity, homogeneity, and uniqueness properties of a norm are immediate. The Triangle Inequality follows from basic properties of suprema, for if f and g are bounded functions on X, then

$$\|f + g\|_u = \sup_{x \in X} |f(x) + g(x)| \leq \sup_{x \in X} \big(|f(x)| + |g(x)|\big)$$

$$\leq \left(\sup_{x \in X} |f(x)|\right) + \left(\sup_{x \in X} |g(x)|\right)$$

$$= \|f\|_u + \|g\|_u.$$

Therefore $\| \cdot \|_u$ is a norm on $\mathcal{F}_b(X)$. $\square$

As stated next, convergence with respect to the norm $\|\cdot\|_u$ is called *uniform convergence*.

Definition 3.5.5 (Uniform Convergence). Let X be a set, and let functions $f_n, f \colon X \to \mathbb{F}$ be given. Then we say that f_n *converges uniformly to f on X*, and write $f_n \to f$ *uniformly*, if

$$\lim_{n \to \infty} \|f - f_n\|_u = \lim_{n \to \infty} \left(\sup_{x \in X} |f(x) - f_n(x)|\right) = 0.$$

Similarly, a sequence $\{f_n\}_{n \in \mathbb{N}}$ that is Cauchy with respect to the uniform norm is said to be a *uniformly Cauchy sequence*. $\diamond$

Suppose that f_n converges uniformly to f. Then, by the definition of a supremum, for each individual point $x \in X$ we have

$$|f(x) - f_n(x)| \leq \|f - f_n\|_u \to 0 \quad \text{as } n \to \infty. \tag{3.19}$$

Thus $f_n(x) \to f(x)$ for each individual x. This shows that *uniform convergence implies pointwise convergence*. However, the converse need not hold, i.e., pointwise convergence does not imply uniform convergence in general. In fact, the sequence of functions defined in Example 3.5.2 gives us an example.

Example 3.5.6 (Shrinking Triangles Revisited). The "Shrinking Triangles" of Example 3.5.2 converge pointwise to the zero function on $[0, 1]$. However, for each $n \in \mathbb{N}$ we have

$$\|0 - f_n\|_u = \sup_{t \in [0,1]} |0 - f_n(t)| = 1,$$

which does not converge to zero as $n \to \infty$. Therefore f_n does not converge uniformly to the zero function. In fact, $\{f_n\}_{n \in \mathbb{N}}$ is not uniformly Cauchy (why?), so there is *no function f* that f_n converges to uniformly. What happens with other norms? Do the Shrinking Triangles converge if we replace the uniform norm with the L^p-norm on $C[0, 1]$? $\diamond$

Next we will show that the space of bounded functions on X is *complete* with respect to the uniform norm.

Theorem 3.5.7 ($\mathcal{F}_b(X)$ Is a Banach Space). *Let X be a set. If $\{f_n\}_{n\in\mathbb{N}}$ is a sequence in $\mathcal{F}_b(X)$ that is Cauchy with respect to $\|\cdot\|_u$, then there exists a function $f \in \mathcal{F}_b(X)$ such that f_n converges uniformly to f. Consequently, $\mathcal{F}_b(X)$ is a Banach space with respect to the uniform norm.*

Proof. Suppose that $\{f_n\}_{n\in\mathbb{N}}$ is a uniformly Cauchy sequence in $\mathcal{F}_b(X)$. If we fix any particular point $x \in X$, then for all m and n we have

$$|f_m(x) - f_n(x)| \leq \|f_m - f_n\|_u.$$

Hence, for this fixed x we see that $\big(f_n(x)\big)_{n\in\mathbb{N}}$ is a *Cauchy sequence of scalars.* Since $\mathbb{F}$ is complete, this sequence of scalars must converge. Define

$$f(x) = \lim_{n\to\infty} f_n(x), \qquad x \in X.$$

We will show that $f \in \mathcal{F}_b(X)$ and $f_n \to f$ uniformly.

Choose any $\varepsilon > 0$. Since $\{f_n\}_{n\in\mathbb{N}}$ is uniformly Cauchy, there exists an N such that $\|f_m - f_n\|_u < \varepsilon$ for all $m, n \geq N$. If $n \geq N$, then for every $x \in X$ we have

$$|f(x) - f_n(x)| = \lim_{m\to\infty} |f_m(x) - f_n(x)| \leq \lim_{m\to\infty} \|f_m - f_n\|_u \leq \varepsilon.$$

Hence for each $n \geq N$ we see that

$$\|f - f_n\|_u = \sup_{x\in X} |f(x) - f_n(x)| \leq \varepsilon.$$

This shows that $f_n \to f$ uniformly as $n \to \infty$. Also, since $\|f - f_n\|_u \leq \varepsilon$, the function $g_n = f - f_n$ is bounded. Since f_n and g_n are bounded, their sum, which is f, is bounded. Thus $f \in \mathcal{F}_b(X)$, so $\mathcal{F}_b(X)$ is complete. $\square$

We will focus more closely now on the case that X is a metric space. In this setting we can distinguish between continuous and discontinuous functions on X. We introduce two spaces of continuous functions on a metric space.

Definition 3.5.8. If X is a metric space, then we let $C(X)$ be the space of all continuous functions $f\colon X \to \mathbb{F}$, and we let $C_b(X)$ be the subspace of all continuous bounded functions on X. That is,

$$C(X) = \{f\colon X \to \mathbb{F} : f \text{ is continuous}\},$$

and

$$C_b(X) = C(X) \cap \mathcal{F}_b(X) = \{f \in C(X) : f \text{ is bounded}\}. \qquad \Diamond$$

Problem 2.9.13 shows that $C_b(X)$ is a subspace of $\mathcal{F}_b(X)$, and therefore $C_b(X)$ is a normed space with respect to the uniform norm. Unless we explicitly specify otherwise, we always assume that the norm on $C_b(X)$ is the uniform norm. Note that if X is compact, then every function on X is bounded by Corollary 2.9.3, so $C_b(X) = C(X)$ when X is compact.

According to the following lemma, the uniform limit of a sequence of continuous functions is continuous.

Theorem 3.5.9. *Let X be a metric space. If $f_n \in C_b(X)$ for $n \in \mathbb{N}$ and $f : X \to \mathbb{F}$ is a function on X such that f_n converges uniformly to f, then $f \in C_b(X)$.*

Proof. Fix any $\varepsilon > 0$. Then, by the definition of uniform convergence, there exists some integer $n > 0$ such that

$$\|f - f_n\|_u < \frac{\varepsilon}{3}. \tag{3.20}$$

In fact, equation (3.20) will be satisfied for all large enough n, but we need only one particular n for this proof. Choose any point $x \in X$. Since f_n is continuous, there is a $\delta > 0$ such that for all $y \in X$ we have

$$d(x, y) < \delta \quad \Longrightarrow \quad |f_n(x) - f_n(y)| < \frac{\varepsilon}{3}.$$

Consequently, if $y \in X$ satisfies $|x - y| < \delta$, then

$$
\begin{aligned}
|f(x) - f(y)| &\leq |f(x) - f_n(x)| + |f_n(x) - f_n(y)| + |f_n(y) - f(y)| \\
&\leq \|f - f_n\|_u + \frac{\varepsilon}{3} + \|f_n - f\|_u \\
&< \frac{\varepsilon}{3} + \frac{\varepsilon}{3} + \frac{\varepsilon}{3} = \varepsilon.
\end{aligned} \tag{3.21}
$$

Applying Lemma 2.9.4, this proves that f is continuous. Since f_n and $f - f_n$ are both bounded, their sum, which is f, is also bounded. Therefore f belongs to $C_b(X)$. □

Now we prove that $C_b(X)$ is complete.

Theorem 3.5.10 ($C_b(X)$ Is a Banach Space). *$C_b(X)$ is a closed subspace of $\mathcal{F}_b(X)$ with respect to the uniform norm, and therefore is a Banach space with respect to $\| \cdot \|_u$.*

Proof. Suppose functions $f_n \in C_b(X)$ and $f \in \mathcal{F}_b(X)$ are given such that $f_n \to f$ uniformly. If we can show that f belongs to $C_b(X)$, then Theorem 2.6.2 will imply that $C_b(X)$ is closed. But the work for this is already done— we just apply Theorem 3.5.9 and conclude that $f \in C_b(X)$. Consequently, $C_b(X)$ is a closed subspace of $\mathcal{F}_b(X)$ with respect to the uniform norm, and therefore it is complete, since $\mathcal{F}_b(X)$ is complete. □

It is not as easy to visualize the appearance of an open ball in $C_b(X)$ as it is in $\mathbb{R}^2$ or $\mathbb{R}^3$, but we can draw a picture that illustrates the idea. Suppose

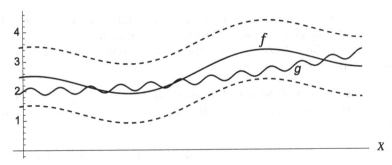

Fig. 3.5 The function g belongs to the open ball $B_1(f)$ centered at f with radius 1, because $\|f - g\|_u < 1$. The dashed curves are the graphs of $f(x) + 1$ and $f(x) - 1$.

that I is an interval in $\mathbb{R}$, and f is some particular function in $C_b(I)$. By definition, the open ball $B_r(f)$ of radius r centered at f is

$$B_r(f) = \{g \in C_b(I) : \|f - g\|_u < r\}.$$

That is, $B_r(f)$ consists of all functions g whose distance from f, as measured by the uniform norm, is strictly less than r. Since

$$\|f - g\|_u = \sup_{x \in I} |f(x) - g(x)|,$$

a function g belongs to $B_r(f)$ if and only if there exists a number $0 \le d < r$ such that

$$|f(x) - g(x)| \le d \quad \text{for every } x \in I.$$

Thus if $g \in B_r(f)$, then $g(x)$ can never stray from $f(x)$ by more than $d < r$ units. For the case of real scalars (i.e., $\mathbb{F} = \mathbb{R}$), Figure 3.5 depicts a function g that belongs to the open ball $B_1(f)$ for a particular f. For the functions f and g in Figure 3.5, there is some $d < 1$ such that $|f(x) - g(x)| \le d$ for every x. The ball $B_1(f)$ consists of all such continuous functions g whose graphs lie between the dashed lines in Figure 3.5.

For functions whose domain is an interval I in the real line, it is often important to consider differentiability as well as continuity. For simplicity of presentation we will take the domain of our functions to be an open interval I. However, only small modifications need to be made if we want to consider functions whose domain is a different type of interval (essentially, one-sided instead of two-sided limits need to be used at each boundary point of the interval).

First, however, we must discuss a technical detail that complicates the issue of which space of differentiable functions we should consider. The problem is that it is not true that a differentiable function must have a continuous derivative. For example, set

$$f(t) = \begin{cases} t^2 \sin(1/t), & t \neq 0, \\ 0, & t = 0. \end{cases} \tag{3.22}$$

According to Problem 3.5.24, this function has the following properties:

- f is continuous everywhere on $\mathbb{R}$,
- $f'(t)$ exists for every point $t \in \mathbb{R}$, but
- f' is discontinuous at $t = 0$.

The graphs of f and f' are shown in Figure 3.6.

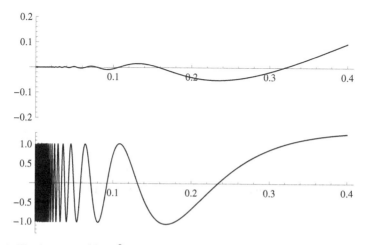

Fig. 3.6 The function $f(t) = t^2 \sin(1/t)$ (top), and its derivative f' (bottom), for $0 < t <$ 0.4. Note the differing vertical scales on the two graphs.

Thus there exist differentiable functions whose derivatives are not very "nice." In order to create a space of differentiable functions on which we can define a good norm, we usually must restrict our attention somewhat. Instead of considering *all* differentiable functions, we often consider the space of differentiable functions *that have a continuous derivative*. In fact, in order to be able to define a norm, we restrict further to functions f such that both f and f' are bounded. This leads us to the following definition.

Definition 3.5.11. Let I be an open interval in the real line (either finite or infinite). Then we define $C_b^1(I)$ to be the set of all bounded differentiable functions on I that have a bounded continuous derivative:

$$C_b^1(I) = \{f \in C_b(I) : f \text{ is differentiable and } f' \in C_b(I)\}. \qquad \diamondsuit$$

The uniform norm $\|\cdot\|_u$ is a norm on $C_b^1(I)$, but $C_b^1(I)$ is not complete with respect to the uniform norm. Instead, we define a norm that takes into account both f and f':

$$\|f\|_{C_b^1} = \|f\|_u + \|f'\|_u = \sup_{t \in I} |f(t)| + \sup_{t \in I} |f'(t)|. \qquad (3.23)$$

The following theorem (whose proof is Problem 3.5.25) states that $C_b^1(I)$ is a Banach space with respect to this norm.

Theorem 3.5.12. *If I is an open interval in $\mathbb{R}$, then $\|\cdot\|_{C_b^1}$ is a norm on $C_b^1(I)$, and $C_b^1(I)$ is complete with respect to this norm.* $\diamond$

We can extend these ideas to higher derivatives. Given an integer $m \geq 1$, we define the following space of m-times differentiable functions:

$$C_b^m(\mathbb{R}) = \{f \in C_b(\mathbb{R}) : f, f', \dots, f^{(m)} \in C_b(\mathbb{R})\}.$$

This is a Banach space with respect to the norm

$$\|f\|_{C_b^m} = \|f\|_u + \|f'\|_u + \cdots + \|f^{(m)}\|_u.$$

Problems

3.5.13. Given $n \in \mathbb{N}$, let $p_n(t) = t^n$ for $t \in [0,1]$. Prove that the sequence $\{p_n\}_{n \in \mathbb{N}}$ converges pointwise to a *discontinuous* function, but p_n does not converge *uniformly* to that function.

3.5.14. For the following choices of functions g_n, determine whether the sequence $\{g_n\}_{n \in \mathbb{N}}$ is pointwise convergent, uniformly convergent, uniformly Cauchy, or bounded with respect to the uniform norm.

(a) $g_n(t) = t^n/n$ for $t \in [0,1]$.

(b) $g_n(t) = \sin 2\pi nt$ for $t \in [0,1]$.

(c) $g_n(t) = nf_n(t)$ for $t \in [0,1]$, where f_n is the Shrinking Triangle from Example 3.5.2.

(d) $g_n(t) = e^{-n|t|}$ for $t \in \mathbb{R}$.

(e) $g_n(t) = te^{-n|t|}$ for $t \in \mathbb{R}$.

(f) $g_n(t) = \dfrac{nt}{1 + n^2 t^2}$ for $t \in \mathbb{R}$.

3.5.15. Let X be a set. Suppose that f_n, f, g_n, $g \in \mathcal{F}_b(X)$ are such that $f_n \to f$ and $g_n \to g$ uniformly. Prove the following statements.

(a) $f_n + g_n \to f + g$ uniformly.

(b) $f_n g_n \to fg$ uniformly.

(c) If there is a $\delta > 0$ such that $|g_n(t)| \geq \delta$ for every $t \in X$ and $n \in \mathbb{N}$, then $f_n/g_n \to f/g$ uniformly.

3.5.16. Show that if $f \in C(\mathbb{R})$ is uniformly continuous and for each $n \in \mathbb{N}$ we set $f_n(t) = f(t - \frac{1}{n})$ for $t \in \mathbb{R}$, then $f_n \to f$ uniformly on $\mathbb{R}$ as $n \to \infty$. Show by example that this can fail if $f \in C(\mathbb{R})$ is not uniformly continuous.

3.5.17. (a) Let X be a set and let Y be a metric space. Given functions f_n, $f \colon X \to Y$, what should it mean to say that "f_n converges to f uniformly"? Formulate a definition that is appropriate for this setting.

(b) Suppose that X and Y are both metric spaces, and let points x_n, $x \in X$ and functions $f_n, f \colon X \to Y$ be given. Prove that if $f_n \to f$ uniformly and $x_n \to x$ in X, then $f_n(x_n) \to f(x)$ in Y.

3.5.18. Let $\{f_n\}_{n \in \mathbb{N}}$ be the sequence of Shrinking Triangles from Example 3.5.2. Prove the following statements.

(a) $\{f_n\}_{n \in \mathbb{N}}$ is a bounded sequence in $C[0,1]$ that contains no convergent subsequences.

(b) The closed unit disk $D = \{f \in C[0,1] : \|f\|_u \le 1\}$ is a closed and bounded subset of $C[0,1]$ that is not sequentially compact.

3.5.19. The space of continuous functions that *vanish at infinity* is

$$C_0(\mathbb{R}) = \left\{ f \in C(\mathbb{R}) : \lim_{t \to \pm\infty} f(t) = 0 \right\}.$$

Prove the following statements.

(a) $C_0(\mathbb{R})$ is a closed subspace of $C_b(\mathbb{R})$ with respect to the uniform norm, and therefore it is a Banach space with respect to $\| \cdot \|_u$.

(b) Every function in $C_0(\mathbb{R})$ is uniformly continuous on $\mathbb{R}$, but there exist functions in $C_b(\mathbb{R})$ that are not uniformly continuous.

(c) If $f \in C_0(\mathbb{R})$ and $g_a(t) = f(t-a)$, where $a \in \mathbb{R}$ is fixed, then $g_a \in C_0(\mathbb{R})$ and $\|g_a\|_u = \|f\|_u$. Therefore $\{g_k\}_{k \in \mathbb{N}}$, where $g_k(t) = f(t-k)$, is a bounded sequence in $C_0(\mathbb{R})$.

(d) The closed unit disk $D = \{f \in C_0(\mathbb{R}) : \|f\|_u \le 1\}$ is a closed and bounded subset of $C_0(\mathbb{R})$ that is not sequentially compact.

3.5.20. A continuous function $f \colon \mathbb{R} \to \mathbb{F}$ has *compact support* if it is identically zero outside of some finite interval (see Definition 2.9.7). Let

$$C_c(\mathbb{R}) = \{f \in C(\mathbb{R}) : f \text{ has compact support}\}.$$

Prove that $C_c(\mathbb{R})$ is a dense but proper subspace of $C_0(\mathbb{R})$ with respect to the uniform norm.

3.5.21. Let $f_n(t) = \sqrt{t^2 + \frac{1}{n^2}}$ for $t \in [-1, 1]$. Prove the following statements.

(a) f_n converges uniformly on $[-1, 1]$ to the function $f(t) = |t|$, which is not differentiable on that domain.

(b) f_n' converges pointwise to some function g on $[-1,1]$, i.e., $g(t) = \lim_{n \to \infty} f_n'(t)$ exists for each $t \in [-1,1]$.

(c) g is not continuous.

(d) $\{f_n'\}_{n \in \mathbb{N}}$ does not converge uniformly on $[-1,1]$.

Thus, even though f_n converges uniformly to f and each f_n is differentiable, the limit f is not differentiable.

3.5.22. Let $f_n(t) = \dfrac{t}{1 + n^2 t^2}$ for $t \in \mathbb{R}$. Prove the following statements.

(a) f_n converges uniformly to the zero function on $\mathbb{R}$.

(b) f_n' converges pointwise, i.e., $g(t) = \lim_{n \to \infty} f_n'(t)$ exists for each $t \in \mathbb{R}$.

(c) g is not identically zero.

(d) $\{f_n'\}_{n \in \mathbb{N}}$ does not converge uniformly on $[-1,1]$.

Thus, even though f_n converges uniformly to zero and each f_n is differentiable, there are points where the limit of the derivatives is not equal to the derivative of the limit, i.e., points where $\left(\lim_{n \to \infty} f_n\right)'(t) \neq \lim_{n \to \infty} f_n'(t)$.

3.5.23. Suppose that functions f_n are differentiable on $\mathbb{R}$, and for each n the derivative f_n' is continuous on $\mathbb{R}$. Suppose further that there exist functions f and g such that:

(a) f_n converges uniformly to f, and

(b) f_n' converges uniformly to g.

Prove that f is differentiable, f' is continuous, and $f' = g$. How does this relate to Problems 3.5.21 and 3.5.23?

3.5.24. Let f be the function defined in equation (3.22). Prove that f is continuous on $\mathbb{R}$ and is differentiable at every point of $\mathbb{R}$, but f' is not continuous at $t = 0$.

Hint: For differentiability, consider $t = 0$ and $t \neq 0$ separately. For $t = 0$, directly use the definition of the derivative to prove that $f'(0)$ exists.

3.5.25. Let I be an open interval in the real line (either finite or infinite). Let $C_b^1(I)$ be the set of all bounded differentiable functions on I that have a bounded continuous derivative:

$$C_b^1(I) = \{f \in C_b(I) : f \text{ is differentiable and } f' \in C_b(I)\}.$$

Prove the following statements.

(a) $\| \cdot \|_u$ is a norm on $C_b^1(I)$, but $C_b^1(I)$ is not complete with respect to this norm.

(b) $\|f\| = \|f'\|_u$ defines a seminorm on $C_b^1(I)$, but it is not a norm.

(c) The following is a norm on $C_b^1(I)$:

$$\|f\|_{C_b^1} = \|f\|_u + \|f'\|_u, \qquad f \in C_b^1(I).$$

(d) $C_b^1(I)$ is complete with respect to the norm given in part (c).

Remark: If I is a closed or half-open interval, then the same results hold if *differentiable on I* means that f is differentiable at every point in the interior and f is differentiable from the right or left (as appropriate) at each endpoint of I.

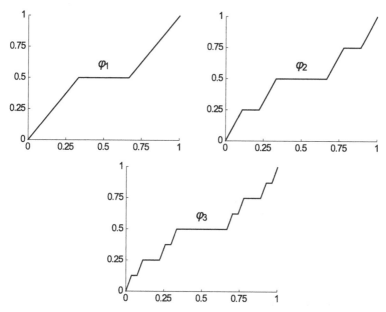

Fig. 3.7 First stages in the construction of the Cantor–Lebesgue function in Problem 3.5.26.

3.5.26. It may be helpful to review Example 2.7.1 before working this problem. Let φ_1, φ_2, φ_3 be the functions pictured in Figure 3.7. The function φ_1 takes the constant value $\frac{1}{2}$ on the interval $(\frac{1}{3}, \frac{2}{3})$ that is removed from $[0, 1]$ in the first stage of the construction of the Cantor set C, and it is linear on the remaining subintervals of $[0, 1]$. The function φ_2 takes the same constant $\frac{1}{2}$ on the interval $(\frac{1}{3}, \frac{2}{3})$ but additionally is constant with values $\frac{1}{4}$ and $\frac{3}{4}$ on the two intervals that are removed during the second stage of the construction of the Cantor set. Continue this process and define $\varphi_3, \varphi_4, \ldots$ in a similar fashion. Each φ_n is continuous, and φ_n is constant on each of the open intervals that are removed during the nth stage of the construction of the Cantor set. Prove the following statements.

(a) For each $n \in \mathbb{N}$,

$$\|\varphi_{n+1} - \varphi_n\|_u = \sup_{t \in [0,1]} |\varphi_{n+1}(t) - \varphi_n(t)| \leq 2^{-n}.$$

(b) $\{\varphi_n\}_{n \in \mathbb{N}}$ is a uniformly Cauchy sequence in $C[0,1]$, and therefore there exists a function $\varphi \in C[0,1]$ such that $\varphi_n \to \varphi$ uniformly.

(c) φ is monotone increasing, i.e., if $0 \leq x \leq y \leq 1$, then $\varphi(x) \leq \varphi(y)$.

(d) If $t \in [0,1] \setminus C$, then φ is differentiable at t and $\varphi'(t) = 0$.

(e) φ is not Lipschitz on $[0,1]$. This continuous function φ is called the *Cantor–Lebesgue function* or the *Devil's staircase* (see Figure 3.8).

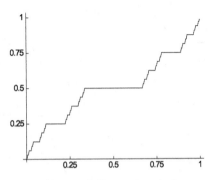

Fig. 3.8 The Devil's staircase (Cantor–Lebesgue function).

3.6 Equivalent Norms

A single vector space X can have many different norms. Each norm $\|\cdot\|$ on X gives us a corresponding *convergence criterion*:

$$x_n \to x \text{ with respect to } \|\cdot\| \qquad means \qquad \lim_{n \to \infty} \|x - x_n\| = 0. \qquad (3.24)$$

In general, if a sequence converges with respect to one norm, it need not converge with respect to a different norm. Even so, it is *possible* for two different norms to determine the same convergence criterion. That is, it sometimes happens that even though $\|\cdot\|_a$ and $\|\cdot\|_b$ are different norms on X, we still have

$$\lim_{n \to \infty} \|x - x_n\|_a = 0 \quad \Longleftrightarrow \quad \lim_{n \to \infty} \|x - x_n\|_b = 0.$$

In this case, a sequence converges with respect to one norm if and only if it converges with respect to the other norm. We will prove in Theorem 3.6.2

that this happens exactly when the two norms are *equivalent* in the sense of the following definition.

Definition 3.6.1 (Equivalent Norms). Suppose that $\|\cdot\|_a$ and $\|\cdot\|_b$ are each norms on a vector space X. We say that these two norms are *equivalent* if there exist constants C_1, $C_2 > 0$ such that

$$\forall\, x \in X, \quad C_1 \|x\|_a \leq \|x\|_b \leq C_2 \|x\|_a. \qquad \diamond \qquad (3.25)$$

For example, according to Problem 3.6.4, any two of the norms $\|\cdot\|_p$ on $\mathbb{F}^d$ defined in Theorem 3.2.5 are equivalent. In contrast, Problem 3.6.7 shows that the ℓ^1 and ℓ^p norms are not equivalent on the infinite-dimensional space ℓ^1 when $p \neq 1$.

The next theorem shows that equivalent norms determine the same convergence criterion and the same topology on a vector space X.

Theorem 3.6.2. *If $\|\cdot\|_a$ and $\|\cdot\|_b$ are two norms on a vector space X, then the following three statements are equivalent.*

(a) $\|\cdot\|_a$ *and* $\|\cdot\|_b$ *are equivalent norms.*

(b) $\|\cdot\|_a$ *and* $\|\cdot\|_b$ *induce the same topologies on X. That is, if $U \subseteq X$, then*

$$U \text{ is open with respect to } \|\cdot\|_a \quad \Longleftrightarrow \quad U \text{ is open with respect to } \|\cdot\|_b.$$

(c) $\|\cdot\|_a$ *and* $\|\cdot\|_b$ *define the same convergence criterion on X. That is, if $\{x_n\}_{n\in\mathbb{N}}$ is a sequence of vectors in X and x is a vector in X, then*

$$\lim_{n\to\infty} \|x - x_n\|_a = 0 \quad \Longleftrightarrow \quad \lim_{n\to\infty} \|x - x_n\|_b = 0. \qquad (3.26)$$

Proof. (a) $\Rightarrow$ (c). The reader should verify that this implication follows directly from the definition of equivalent norms.

(c) $\Rightarrow$ (b). Assume that statement (c) holds, and suppose that F is a subset of X that is closed with respect to the norm $\|\cdot\|_a$. We will show that F is closed with respect to the norm $\|\cdot\|_b$. To do this, suppose that $x_n \in F$, $x \in X$, and $\|x - x_n\|_b \to 0$. Since we have assumed that statement (c) holds, it follows that $\|x - x_n\|_a \to 0$. Since F is closed with respect to $\|\cdot\|_a$, Theorem 2.4.2 therefore implies that $x \in F$. In summary, we have shown that if $x_n \in F$ and $x_n \to x$ with respect to $\|\cdot\|_b$, then $x \in F$. Applying Theorem 2.4.2 again, this implies that F is closed with respect to $\|\cdot\|_b$.

The above work shows that every subset of X that is closed with respect to $\|\cdot\|_a$ is closed with respect to $\|\cdot\|_b$. A symmetric argument shows that every set that is closed with respect to $\|\cdot\|_b$ is closed with respect to $\|\cdot\|_a$. Taking complements, we conclude that statement (b) holds.

(b) $\Rightarrow$ (a). Assume that statement (b) holds, and let $B_r^a(x)$ and $B_r^b(x)$ denote the open balls of radius r centered at $x \in X$ with respect to the norms $\|\cdot\|_a$ and $\|\cdot\|_b$, respectively. Since $B_1^a(0)$ is open with respect to $\|\cdot\|_a$, statement (b) implies that $B_1^a(0)$ is open with respect to $\|\cdot\|_b$. Therefore, since $0 \in B_1^a(0)$, there must exist some radius $r > 0$ such that $B_r^b(0) \subseteq B_1^a(0)$.

Choose any vector $x \in X$ and fix $0 < t < r$. Then

$$\frac{t}{\|x\|_b} x \in B_r^b(0) \subseteq B_1^a(0).$$

By the definition of an open ball, this tells us that

$$\left\| \frac{tx}{\|x\|_b} \right\|_a < 1.$$

Rearranging, we see that $t\|x\|_a < \|x\|_b$, and therefore by letting $t \to r$ we obtain $r\|x\|_a \leq \|x\|_b$. This is true for all $x \in X$, so we can take $C_1 = r$. A symmetric argument, interchanging the roles of the two norms, completes the proof. $\square$

Problems

3.6.3. Let $\|\cdot\|_a$ and $\|\cdot\|_b$ be equivalent norms on a vector space X, and let E be a subset of X. For each of the following statements, prove that the statement is true with respect to $\|\cdot\|_a$ if and only if it is true with respect to $\|\cdot\|_b$.

(a) E is closed.

(b) E is compact.

(c) E is bounded.

(d) $x \in X$ is an accumulation point of E.

(e) E is complete (every Cauchy sequence in E converges to a point of E).

(f) E is convex.

3.6.4. Prove that any two of the norms on $\mathbb{F}^d$ defined in Theorem 3.2.5 are equivalent.

3.6.5. Let X be a finite-dimensional vector space, and let $\mathcal{B} = \{e_1, \ldots, e_d\}$ be a Hamel basis for X (see Section 1.11). Then for each $x \in X$ there exist unique scalars $c_1(x), \ldots, c_d(x)$ such that $x = \sum_{k=1}^d c_k(x)\, e_k$. Define

$$\|x\|_p = \begin{cases} \left(|c_1(x)|^p + \cdots + |c_d(x)|^p \right)^{1/p}, & \text{if } 1 \leq p < \infty, \\ \max\{|c_1(x)|, \ldots, |c_d(x)|\}, & \text{if } p = \infty. \end{cases} \qquad (3.27)$$

Prove that:

(a) $\|\cdot\|_p$ is a norm on X,

(b) X is a Banach space with respect to $\|\cdot\|_p$, and

(c) any two of these norms are equivalent.

3.6.6. (a) Given x_n, $x \in \ell^1$, prove that

$$\lim_{n\to\infty} \|x - x_n\|_1 = 0 \quad \Longrightarrow \quad \lim_{n\to\infty} \|x - x_n\|_\infty = 0.$$

(b) Show by example that the converse implication fails, i.e., exhibit vectors x_n, $x \in \ell^1$ such that $x_n \to x$ with respect to $\|\cdot\|_\infty$ but $x_n \not\to x$ with respect to $\|\cdot\|_1$.

3.6.7. Since ℓ^1 is a subspace of both ℓ^2 and ℓ^∞, if we restrict the ℓ^2-norm or the ℓ^∞-norm to the space ℓ^1, then we obtain a norm on ℓ^1. Prove that no two of the norms $\|\cdot\|_1$, $\|\cdot\|_2$, and $\|\cdot\|_\infty$ are equivalent on ℓ^1.

3.6.8. Let X and Y be normed vector spaces with norms $\|\cdot\|_X$ and $\|\cdot\|_Y$, respectively. Given vectors $x \in X$ and $y \in Y$, if $1 \le p < \infty$, then set

$$\|(x, y)\|_p = \left(\|x\|_X^p + \|y\|_Y^p \right)^{1/p},$$

and for $p = \infty$ let $\|(x, y)\|_\infty = \max\{\|x\|_X, \|y\|_Y\}$. Prove the following statements.

(a) $\|\cdot\|_p$ is a norm on the Cartesian product $X \times Y$.

(b) Any two of these norms on $X \times Y$ are equivalent.

(c) If X and Y are both Banach spaces, then $X \times Y$ is a Banach space with respect to $\|\cdot\|_p$.

3.6.9. Suppose that $\|\cdot\|_a$ and $\|\cdot\|_b$ are two norms on a vector space X. Let $B_r^a(x)$ and $B_r^b(x)$ denote the open balls of radius r centered at $x \in X$ with respect to the norms $\|\cdot\|_a$ and $\|\cdot\|_b$, respectively. Prove that $\|\cdot\|_a$ and $\|\cdot\|_b$ are equivalent norms if and only if there exists some $r > 0$ such that

$$B_{1/r}^a(0) \subseteq B_1^b(0) \subseteq B_r^a(0).$$

3.6.10. Fix $1 \le p < \infty$. Prove that the L^p-norm on $C[a, b]$ (the norm $\|\cdot\|_p$ defined in Theorem 3.2.6) is not equivalent to the uniform norm $\|\cdot\|_u$.

3.6.11. (a) Let $w \in C[a, b]$ be any function such that $w(t) > 0$ for every t. Prove that

$$\|f\|_w = \int_a^b |f(t)|\, w(t)\, dt, \qquad f \in C[a, b], \tag{3.28}$$

defines a norm on $C[a, b]$ that is equivalent to the norm $\|\cdot\|_1$ defined in equation (3.1).

(b) Suppose that $w \in C[a, b]$ satisfies $w \ge 0$ with $w(t_0) = 0$ for only a *single* point $t_0 \in [a, b]$. Prove that equation (3.28) defines a norm on $C[a, b]$, but in this case $\|\cdot\|_1$ and $\|\cdot\|_w$ are not equivalent.

3.7 Norms on Finite-Dimensional Spaces

We will use the properties of compact sets to prove that *all norms on a finite-dimensional vector space X are equivalent.* If X is the *trivial space* $X = \{0\}$ then the only norm on X is the zero function, so we will focus our attention on nontrivial vector spaces.

A nontrivial finite-dimensional vector space X, say with dimension d, has a Hamel basis $\mathcal{B} = \{e_1, \ldots, e_d\}$. Each vector $x \in X$ has a unique representation

$$x = c_1(x) e_1 + \cdots + c_d(x) e_d, \tag{3.29}$$

and it is easy to see that

$$\|x\|_\infty = \max\{|c_1(x)|, \ldots, |c_d(x)|\} \tag{3.30}$$

is a norm on X (indeed, this is part of Problem 3.6.5). The "ℓ^∞-unit circle" corresponding to this norm is

$$S = \{x \in X : \|x\|_\infty = 1\}. \tag{3.31}$$

The following lemma shows that this unit circle is a compact subset of X.

Lemma 3.7.1. *If X is a nontrivial finite-dimensional vector space, then the unit circle S defined in equation (3.31) is a compact set with respect to the norm $\|\cdot\|_\infty$ defined in equation (3.30).*

Proof. Each $x \in X$ can be uniquely written as $x = c_1(x) e_1 + \cdots + c_d(x) e_d$. By the definition of the norm $\|\cdot\|_\infty$, we have

$$x \in S \quad \Longleftrightarrow \quad |c_k(x)| = 1 \text{ for some } 1 \le k \le d.$$

We will show that S is sequentially compact. To do this, let $\{x_n\}_{n\in\mathbb{N}}$ be any sequence of vectors contained in S. Then for each $n \in \mathbb{N}$ there exists some index $1 \le k_n \le d$ such that $|c_{k_n}(x_n)| = 1$. Since there are infinitely many n but only finitely many choices for k_n, there must be some single index $1 \le k \le d$ such that $|c_k(x_n)| = 1$ for infinitely many distinct values of n. Let $\{y_n\}_{n\in\mathbb{N}}$ be a subsequence of $\{x_n\}_{n\in\mathbb{N}}$ such that $|c_k(y_n)| = 1$ for every n.

Next, let

$$D = \{c \in \mathbb{F} : |c| \le 1\}.$$

That is, if $\mathbb{F} = \mathbb{R}$ then D is the interval $[-1, 1]$, while if $\mathbb{F} = \mathbb{C}$ then D is the closed unit disk in the complex plane. Since D is a closed and bounded subset of $\mathbb{F}$, the Heine–Borel Theorem implies that D is compact, and therefore D is sequentially compact by Theorem 2.8.9. Since $\big(c_1(y_n)\big)_{n\in\mathbb{N}}$ is an infinite sequence of scalars contained in D, it follows that there must exist a subsequence $\{y_n'\}_{n\in\mathbb{N}}$ of $\{y_n\}_{n\in\mathbb{N}}$ such that $\big(c_1(y_n')\big)_{n\in\mathbb{N}}$ converges. Let b_1 be the limit of this sequence, and note that $|b_1| \le 1$.

Then we repeat the same argument over and over. That is, we take $j = 2$ and find a subsequence $\{y_n''\}_{n \in \mathbb{N}}$ of $\{y_n'\}_{n \in \mathbb{N}}$ such that $\left(c_2(y_n'')\right)_{n \in \mathbb{N}}$ converges, say to b_2. Then we take $j = 3$ and find b_3 and a subsequence $\{y_n'''\}_{n \in \mathbb{N}}$, and so forth. Repeating this for each coordinate j in turn, we eventually pass to a subsequence $\{y_n^{'''\cdots'}\}_{n \in \mathbb{N}}$, which for simplicity we will just call $\{w_n\}_{n \in \mathbb{N}}$, such that

$$b_j = \lim_{n \to \infty} c_j(w_n) \text{ exists for each } j = 1, \ldots, d. \tag{3.32}$$

Each scalar b_j satisfies $|b_j| \leq 1$, and for the index k we have $|b_k| = 1$. Hence the vector

$$x = b_1 e_1 + \cdots + b_d e_d$$

belongs to S. It follows from equation (3.32) and the definition of $\|\cdot\|_\infty$ that $\|x - w_n\|_\infty \to 0$ as $n \to \infty$. Hence S is sequentially compact, so Theorem 2.8.9 implies that S is compact. $\square$

Now we use the compactness of the set S to prove that all norms on a finite-dimensional vector space are equivalent.

Theorem 3.7.2. *If X is a finite-dimensional vector space, then any two norms on X are equivalent.*

Proof. As before, choose a basis $\mathcal{B} = \{e_1, \ldots, e_d\}$ for X, write vectors uniquely as $x = c_1(x) e_1 + \cdots + c_d(x) e_d$, and define $\|x\|_\infty = \max |c_k(x)|$.

Let $\|\cdot\|$ be any norm on X. We will prove that $\|\cdot\|$ is equivalent to $\|\cdot\|_\infty$. To do this, we must show that there exist strictly positive constants C_1, $C_2 > 0$ such that $C_1 \|x\|_\infty \leq \|x\| \leq C_2 \|x\|_\infty$ for every $x \in X$. The upper inequality follows easily by applying the Triangle Inequality:

$$\begin{aligned}
\|x\| &= \left\| \sum_{k=1}^{d} c_k(x) e_k \right\| \\
&\leq \sum_{k=1}^{d} |c_k(x)| \|e_k\| \\
&\leq \left(\sum_{k=1}^{d} \|e_k\| \right) \left(\max_k |c_k(x)| \right) \\
&= C_2 \|x\|_\infty, \tag{3.33}
\end{aligned}$$

where $C_2 = \sum_{k=1}^{d} \|e_k\|$. Each vector e_k is nonzero (since $\mathcal{B}$ is a basis), so C_2 is a positive (and finite) constant.

To establish the lower inequality, set $S = \{x \in X : \|x\|_\infty = 1\}$ and define

$$C_1 = \inf_{x \in S} \|x\|. \tag{3.34}$$

Since $x \in S$ if and only if $\|x\|_\infty = 1$, it follows that $C_1 \|x\|_\infty \leq \|x\|$ for every vector $x \in X$. Our task is to show that $C_1 > 0$.

If we had $C_1 = 0$, then there would exist vectors $x_n \in S$ such that $\|x_n\| \to 0$ as $n \to \infty$. Lemma 3.7.1 showed that S is compact with respect to $\|\cdot\|_\infty$, so $\{x_n\}_{n \in \mathbb{N}}$ must have a convergent subsequence. That is, there exist some subsequence $\{x_{n_k}\}_{k \in \mathbb{N}}$ and some vector $x \in S$ such that

$$\lim_{k \to \infty} \|x - x_{n_k}\|_\infty = 0.$$

But $\|x - x_{n_k}\| \leq C_2 \|x - x_{n_k}\|_\infty$ by equation (3.33), so we conclude that

$$\lim_{k \to \infty} \|x - x_{n_k}\| = 0.$$

Hence $x_{n_k} \to x$ with respect to the norm $\|\cdot\|$. The continuity of the norm (Lemma 3.3.3) therefore implies that $\|x_{n_k}\| \to \|x\|$. Since $\|x_{n_k}\| \to 0$, we conclude that $x = 0$, which contradicts the fact that $x \in S$. Therefore C_1 must be strictly positive (see Problem 3.7.7 for an alternative derivation of this fact).

In summary, we have shown that every norm $\|\cdot\|$ is equivalent to the particular norm $\|\cdot\|_\infty$, and it follows from this (why?) that any two norms on X are equivalent to each other. $\square$

Problem 3.7.4 shows how to derive the following result by applying Theorem 3.7.2.

Corollary 3.7.3. *If X is a normed space and M is a finite-dimensional subspace of X, then M is closed.* $\diamondsuit$

An infinite-dimensional subspace of a normed space need not be closed. For example, Problem 2.6.18 showed that c_{00} is a dense subspace of ℓ^1, i.e., its closure is all of ℓ^1. Therefore c_{00} is not equal to its closure, so it is not a closed set.

Problems

3.7.4. Let M be a finite-dimensional subspace of a normed space X. Suppose that $x_n \in M$ and $x_n \to y \in X$, but $y \notin M$. Define

$$M_1 = M + \mathrm{span}\{y\} = \{m + cy : m \in M, c \in \mathbb{F}\}.$$

(a) Prove that if $x \in M_1$, then there exist a *unique* vector $m_x \in M$ and a *unique* scalar $c_x \in \mathbb{F}$ such that $x = m_x + c_x y$.

(b) Given $x \in M_1$, let $x = m_x + c_x y$ as in part (a) and set

$$\|x\|_{M_1} = \|m_x\| + |c_x|.$$

Prove that $\| \cdot \|_{M_1}$ is a norm on M_1.

(c) Prove that $\|y - x_n\|_{M_1} \to 0$ as $n \to \infty$ (note that M_1 is finite-dimensional, so all norms on M_1 are equivalent).

(d) Derive a contradiction, and use this to show that M is closed.

3.7.5. Given a finite-dimensional normed space X, prove the following statements.

(a) X is separable.

(b) Every closed and bounded subset of X is compact.

(c) No proper subspace of X is dense in X.

3.7.6. Assume that X is a Banach space, and M is an infinite-dimensional subspace of X that has a countable Hamel basis. Prove that M is a meager subset of X.

3.7.7. This problem will give an alternative proof that the quantity $C_1 = \inf_{x \in S} \|x\|$ defined in equation (3.34) is strictly positive.

(a) Prove that the function $\rho(x) = \|x\|$ is continuous *with respect to the norm* $\| \cdot \|_\infty$.

(b) Use the fact that S is compact with respect to $\| \cdot \|_\infty$ to show that $C_1 > 0$.

Chapter 4
Further Results on Banach Spaces

In this chapter we will take a closer look at normed and Banach spaces. In a generic metric space there need not be any way to "add" points in the space, but a normed space is a vector space, so we can form linear combinations of vectors. Moreover, because we also have a notion of *limits* we can go even further and define *infinite series* of vectors (which are *limits* of *partial sums* of the series). We study infinite series and their applications in Sections 4.1–4.5. In Sections 4.6–4.9, we focus specifically on spaces of continuous functions, considering issues related to approximation and compactness.

4.1 Infinite Series in Normed Spaces

Suppose that $x_1, x_2, \ldots$ are infinitely many vectors in a vector space X. Does it make sense to sum all of these vectors, i.e., to form the *infinite series*

$$\sum_{n=1}^{\infty} x_n \;=\; x_1 + x_2 + \cdots ? \tag{4.1}$$

Since a vector space has an operation of *vector addition*, we do have a way to add two vectors together, and hence by induction we can compute the sum of *finitely many* vectors. That is, we can define the *partial sums*

$$s_N \;=\; \sum_{n=1}^{N} x_n \;=\; x_1 + x_2 + \cdots + x_N$$

for any finite integer $N = 1, 2, \ldots$. However, this does not by itself tell us how to compute a sum of *infinitely many* vectors. To make sense of this we have to see what happens to the partial sums s_N as N increases. And to do this we need to be able to compute the *limit* of the partial sums, which requires that we have a notion of *convergence* in our space. If all we know is

© Springer International Publishing AG, part of Springer Nature 2018
C. Heil, *Metrics, Norms, Inner Products, and Operator Theory*, Applied and
Numerical Harmonic Analysis, https://doi.org/10.1007/978-3-319-65322-8_4

that X is a vector space, then we do not have any way to measure distance or to determine convergence in X. But if X is a *normed space*, then we do know what convergence and limits mean, and so we can determine whether the partial sums $s_N = x_1 + \cdots + x_N$ converge to some limit as $N \to \infty$. If they do, then this is what it means to form an infinite sum. We make all of this precise in the following definition.

Definition 4.1.1 (Convergent Series). Let $\{x_n\}_{n \in \mathbb{N}}$ be a sequence of vectors in a normed space X. We say that the infinite series $\sum_{n=1}^{\infty} x_n$ *converges* if there is a vector $x \in X$ such that the *partial sums* $s_N = \sum_{n=1}^{N} x_n$ converge to x, i.e., if

$$\lim_{N \to \infty} \|x - s_N\| = \lim_{N \to \infty} \left\| x - \sum_{n=1}^{N} x_n \right\| = 0. \qquad (4.2)$$

In this case, we write $x = \sum_{n=1}^{\infty} x_n$, and we also use the shorthand $x = \sum x_n$ or $x = \sum_n x_n$. $\Diamond$

In order for an infinite series to converge in X, the *norm of the difference* between x and the partial sum s_N must converge to zero. If we wish to emphasize which norm we are referring to, we may write that $x = \sum x_n$ *converges with respect to* $\| \cdot \|$, or we may say that $x = \sum x_n$ *converges in* X.

Example 4.1.2. Let $X = \ell^\infty$, and let $\{\delta_n\}_{n \in \mathbb{N}}$ be the sequence of standard basis vectors. Does the series $\sum_{n=1}^{\infty} \delta_n$ converge in ℓ^∞? The Nth partial sum of this series is

$$s_N = \sum_{n=1}^{N} \delta_n = (1, \ldots, 1, 0, 0, \ldots),$$

where the 1 is repeated N times. It may appear that s_N converges to the constant vector $x = (1, 1, \ldots)$. However, while s_N does converge *componentwise* to x, *it does not converge in* ℓ^∞-*norm*, because

$$\|x - s_N\|_\infty = \|(0, \ldots, 0, 1, 1, \ldots)\|_\infty = 1 \nrightarrow 0.$$

In fact, the sequence of partial sums $\{s_N\}_{N \in \mathbb{N}}$ is not even Cauchy in ℓ^∞ (why not?), so the partial sums do not converge to any vector. Therefore $\sum_{n=1}^{\infty} \delta_n$ is not a convergent series in ℓ^∞. Does this series converge in ℓ^p for any finite value of p? $\Diamond$

Example 4.1.3. (a) Consider the series $\sum_{n=1}^{\infty} \frac{(-1)^n}{n} \delta_n$ in the space $X = \ell^1$. The partial sums of this series are

$$s_N = \sum_{n=1}^{N} \frac{(-1)^n}{n} \delta_n = \left(-1, \tfrac{1}{2}, -\tfrac{1}{3}, \ldots, \frac{(-1)^N}{N}, 0, 0, \ldots \right).$$

These partial sums converge componentwise to

$$x = \left(\frac{(-1)^n}{n}\right)_{n \in \mathbb{N}} = \left(-1, \tfrac{1}{2}, -\tfrac{1}{3}, \dots\right), \qquad (4.3)$$

but this vector does not belong to ℓ^1. The partial sums s_N do not converge to x in ℓ^1-norm, or to any vector in ℓ^1, because convergent sequences must be bounded but the s_N are not:

$$\sup_{N \in \mathbb{N}} \|s_N\|_1 = \sup_{N \in \mathbb{N}} \sum_{n=1}^{N} \left|\frac{(-1)^n}{n}\right| = \sup_{N \in \mathbb{N}} \sum_{n=1}^{N} \frac{1}{n} = \infty.$$

Consequently $\sum_{n=1}^{\infty} \frac{(-1)^n}{n} \delta_n$ is not a convergent series in ℓ^1.

(b) Again consider $\sum_{n=1}^{\infty} \frac{(-1)^n}{n} \delta_n$, but this time in ℓ^p where $1 < p < \infty$. In this case the vector x defined in equation (4.3) does belong to ℓ^p, because $\sum_{n=1}^{\infty} \frac{1}{n^p} < \infty$. Furthermore,

$$x - s_N = \left(0, \dots, 0, \frac{(-1)^{N+1}}{N+1}, \frac{(-1)^{N+2}}{N+2}, \dots\right),$$

so by applying Lemma 1.8.4 we see that

$$\lim_{N \to \infty} \|x - s_N\|_1 = \lim_{N \to \infty} \sum_{n=N+1}^{\infty} \left|\frac{(-1)^n}{n}\right|^p = \lim_{N \to \infty} \sum_{n=N+1}^{\infty} \frac{1}{n^p} = 0.$$

Thus $\sum_{n=1}^{\infty} \frac{(-1)^n}{n} \delta_n$ is a convergent series in ℓ^p when $1 < p < \infty$. Do the $(-1)^n$ factors play any role in this example, i.e., does anything change if we consider the series $\sum_{n=1}^{\infty} \frac{1}{n} \delta_n$ instead? $\diamond$

The following result (whose proof is Problem 4.1.7) gives a complete characterization of the scalars c_n such that $\sum_{n=1}^{\infty} c_n \delta_n$ converges in ℓ^p. Why is there a difference between the finite p and $p = \infty$ cases?

Lemma 4.1.4. *Assume c_n is a scalar for $n \in \mathbb{N}$, and let $c = (c_n)_{n \in \mathbb{N}}$.*
(a) If $1 \le p < \infty$, then $\sum_{n=1}^{\infty} c_n \delta_n$ converges in ℓ^p if and only if $c \in \ell^p$.
(b) The series $\sum_{n=1}^{\infty} c_n \delta_n$ converges in ℓ^∞ if and only if $c \in c_0$. $\diamond$

Now we give an example of a convergent series in $C_b(\mathbb{R})$, the space of all bounded continuous functions $f \colon \mathbb{R} \to \mathbb{F}$. Recall from Section 3.5 that $C_b(\mathbb{R})$ is a Banach space with respect to the uniform norm

$$\|f\|_u = \sup_{t \in \mathbb{R}} |f(t)|, \qquad f \in C_b(\mathbb{R}).$$

Choose your favorite bounded continuous function $f \in C_b(\mathbb{R})$, and consider the series

$$g(t) = \sum_{n=0}^{\infty} \frac{f(t)^n}{n!}, \qquad t \in \mathbb{R}. \qquad (4.4)$$

If we fix a particular point $t \in \mathbb{R}$ and think of equation (4.4) as a sum of the scalars $f(t)^n/n!$, then we know from basic calculus facts that the series converges. In fact, it converges to the number $g(t) = e^{f(t)}$ (see Lemma 1.8.1). Because the series converges for each individual t, we say that it converges *pointwise*.

However, we can also think of equation (4.4) as being a *sum of functions*. That is, if we let $f_n(t) = f(t)^n/n!$ then each function f_n is a vector in $C_b(\mathbb{R})$, and we can ask whether the infinite series

$$g = \sum_{n=0}^{\infty} f_n$$

converges *in the space* $C_b(\mathbb{R})$. Since the norm on $C_b(\mathbb{R})$ is the uniform norm, to answer this question we must determine whether the partial sums

$$s_N(t) = \sum_{n=0}^{N} f_n(t) = \sum_{n=0}^{N} \frac{f(t)^n}{n!}$$

converge *with respect to the uniform norm* rather than pointwise. A series whose partial sums converge uniformly is said to be a *uniformly convergent series*. For the particular series given in equation (4.4), the easy way to prove uniform convergence is to use results from Section 4.2, but since we have not covered that material yet, we will provide a direct (but longer) proof.

Lemma 4.1.5. *If $f \in C_b(\mathbb{R})$, then the series in equation (4.4) converges uniformly, and the function in $C_b(\mathbb{R})$ that it converges to is $g(t) = e^{f(t)}$.*

Proof. We observed above that equation (4.4) converges for each individual $t \in \mathbb{R}$, i.e., for each t we have $s_N(t) \to g(t) = e^{f(t)}$ as $N \to \infty$. However, pointwise convergence is not enough. We have to prove that the sequence of partial sums converges *uniformly* to g. We break this into two steps. First we will show that the series converges uniformly to *some* function h, and then we will prove that h equals the function $g(t) = e^{f(t)}$.

Step 1. Since $C_b(\mathbb{R})$ is a Banach space, we can prove that a sequence converges by showing that it is Cauchy. So, our goal is to show that $\{s_N\}_{N \in \mathbb{N}}$ is Cauchy with respect to the uniform norm (i.e., using the terminology of Definition 3.5.5, we want to show that $\{s_N\}_{N \in \mathbb{N}}$ is a *uniformly Cauchy sequence*).

For simplicity of notation, let $R = \|f\|_u$ be the uniform norm of our function f. By definition,

$$|f(t)| \leq R, \qquad \text{for all } t \in \mathbb{R}. \tag{4.5}$$

Fix now any $\varepsilon > 0$. Since $\sum_{n=0}^{\infty} \frac{R^n}{n!} = e^R < \infty$, Lemma 1.8.4 implies that there exists some integer K such that

$$\sum_{n>K} \frac{R^n}{n!} < \varepsilon. \tag{4.6}$$

We will show that $\|s_N - s_M\|_u < \varepsilon$ whenever M and N are larger than K. One of M and N must be larger, so without loss of generality we may assume that $N > M > K$. Then

$$\|s_N - s_M\|_u = \left\| \sum_{n=0}^{N} \frac{f(t)^n}{n!} - \sum_{n=0}^{M} \frac{f(t)^n}{n!} \right\|_u \qquad \text{(definition of } s_N \text{ and } s_M\text{)}$$

$$= \left\| \sum_{n=M+1}^{N} \frac{f(t)^n}{n!} \right\|_u \qquad \text{(collect terms)}$$

$$\leq \sum_{n=M+1}^{N} \left\| \frac{f(t)^n}{n!} \right\|_u \qquad \text{(Triangle Inequality)}$$

$$\leq \sum_{n=K+1}^{\infty} \frac{R^n}{n!} \qquad \text{(by equation (4.5))}$$

$$< \varepsilon \qquad \text{(by equation (4.6))}.$$

Thus $\{s_N\}_{N \in \mathbb{N}}$ is Cauchy in $C_b(\mathbb{R})$, and therefore it must converge to some function $h \in C_b(\mathbb{R})$.

Step 2. Now we determine what the function h is. We observed before that $s_N(t) \to g(t) = e^{f(t)}$ for each individual t. Now, since $s_N \to h$ uniformly and since uniform convergence implies pointwise convergence, we also have that $s_N(t) \to h(t)$ for each t. Consequently, by the uniqueness of limits, $h(t) = g(t) = e^{f(t)}$ for each t. $\square$

Problems

4.1.6. (a) Suppose that $\sum x_n$ and $\sum y_n$ are convergent series in a normed space X. Show that $\sum (x_n + y_n)$ is convergent and equals $\sum x_n + \sum y_n$.

(b) Show by example that $\sum (x_n + y_n)$ can converge even if $\sum x_n$ and $\sum y_n$ do not converge.

(c) If $\sum x_n$ converges but $\sum y_n$ does not, is it possible that $\sum (x_n + y_n)$ can converge?

4.1.7. Prove Lemma 4.1.4.

4.1.8. This problem will characterize the functions $f \in C[a, b]$ for which the infinite series $\sum_{n=1}^{\infty} f(t)^n$ converges uniformly. Let $s_N(t) = \sum_{n=1}^{N} f(t)^n$ denote the Nth partial sum of the series, and prove the following statements.

(a) If $\|f\|_u < 1$, then the partial sums $\{s_N\}_{N \in \mathbb{N}}$ are uniformly Cauchy, and therefore $\sum_{n=1}^{\infty} f(t)^n$ converges in the space $C[a, b]$.

(b) If $\|f\|_u \geq 1$, then there is some $t \in [a, b]$ such that $\sum_{n=1}^{\infty} f(t)^n$ does not converge. Explain why this implies that the infinite series $\sum_{n=1}^{\infty} f(t)^n$ does not converge uniformly.

4.1.9. Given $f \in C_b(\mathbb{R})$, prove that the infinite series $\sum_{n=1}^{\infty} f(t)^n$ converges uniformly if and only if $\|f\|_u < 1$.

Note: This problem is more difficult than Problem 4.1.8, because there exist functions $f \in C_b(\mathbb{R})$ that satisfy $\|f\|_u = 1$ even though there is no point t such that $|f(t)| = 1$.

4.1.10. Recall from Theorem 3.5.7 that the space $\mathcal{F}_b[0, \infty)$ of all bounded functions on $[0, \infty)$ is a Banach space with respect to the uniform norm.

(a) Let f be the *sinc function*, defined by the rule

$$f(t) = \frac{\sin 2\pi t}{2\pi t}, \qquad \text{for } t \neq 0,$$

and $f(0) = 1$. For each integer $n \geq 0$, define $f_n(t) = f(t)$ if $t \in [n, n+1)$, and $f_n(t) = 0$ otherwise. Prove that the series $\sum_{n=0}^{\infty} f_n$ converges uniformly to f.

(b) Does part (a) hold for every function f in $\mathcal{F}_b[0, \infty)$? Can you explicitly characterize those functions f for which part (a) holds?

4.2 Absolute Convergence of Series

Suppose that $x = \sum x_n$ is a convergent infinite series in a normed space X. Then the partial sums s_N of the series converge to x in norm, so by applying Lemma 3.3.3 and the Triangle Inequality we see that

$$\left\| \sum_{n=1}^{\infty} x_n \right\| = \|x\| = \lim_{N \to \infty} \|s_N\| \qquad \text{(continuity of the norm)}$$

$$= \lim_{N \to \infty} \left\| \sum_{n=1}^{N} x_n \right\| \qquad \text{(definition of } s_N)$$

$$\leq \lim_{N \to \infty} \sum_{n=1}^{N} \|x_n\| \qquad \text{(Triangle Inequality)}$$

$$= \sum_{n=1}^{\infty} \|x_n\| \qquad \text{(definition of infinite series)}.$$

We summarize this in the following lemma.

Lemma 4.2.1. *If* $\sum x_n$ *is a convergent infinite series in a normed space* X, *then*

$$\left\| \sum_{n=1}^{\infty} x_n \right\| \leq \sum_{n=1}^{\infty} \|x_n\|. \quad \Diamond$$

Lemma 4.2.1 is a kind of infinite series version of the Triangle Inequality, and as such it can be very useful. However, it has some important limitations. First, the lemma applies only to series that we know are convergent; we cannot use it to prove that a given series converges. Second, even if we know that $\sum x_n$ converges, Lemma 4.2.1 may not give us any useful information, because it is entirely possible that we might have $\sum \|x_n\| = \infty$. For example, if $1 < p < \infty$, then the argument of Example 4.1.3(b) shows that the series $\sum \frac{1}{n} \delta_n$ converges in ℓ^p, yet we have

$$\sum_{n=1}^{\infty} \left\| \frac{1}{n} \delta_n \right\|_p = \sum_{n=1}^{\infty} \frac{1}{n} \|\delta_n\|_p = \sum_{n=1}^{\infty} \frac{1}{n} = \infty. \tag{4.7}$$

Even so, Lemma 4.2.1 suggests that there is some kind of connection between an infinite series of vectors $\sum x_n$ and the infinite series of scalars $\sum \|x_n\|$. We will explore this connection, but first we introduce some terminology.

Definition 4.2.2 (Absolute Convergence). Let $\{x_n\}_{n \in \mathbb{N}}$ be a sequence in a normed space X. We say that the series $\sum_{n=1}^{\infty} x_n$ is *absolutely convergent* if

$$\sum_{n=1}^{\infty} \|x_n\| < \infty. \quad \Diamond$$

Even though the word "convergence" appears in the name "absolute convergence," this does not mean that an absolutely convergent series must be a convergent series, nor is it true that a convergent series must be absolutely convergent. Absolute convergence just means that the series of scalars $\sum \|x_n\|$ converges.

Here is an example of an infinite series that converges, but does not converge absolutely.

Example 4.2.3. As we noted above, Example 4.1.3(b) shows that the series $\sum \frac{1}{n} \delta_n$ converges in ℓ^p-norm whenever $1 < p < \infty$. However, equation (4.7) tells us that the series $\sum \frac{1}{n} \delta_n$ does not converge absolutely in ℓ^p. $\Diamond$

Here is an example of a series that converges absolutely, but does not converge.

Example 4.2.4. For this example we take our space to be c_{00}, the set of all "finite sequences," and we take the norm on c_{00} to be the ℓ^1-norm. Consider the series

$$\sum_{n=1}^{\infty} 2^{-n} \delta_n.$$

This series converges absolutely, because

$$\sum_{n=1}^{\infty} \left\| 2^{-n} \delta_n \right\|_1 = \sum_{n=1}^{\infty} 2^{-n} \left\| \delta_n \right\|_1 = \sum_{n=1}^{\infty} 2^{-n} = 1 < \infty. \qquad (4.8)$$

However, the partial sums of the series $\sum 2^{-n} \delta_n$ are

$$s_N = \sum_{n=1}^{N} 2^{-n} \delta_n = \left(\tfrac{1}{2}, \tfrac{1}{4}, \dots, \tfrac{1}{2^n}, 0, 0, \dots \right), \qquad (4.9)$$

and these do not converge to any vector in c_{00} (why not?). Thus, even though the series $\sum 2^{-n} \delta_n$ converges absolutely, it does not converge in c_{00}. $\diamond$

Convergence of a series depends both on what space we are considering and on the norm we choose for that space. To illustrate, we consider next what happens if we change the space in Example 4.2.4 from c_{00} to ℓ^1.

Example 4.2.5. Now we take our space to be ℓ^1, under the ℓ^1-norm. The computation in equation (4.8) shows that the infinite series $\sum 2^{-n} \delta_n$ converges absolutely with respect to this norm. The partial sums of the series are precisely the vectors s_N given in equation (4.9). These vectors converge componentwise to the vector

$$s = \left(2^{-n} \right)_{n \in \mathbb{N}} = \left(\tfrac{1}{2}, \tfrac{1}{4}, \tfrac{1}{8}, \dots \right),$$

and the reader should show that we actually have ℓ^1-norm convergence, i.e., $\| s - s_N \|_1 \to 0$. Therefore the series $\sum 2^{-n} \delta_n$ converges in the space ℓ^1, and it also converges absolutely in that space. $\diamond$

Since ℓ^1 is complete with respect to the norm $\| \cdot \|_1$ while c_{00} is not, the two preceding examples suggest that *completeness* may be an important consideration in dealing with absolute convergence. Indeed, we prove in the next theorem that if X is complete, then every absolutely convergent series in X converges, while if X is not complete, then there will exist vectors $x_n \in X$ such that $\sum \| x_n \| < \infty$ yet $\sum x_n$ does not converge.

Theorem 4.2.6. *If X is a normed space, then the following two statements are equivalent.*

(a) *X is complete (i.e., X is a Banach space).*

(b) *Every absolutely convergent series in X converges in X. That is, if $\{x_n\}_{n \in \mathbb{N}}$ is a sequence in X such that $\sum \| x_n \| < \infty$, then the series $\sum x_n$ converges in X.*

Proof. (a) $\Rightarrow$ (b). Assume that X is complete, and suppose that $\sum \| x_n \| < \infty$. Set

$$s_N = \sum_{n=1}^{N} x_n \qquad \text{and} \qquad t_N = \sum_{n=1}^{N} \|x_n\|.$$

Given any $N > M$, we have

$$\|s_N - s_M\| = \left\| \sum_{n=M+1}^{N} x_n \right\| \le \sum_{n=M+1}^{N} \|x_n\| = |t_N - t_M|.$$

Since $\{t_N\}_{N \in \mathbb{N}}$ is a Cauchy sequence of scalars, this implies that $\{s_N\}_{N \in \mathbb{N}}$ is a Cauchy sequence of vectors in X. Since X is complete, the partial sums s_N must therefore converge. Hence, by the definition of an infinite series, $\sum x_n$ converges in X.

(b) $\Rightarrow$ (a). Suppose that every absolutely convergent series in X is convergent. Let $\{x_n\}_{n \in \mathbb{N}}$ be a Cauchy sequence in X. Appealing to Problem 2.2.23, there exists a subsequence $\{x_{n_k}\}_{k \in \mathbb{N}}$ such that for each $k \in \mathbb{N}$ we have

$$\|x_{n_{k+1}} - x_{n_k}\| < 2^{-k}.$$

Consequently, the series $\sum_{k=1}^{\infty} (x_{n_{k+1}} - x_{n_k})$ is absolutely convergent, because

$$\sum_{k=1}^{\infty} \|x_{n_{k+1}} - x_{n_k}\| \le \sum_{k=1}^{\infty} 2^{-k} = 1 < \infty.$$

By hypothesis, the series $\sum_{k=1}^{\infty} (x_{n_{k+1}} - x_{n_k})$ must therefore converge in X. Let

$$x = \sum_{k=1}^{\infty} (x_{n_{k+1}} - x_{n_k}).$$

By definition, this means that the partial sums

$$s_M = \sum_{k=1}^{M} (x_{n_{k+1}} - x_{n_k}) = x_{n_{M+1}} - x_{n_1}$$

converge to x as $M \to \infty$. Let $y = x + x_{n_1}$. Then, since n_1 is fixed,

$$x_{n_M} = s_{M-1} + x_{n_1} \to x + x_{n_1} = y \qquad \text{as } M \to \infty.$$

Thus $\{x_n\}_{n \in \mathbb{N}}$ is a Cauchy sequence that has a subsequence $\{x_{n_k}\}_{k \in \mathbb{N}}$ that converges to the vector y. Appealing now to Problem 2.2.21, this implies that $x_n \to y$. Hence every Cauchy sequence in X converges, so X is complete. $\square$

To illustrate, recall that Lemma 4.1.5 showed that if f is a function in $C_b(\mathbb{R})$, then the series

$$g(t) = \sum_{n=0}^{\infty} \frac{f(t)^n}{n!}$$

converges uniformly. We can give a simpler proof of this fact by applying
Theorem 4.2.6. Since $|f(t)| \leq \|f\|_u$ for every t, the uniform norm of the nth
term $f_n(t) = f(t)^n/n!$ satisfies

$$\|f_n\|_u = \sup_{t \in \mathbb{R}} \left| \frac{f(t)^n}{n!} \right| = \sup_{t \in \mathbb{R}} \frac{|f(t)|^n}{n!} \leq \frac{\|f\|_u^n}{n!}.$$

Therefore

$$\sum_{n=0}^{\infty} \|f_n\|_u \leq \sum_{n=0}^{\infty} \frac{\|f\|_u^n}{n!} = e^{\|f\|_u} < \infty.$$

This shows that the series $\sum_{n=0}^{\infty} f_n$ converges absolutely in the Banach space
$C_b(\mathbb{R})$. Theorem 4.2.6 therefore implies that $\sum_{n=0}^{\infty} f_n$ converges with respect
to the norm of that space, i.e., it converges uniformly.

Problems

4.2.7. Exhibit a convergent series of real numbers $\sum_{k=1}^{\infty} x_k$ and real scalars
c_k with $0 < |c_k| < 1$ for every k such that $\sum_{k=1}^{\infty} c_k x_k$ converges, but

$$\left| \sum_{k=1}^{\infty} c_k x_k \right| > \left| \sum_{k=1}^{\infty} x_k \right|.$$

4.2.8. Let X be a Banach space.

(a) Suppose that a series $\sum_{n=1}^{\infty} x_n$ converges absolutely in X. Given indices
$n_1 < n_2 < \cdots$, prove that the series $\sum_{k=1}^{\infty} x_{n_k}$ converges in X.

(b) Does the conclusion of part (a) remain valid if we assume only that
$\sum_{n=1}^{\infty} x_n$ converges? Either prove or exhibit a counterexample.

4.2.9. Prove that c_0 is a Banach space with respect to the sup-norm by
showing that every absolutely convergent series in c_0 converges and applying
Theorem 4.2.6.

4.2.10. This problem will show that there can exist convergent series that
are not absolutely convergent, even in a Banach space.

Consider the infinite series

$$f(t) = \sum_{n=1}^{\infty} (-1)^n \frac{t^2 + n}{n^2} \tag{4.10}$$

in the Banach space $C[0,1]$ (with respect to the uniform norm, as usual).
Prove the following statements.

(a) The series converges pointwise (i.e., it converges for each individual t).

(b) The series does not converge absolutely in $C[0,1]$.

(c) The partial sums of the series converge uniformly, and therefore the series in equation (4.10) is uniformly convergent.

4.2.11. For each of the following choices of $f_n \in C[0,1]$, determine whether the series $\sum_{n=1}^{\infty} f_n$ converges pointwise, converges uniformly, or converges absolutely in the space $C[0,1]$.

(a) $f_n(t) = \dfrac{t}{1+n^2 t}$,   (b) $f_n(t) = \dfrac{1}{1+n^2 t}$,   (c) $f_n(t) = \dfrac{(-1)^n}{1+n^2 t}$.

Do any answers change if we replace the interval $[0,1]$ with the smaller interval $[\delta, 1]$, where $0 < \delta < 1$?

4.2.12. Let $f_n(t) = \dfrac{e^{-n^2 t^2}}{n^2}$ for $t \in \mathbb{R}$.

(a) Prove that the series $\sum_{n=1}^{\infty} f_n$ converges absolutely in $C_b(\mathbb{R})$ with respect to the uniform norm, and therefore converges, since $C_b(\mathbb{R})$ is a Banach space.

(b) Prove that the series $\sum_{n=1}^{\infty} f_n'$ converges in $C_b(\mathbb{R})$ with respect to the uniform norm, but it does not converge absolutely with respect to that norm.

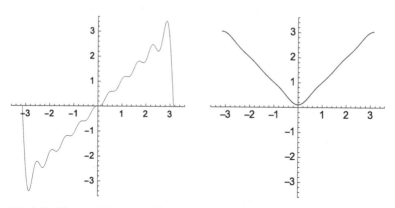

Fig. 4.1 Left: The partial sum s_{10} for the first series in Problem 4.2.13. Right: $\frac{\pi}{2}$ minus the partial sum s_3 for the second series in Problem 4.2.13.

4.2.13. Consider the two series

$$f(t) = \sum_{n=1}^{\infty} \frac{2(-1)^{n+1}}{n} \sin nt, \quad g(t) = \frac{\pi}{2} - \sum_{n=1}^{\infty} \frac{4}{(2n-1)^2 \pi} \cos(2n-1)t$$

in the space $C[-\pi, \pi]$, with respect to the uniform norm. Plot the Nth partial sums of these series for various values of N, and observe that the first series appears to converge to $f(t) = t$ on the interval $(-\pi, \pi)$, while the second appears to converge to $g(t) = |t|$ on $[-\pi, \pi]$ (see Figure 4.1). Assume for this problem that the series do converge *pointwise* to these functions (at the endpoints, take $f(\pm\pi) = 0$).

(a) Prove that the first series does not converge absolutely in $C[-\pi, \pi]$. Does this imply that the series does not converge in $C[-\pi, \pi]$? Let s_N be the Nth partial sum of the series for f, and prove that $\|f - s_N\|_u \geq 1$ for every N. What does this imply?

(b) Prove that the second series converges absolutely in $C[-\pi, \pi]$, and therefore converges uniformly.

4.2.14. Let I be an interval in the real line (of any type). We say that a function $f: I \to \mathbb{C}$ is *Lipschitz continuous* on I if there exists a constant $K > 0$ such that

$$|f(x) - f(y)| \leq K |x - y|, \qquad \text{for all } x, y \in I.$$

We let $\text{Lip}(I)$ denote the space of bounded functions that are Lipschitz on I:

$$\text{Lip}(I) = \{f \in \mathcal{F}_b(I) : f \text{ is Lipschitz}\}.$$

Prove the following statements, which show that the Lipschitz condition is something "strictly between" continuity and differentiability.

(a) Every Lipschitz continuous function is uniformly continuous on I.

(b) We have
$$C_b^1(I) \subsetneqq \text{Lip}(I) \subsetneqq C_b(I),$$

where $C_b^1(I)$ is the space introduced in Problem 3.5.25,

Hint: If $\mathbb{F} = \mathbb{R}$, then use the Mean Value Theorem to prove the first inclusion. However, recall that the MVT does not hold for complex-valued functions (see Problem 1.9.2). Therefore, if $\mathbb{F} = \mathbb{C}$, then consider the real and imaginary parts of f separately.

(c) If $a, b > 0$ and

$$f(t) = \begin{cases} |t|^a \sin |t|^{-b}, & t \neq 0, \\ 0, & t = 0, \end{cases}$$

then $f \in \text{Lip}(-1, 1)$ if $a \geq 1 + b$, and $f \in C_b^1(-1, 1)$ if $a > 1 + b$.

4.2.15. Given an interval I, let $\text{Lip}(I)$ be the space of Lipschitz functions introduced in Problem 4.2.14. Prove that the following is a norm on $\text{Lip}(I)$:

$$\|f\|_{\text{Lip}} = \|f\|_u + \sup_{x \neq y} \frac{|f(x) - f(y)|}{|x - y|}.$$

Then prove that $\text{Lip}(I)$ is a Banach space with respect to this norm by showing that every absolutely convergent series in $\text{Lip}(I)$ converges.

4.2.16. Let I be an interval, and prove the following statements.

(a) For each $n \in \mathbb{N}$,

$$F_n = \{f \in C(I) : |f(x) - f(y)| \leq n\,|x - y| \text{ for all } x, y \in I\}$$

is a closed subset of $C(I)$ (with respect to the uniform norm).

(b) If $f \in F_n$ and $\varepsilon > 0$, then there exists a function $g \in C(I)$ such that $\|f - g\|_{\mathrm{u}} < \varepsilon$ but $g \notin F_n$.

(c) $\text{Lip}(I)$ is a meager subset of $C(I)$ (with respect to $\|\cdot\|_{\mathrm{u}}$).

Remark: Since $C_b^1(I) \subseteq \text{Lip}(I)$, this implies that $C_b^1(I)$ is only a meager subset of $C(I)$, and hence is just a "small" subset of $C(I)$ in some sense. Thus "most" continuous functions on I do not belong to $C_b^1(I)$. If we let S be the set of all continuous functions on I that are differentiable at even just one point, then a similar but more delicate analysis shows that S is still only a meager subset of $C_b(I)$. Hence "most" continuous functions are not differentiable at any point—even though it is not easy to explicitly construct such functions!

4.3 Unconditional Convergence of Series

The ordering of the vectors in a convergent series may be important. If we reorder a series, or in other words consider a new series $\sum_{n=1}^{\infty} x_{\sigma(n)}$, where $\sigma \colon \mathbb{N} \to \mathbb{N}$ is a bijection, then there is no guarantee that this reordered series will still converge (a bijection $\sigma \colon \mathbb{N} \to \mathbb{N}$ is also sometimes called a *permutation of the natural numbers*). We introduce some terminology for this situation.

Definition 4.3.1 (Unconditional Convergence). Let $\{x_n\}_{n \in \mathbb{N}}$ be a sequence of vectors in a normed space X. If $\sum_{n=1}^{\infty} x_{\sigma(n)}$ converges for every bijection $\sigma \colon \mathbb{N} \to \mathbb{N}$, then we say that the infinite series $\sum_{n=1}^{\infty} x_n$ *converges unconditionally.* A series that converges but does not converge unconditionally is said to be *conditionally convergent.* ◇

That is, a series $\sum_{n=1}^{\infty} x_n$ converges unconditionally if and only if for each permutation σ the partial sums $\sum_{n=1}^{N} x_{\sigma(n)}$ converge to some vector as $N \to \infty$. This definition does not require that the partial sums converge to the *same* limit for each bijection σ. However, the following result states that if a series converges unconditionally, then every rearrangement of the series converges to the same limit (for one proof, see [Heil11, Cor. 3.11]).

Lemma 4.3.2. *If X is a normed space and the series $\sum_{n=1}^{\infty} x_n$ converges unconditionally in X, then there is a vector $x \in X$ such that $\sum_{n=1}^{\infty} x_{\sigma(n)} = x$ for every bijection $\sigma \colon \mathbb{N} \to \mathbb{N}$.* ◇

Unconditional convergence of an infinite series is highly desirable, because it says that the particular choice of ordering of the index set is irrelevant—if a series converges unconditionally, then it will converge no matter how we choose to order the terms, and each reordering of the series will converge to the same limit. Series that converge *conditionally* can be very difficult to work with.

Before considering series in general Banach spaces, we first look at what happens in the scalar field $X = \mathbb{F}$. The next result shows that *for series of scalars*, unconditional and absolute convergence are equivalent.

Theorem 4.3.3. *If $(c_n)_{n \in \mathbb{N}}$ is a sequence of scalars, then*

$$\sum_{n=1}^{\infty} |c_n| < \infty \quad \Longleftrightarrow \quad \sum_{n=1}^{\infty} c_n \text{ converges unconditionally.}$$

Proof. $\Rightarrow$. Suppose that $\sum |c_n| < \infty$, and let σ be any permutation of $\mathbb{N}$. We must show that $\sum c_{\sigma(n)}$ converges. To do this, let s_N denote the Nth partial sum of this series, i.e.,

$$s_N = \sum_{n=1}^{N} c_{\sigma(n)}.$$

We will show that the series $\sum c_{\sigma(n)}$ converges by proving that $\{s_N\}_{N \in \mathbb{N}}$ is a Cauchy sequence in X.

To do this, fix any $\varepsilon > 0$. We must show that $|s_N - s_M| < \varepsilon$ for all large enough M and N. Now, since $\sum |c_n| < \infty$, there is some number K such that

$$\sum_{n=K+1}^{\infty} |c_n| < \varepsilon.$$

And since $\sigma(1), \sigma(2), \ldots$ is a listing of *all* of the positive integers, there must exist some number L such that

$$\sigma(1), \sigma(2), \ldots, \sigma(L) \quad \text{includes} \quad 1, \ldots, K$$

(and might include other integers as well). Suppose that $N > M > L$. Then $\sigma(M+1), \ldots, \sigma(N)$ does not include $\sigma(1), \ldots, \sigma(L)$, and therefore does not include $1, \ldots, K$. Instead, $\sigma(M+1), \ldots, \sigma(N)$ are finitely many of the integers from $K + 1$ onward. Therefore

$$|s_N - s_M| = \left| \sum_{n=M+1}^{N} c_{\sigma(n)} \right| \leq \sum_{n=M+1}^{N} |c_{\sigma(n)}| \leq \sum_{n=K+1}^{\infty} |c_n| < \varepsilon.$$

This shows that $\{s_N\}_{N \in \mathbb{N}}$ is a Cauchy sequence, and therefore $\sum c_{\sigma(n)}$ converges, since $\mathbb{F}$ is complete.

$\Leftarrow$. *Case 1: Real scalars.* We will prove the contrapositive statement. Suppose that $\sum c_n$ is a sequence of real scalars that does not converge absolutely. Let (p_n) be the sequence of nonnegative terms of (c_n) in order, and let (q_n) be the sequence of negative terms of (c_n) in order (either (p_n) or (q_n) may possibly be a finite sequence). If $\sum p_n$ and $\sum q_n$ both converge, then $\sum |c_n|$ converges and equals $\sum p_n - \sum q_n$, which is a contradiction. Hence at least one of $\sum p_n$ and $\sum q_n$ must diverge. The proof is similar no matter which one of these diverges, so let us assume that $\sum p_n$ diverges. Since each p_n is nonnegative, this means that $\sum p_n = \infty$. Hence the partial sums of this series increase without bound, and so there must exist some number $m_1 > 0$ such that

$$p_1 + \cdots + p_{m_1} > 1.$$

Then there must exist an $m_2 > m_1$ such that

$$p_1 + \cdots + p_{m_1} - q_1 + p_{m_1+1} + \cdots + p_{m_2} > 2,$$

and an $m_3 > m_2$ such that

$$p_1 + \cdots + p_{m_1} - q_1 + p_{m_1+1} + \cdots + p_{m_2} - q_2 + p_{m_2+1} + \cdots + p_{m_3} > 3,$$

and so forth. Continuing in this way, we see that

$$p_1 + \cdots + p_{m_1} - q_1 + p_{m_1+1} + \cdots + p_{m_2} - q_2 + \cdots$$

is a rearrangement of $\sum c_n$ that diverges. Hence $\sum c_n$ cannot converge unconditionally.

Case 2: Complex scalars. Now suppose that the scalars c_n are complex. This time we use a direct argument, breaking the series into real and imaginary parts and applying Case 1.

Suppose that the series $c = \sum c_n$ converges unconditionally. Write $c = a + ib$ and $c_n = a_n + ib_n$, and let σ be any permutation of $\mathbb{N}$. Then $c = \sum c_{\sigma(n)}$ converges by hypothesis. Now, for every complex number z we have

$$|\mathrm{Re}(z)| \leq \sqrt{\mathrm{Re}(z)^2 + \mathrm{Im}(z)^2} = |z|.$$

Therefore

$$\left| a - \sum_{n=1}^{N} a_{\sigma(n)} \right| = \left| \mathrm{Re}\left(c - \sum_{n=1}^{N} c_{\sigma(n)} \right) \right| \leq \left| c - \sum_{n=1}^{N} c_{\sigma(n)} \right| \to 0$$

as $N \to \infty$. Thus the series of real scalars $a = \sum a_{\sigma(n)}$ converges. This is true for every permutation σ, so $\sum a_n$ converges unconditionally. Case 1 therefore implies that $\sum |a_n| < \infty$. An identical argument shows that $\sum b_n$ converges absolutely, so

$$\sum_{n=1}^{\infty} |c_n| = \sum_{n=1}^{\infty} |a_n + ib_n| \le \sum_{n=1}^{\infty} |a_n| + \sum_{n=1}^{\infty} |b_n| < \infty. \quad \square$$

Consider, for example, the alternating harmonic series $\sum_{n=1}^{\infty} \frac{(-1)^n}{n}$, which converges to the value $-\ln 2$. This series does not converge absolutely, so by Theorem 4.3.3 it cannot converge unconditionally. Hence there must exist some permutation σ such that $\sum \frac{(-1)^{\sigma(n)}}{\sigma(n)}$ does not converge. Can you exhibit such a permutation? Even more interestingly, Problem 4.3.6 shows how to prove that if we choose *any* real number x, then there is a permutation σ such that the reordered series converges to x, i.e.,

$$\sum_{n=1}^{\infty} \frac{(-1)^{\sigma(n)}}{\sigma(n)} = \lim_{N \to \infty} \sum_{n=1}^{N} \frac{(-1)^{\sigma(n)}}{\sigma(n)} = x.$$

Theorem 4.3.3 deals only with series of scalars. If X is a *finite-dimensional* normed space, then it can be shown that an infinite series converges absolutely if and only if it converges unconditionally. The next lemma shows that one direction of this implication still holds in every complete normed space: Absolute convergence *always* implies unconditional convergence.

Lemma 4.3.4. *Let X be a Banach space. If a series $\sum x_n$ converges absolutely in X, then it converges unconditionally. In other words,*

$$\sum_{n=1}^{\infty} \|x_n\| < \infty \quad \Longrightarrow \quad \sum_{n=1}^{\infty} x_n \text{ converges unconditionally.} \qquad (4.11)$$

Proof. Assume that $\sum \|x_n\| < \infty$. Then $\sum \|x_n\|$ is an absolutely convergent series of scalars, so Theorem 4.3.3 implies that it converges unconditionally. Hence, if σ is a permutation of $\mathbb{N}$, then $\sum_{n=1}^{\infty} \|x_{\sigma(n)}\|$ converges. Since every term is nonnegative, we therefore have

$$\sum_{n=1}^{\infty} \|x_{\sigma(n)}\| < \infty.$$

This tells us that $\sum x_{\sigma(n)}$ converges absolutely. Hence, since X is a Banach space, the series $\sum x_{\sigma(n)}$ converges (see Theorem 4.2.6). Since this is true for every bijection σ, we have shown that $\sum x_n$ converges unconditionally. $\square$

Unconditional convergence *does not* imply absolute convergence in an infinite-dimensional normed space. For example, if $1 < p \le \infty$, then $\sum \frac{1}{n} \delta_n$ converges unconditionally but not absolutely in ℓ^p (see Problem 4.3.5). Later, we will see in Problem 5.7.3 that if H is any infinite-dimensional *Hilbert space* (a special type of Banach space), then there is a series in H that converges unconditionally but not absolutely. Generic Banach spaces are much more difficult to deal with, but an important (and nontrivial) result known as the *Dvoretzky–Rogers Theorem* implies that *every* infinite-dimensional Banach space contains series that converge unconditionally but not absolutely (for a discussion and proof of the Dvoretzky–Rogers Theorem see [Heil11, Sec. 3.6]).

Problems

4.3.5. Given scalars c_n, set $c = (c_n)_{n \in \mathbb{N}}$ and prove the following statements.

(a) If $1 \le p < \infty$ and $c \in \ell^p$, then $\sum c_n \delta_n$ converges unconditionally in ℓ^p (with respect to the norm $\|\cdot\|_p$).

(b) If $p = \infty$ and $c \in c_0$, then $\sum c_n \delta_n$ converges unconditionally (with respect to the sup-norm).

4.3.6. Assume that $\sum c_n$ is a conditionally convergent series of real scalars, i.e., the series converges but does not converge unconditionally.

(a) Let (p_n) be the sequence of nonnegative terms of (c_n) in order, and let (q_n) be the sequence of negative terms of (c_n) in order. Show that $\sum p_n$ and $\sum q_n$ must both diverge.

(b) This part will show that if x is a real number, then there is a permutation σ of $\mathbb{N}$ such that $\sum c_{\sigma(n)} = x$. Let m_1 be the first integer such that

$$p_1 + \cdots + p_{m_1} > x.$$

Then let n_1 be the first integer such that

$$p_1 + \cdots + p_{m_1} - q_1 - \cdots - q_{n_1} < x.$$

Then let m_2 be the first integer greater than m_1 such that

$$p_1 + \cdots + p_{m_1} - q_1 - \cdots - q_{n_1} + p_{m_1+1} + \cdots + p_{m_2} > x.$$

Continue in this way, and prove that

$$p_1 + \cdots + p_{m_1} - q_1 - \cdots - q_{n_1} + p_{m_1+1} + \cdots + p_{m_2} - q_{n_1+1} - \cdots - q_{n_2} + \cdots = x.$$

4.3.7. Continuing Problem 4.1.10(a), prove that the series $\sum_{n=0}^{\infty} f_n$ in that problem converges unconditionally but not absolutely in the space $\mathcal{F}_b[0, \infty)$ (with respect to the uniform norm).

4.3.8. Prove *Fubini's Theorem for iterated series*: If c_{mn} is a scalar for all m, $n \in \mathbb{N}$ and

$$\sum_{m=1}^{\infty} \left(\sum_{n=1}^{\infty} |c_{mn}| \right) < \infty, \tag{4.12}$$

then for each bijection $\sigma\colon \mathbb{N} \to \mathbb{N} \times \mathbb{N}$ the following series converge and are equal as indicated:

$$\sum_{m=1}^{\infty}\left(\sum_{n=1}^{\infty} c_{mn}\right) = \sum_{k=1}^{\infty} c_{\sigma(k)} = \sum_{n=1}^{\infty}\left(\sum_{m=1}^{\infty} c_{mn}\right). \tag{4.13}$$

Using the terminology of Definition 4.3.1, this says that the series $\sum_{m,n} c_{mn}$ converges unconditionally. (We have included parentheses in equations (4.12) and (4.13) to emphasize the iteration of the series, but they are typically omitted.)

4.3.9. Prove *Tonelli's Theorem for iterated series*: If $c_{mn} \geq 0$ for each m, $n \in \mathbb{N}$, then

$$\sum_{m=1}^{\infty}\sum_{n=1}^{\infty} c_{mn} = \sum_{n=1}^{\infty}\sum_{m=1}^{\infty} c_{mn},$$

in the sense that either both sides converge and are equal, or both sides are infinite.

4.4 Span, Closed Span, and Complete Sequences

The *span* of a set of vectors was introduced in Section 1.11, but for convenience we restate the definition here.

Definition 4.4.1 (Span). If A is a subset of a vector space X, then the *finite linear span* of A, or simply the *span* for short, is the set of all *finite linear combinations* of elements of A:

$$\mathrm{span}(A) = \left\{\sum_{n=1}^{N} c_n x_n : N > 0,\ x_n \in A,\ c_n \in \mathbb{F}\right\}. \tag{4.14}$$

We say that A *spans* X if $\mathrm{span}(A) = X$.

If $A = \{x_n\}_{n\in\mathbb{N}}$ is a sequence, then we usually write $\mathrm{span}\{x_n\}_{n\in\mathbb{N}}$ instead of $\mathrm{span}(\{x_n\}_{n\in\mathbb{N}})$. $\Diamond$

We saw in Example 1.11.2 that if we let $\mathcal{M} = \{t^k\}_{k=0}^{\infty}$, then $\mathrm{span}(\mathcal{M}) = \mathrm{span}\{t^k\}_{k=0}^{\infty} = \mathcal{P}$, the set of all polynomials. Hence the set of monomials $\mathcal{M}$ spans $\mathcal{P}$.

Here is another example of a spanning set.

Example 4.4.2. Let $\mathcal{E} = \{\delta_n\}_{n\in\mathbb{N}}$ be the sequence of standard basis vectors. The span of $\mathcal{E}$ is the set of all finite linear combinations of standard basis vectors, which is

$$\text{span}(\mathcal{E}) = \text{span}\{\delta_n\}_{n \in \mathbb{N}} = \left\{ \sum_{n=1}^{N} c_n \delta_n : N > 0, \, c_n \in \mathbb{F} \right\}$$

$$= \left\{ (c_1, \dots, c_N, 0, 0, \dots) : N > 0, \, c_n \in \mathbb{F} \right\}$$

$$= c_{00}.$$

We can only form *finite* linear combinations when constructing a span, so a sequence such as $x = (1, 1, 1, \dots)$ does not belong to $\text{span}(\mathcal{E})$. The standard basis $\mathcal{E}$ spans c_{00}, but it does not span ℓ^p for any index p. $\Diamond$

If X is a *normed space*, then we can take limits as well as forming finite linear combinations. In particular, the following definition gives a name to the *closure* of the span of a set.

Definition 4.4.3 (Closed Span). If A is a subset of a normed space X, then the *closed linear span* or simply the *closed span* of A is the closure of $\text{span}(A)$. For compactness of notation, we set

$$\overline{\text{span}}(A) = \overline{\text{span}(A)}.$$

If $A = \{x_n\}_{n \in \mathbb{N}}$ is a sequence, then to avoid extra parentheses we usually write $\overline{\text{span}}\{x_n\}_{n \in \mathbb{N}}$ instead of $\overline{\text{span}}(\{x_n\}_{n \in \mathbb{N}})$. $\Diamond$

Let A be an arbitrary subset of a normed space X. Part (b) of Theorem 2.6.2 tells us that $\overline{\text{span}}(A)$ is the set of all limits of elements of the span of A:

$$\overline{\text{span}}(A) = \{ y \in X : \exists \, y_n \in \text{span}(A) \text{ such that } y_n \to y \}. \tag{4.15}$$

For example, suppose that $x_n \in A$, c_n is a scalar, and $x = \sum_{n=1}^{\infty} c_n x_n$ is a convergent infinite series in X. Then the partial sums

$$s_N = \sum_{n=1}^{N} c_n x_n$$

belong to $\text{span}(A)$ and converge to x, so it follows from equation (4.15) that $x \in \overline{\text{span}}(A)$. As a consequence, we have the inclusion

$$\left\{ \sum_{n=1}^{\infty} c_n x_n : x_n \in A, \, c_n \text{ scalar}, \sum_{n=1}^{\infty} c_n x_n \text{ converges} \right\} \subseteq \overline{\text{span}}(A). \tag{4.16}$$

That is, every convergent "infinite linear combination" $\sum_{n=1}^{\infty} c_n x_n$ belongs to the closed span. However, *equality need not hold in equation* (4.16), i.e., it need not be the case that every element of the closed span can be written in the form $\sum_{n=1}^{\infty} c_n x_n$ for some scalars c_n and some vectors $x_n \in A$. We will see an example of such a situation in Section 4.6 (in particular, see the discussion related to Theorem 4.6.3).

We often need to consider sequences whose closed span is the entire space X, so we introduce the following terminology.

Definition 4.4.4 (Complete Sequence). Let $\{x_n\}_{n \in \mathbb{N}}$ be a sequence of vectors in a normed space X. We say that the sequence $\{x_n\}_{n \in \mathbb{N}}$ is *complete* in X if $\mathrm{span}\{x_n\}_{n \in \mathbb{N}}$ is dense in X, i.e., if

$$\overline{\mathrm{span}}\{x_n\}_{n \in \mathbb{N}} = X.$$

Complete sequences are also known as *total* or *fundamental* sequences. ◇

We will see several examples of complete sequences in this chapter, and more in Chapter 5. Note that the meaning of a "complete sequence" (Definition 4.4.4) is quite different from that of a "complete metric space" (Definition 2.2.9). A *complete sequence* is a sequence whose closed span equals the entire space X, whereas X is a *complete space* if every Cauchy sequence in X converges to an element of X.

Problems

4.4.5. Let A be a subset of a normed space X. Show that

$$\overline{\mathrm{span}}(A) = \bigcap \{M : M \text{ is a closed subspace of } X \text{ and } A \subseteq M\}.$$

Explain why this shows that $\overline{\mathrm{span}}(A)$ is the *smallest closed subspace of X that contains the set A.*

4.4.6. Let A be a subset of a normed space X and let y be a vector in X. Either prove or provide a counterexample for each of the following statements.

(a) If $y \in \mathrm{span}(A)$, then $\mathrm{span}(A \cup \{y\}) = \mathrm{span}(A)$.

(b) If $y \in \overline{\mathrm{span}}(A)$, then $\mathrm{span}(A \cup \{y\}) = \mathrm{span}(A)$.

(c) If $y \in \overline{\mathrm{span}}(A)$, then $\overline{\mathrm{span}}(A \cup \{y\}) = \overline{\mathrm{span}}(A)$.

(d) If $y \notin \mathrm{span}(A)$, then $\overline{\mathrm{span}}(A \cup \{y\}) \neq \overline{\mathrm{span}}(A)$.

4.4.7. Let $\{x_n\}_{n \in N}$ be a sequence in a normed space X. Either prove or provide a counterexample for each of the following statements.

(a) If $\{x_n\}_{n \in N}$ is complete, then $\{x_n\}_{n \in N}$ is dense in X.

(b) If $\{x_n\}_{n \in N}$ is dense, then $\{x_n\}_{n \in N}$ is complete in X.

(c) If $\{x_n\}_{n \in N}$ is complete and $m \in \mathbb{N}$ is fixed, then $\{x_n\}_{n \neq m}$ cannot be complete.

4.4.8. Determine whether $\mathrm{span}(\mathcal{F}) = C[0,1]$ for each of the following collections of functions $\mathcal{F}$.

(a) $\mathcal{F} = \{\sin nt\}_{n \in \mathbb{N}}$.   (b) $\mathcal{F} = \{e^{\alpha t}\}_{\alpha \in \mathbb{R}}$.   (c) $\mathcal{F} = \{|t - a|\}_{0 < a < 1}$.

4.4.9. Prove that the following sets do not span ℓ^1:

(a) $A = \{\delta_1 + \delta_n\}_{n \in \mathbb{N}}$. (b) $B = \{\delta_1 + \cdots + \delta_n\}_{n \in \mathbb{N}}$.

Can you find an uncountable collection of sequences that spans ℓ^1? Can you find a *countable* collection that spans ℓ^1?

4.4.10. Suppose that $\{x_n\}_{n \in \mathbb{N}}$ is a countable set of elements of ℓ^1. For each $N \in \mathbb{N}$, let

$$F_N = \operatorname{span}\{x_1, \ldots, x_N\} = \left\{ \sum_{k=1}^{N} c_k x_k : c_1, \ldots, c_N \in \mathbb{F} \right\}.$$

Prove the following statements.

(a) F_N is a closed subset of ℓ^1.

(b) If $x \in F_N$ and $\varepsilon > 0$, then there exists some $y \notin F_N$ such that $\|x - y\|_1 < \varepsilon$.

(c) $F_N^\circ = \varnothing$, i.e., F_N has empty interior.

(d) $\{x_n\}_{n \in \mathbb{N}}$ does not span ℓ^1. Hint: Baire Category Theorem.

4.4.11. Use the Baire Category Theorem to prove the following statements.

(a) There is no norm $\|\cdot\|$ on the set of polynomials $\mathcal{P}$ such that $\mathcal{P}$ is a Banach space with respect to $\|\cdot\|$.
  Hint: Consider the set $\mathcal{P}_n$ of all polynomials of degree at most n.

(b) There is no norm $\|\cdot\|$ on the set of finite sequences c_{00} such that c_{00} is a Banach space with respect to $\|\cdot\|$.

4.5 Independence, Hamel Bases, and Schauder Bases

The definition of linear independence of vectors was given in Section 1.11, but we recall it again here.

Definition 4.5.1 (Finite Independence). Let A be a subset of a vector space X. We say that A is *finitely linearly independent, linearly independent,* or just *independent* for short if for every choice of finitely many distinct vectors $x_1, \ldots, x_N \in A$ and scalars $c_1, \ldots, c_N \in \mathbb{F}$, we have

$$\sum_{n=1}^{N} c_n x_n = 0 \quad \Longleftrightarrow \quad c_1 = \cdots = c_N = 0. \qquad \Diamond$$

For example, the sequence of standard basis vectors $\mathcal{E} = \{\delta_n\}_{n \in \mathbb{N}}$ is finitely linearly independent, and Example 1.11.4 showed that the set of monomials $\{t^k\}_{k \geq 0}$ is linearly independent.

In the field of linear algebra, which deals with general vector spaces, a set that is both linearly independent and spans a vector space X is usually called a *basis* for X. This definition of basis only employs finite linear combinations of elements. We, however, are interested in *normed spaces*, and in a normed space we can work with *infinite series* rather than just finite sums. Later we will use the existence of infinite sums in normed spaces to define a more useful basis concept known as a *Schauder basis*. In order to clearly distinguish between these two types of "bases," we will refer to a set that is a basis in the strict linear-algebraic sense as a *Hamel basis* or a *vector space basis*. The formal definition is as follows.

Definition 4.5.2 (Hamel Basis). Let X be a nontrivial vector space. A *Hamel basis, vector space basis,* or simply a *basis* for X is a set $\mathcal{B} \subseteq X$ such that

(a) $\mathcal{B}$ is linearly independent, and

(b) $\text{span}(\mathcal{B}) = X$.

For the trivial vector space $X = \{0\}$ there is no set $\mathcal{B}$ that both spans X and is independent, but we declare that the empty set $\varnothing$ is a Hamel basis for $X = \{0\}$. $\diamondsuit$

The *Axiom of Choice* is one of the axioms of the standard form of set theory most commonly accepted in mathematics (Zermelo–Fraenkel set theory with the Axiom of Choice, or ZFC). One consequence of the Axiom of Choice is that *every vector space has a Hamel basis* (in fact, this statement is equivalent to the Axiom of Choice). Further, any two Hamel bases for a given vector space X have the same cardinality.

We say that X is *finite-dimensional* if it has a finite Hamel basis. If X is a nontrivial finite-dimensional vector space then X has infinitely many Hamel bases, but each basis has exactly the same number of vectors. We call this number the *dimension* of X and denote it by $\dim(X)$. If $d = \dim(X)$ is the dimension of X and $\mathcal{B} = \{x_1, \ldots, x_d\}$ is a Hamel basis for X, then every vector $x \in X$ can be written

$$x = \sum_{n=1}^{d} c_n(x)\, x_n$$

for a unique choice of scalars $c_1(x), \ldots, c_d(x)$. We declare that the dimension of the trivial vector space $\{0\}$ is zero.

A vector space that does not have a finite Hamel basis is said to be *infinite-dimensional*, and in this case we write $\dim(X) = \infty$. For example, we saw in Section 1.11 that $\{t^k\}_{k \geq 0}$ is a Hamel basis for the set of polynomials $\mathcal{P}$. Since any two Hamel bases must have the same cardinality, it follows that every Hamel basis for $\mathcal{P}$ is infinite. Therefore $\mathcal{P}$ is an infinite-dimensional space. Similarly, c_{00} is infinite-dimensional, since the standard basis $\mathcal{E} = \{\delta_n\}_{n \in \mathbb{N}}$ is a Hamel basis for c_{00}.

Observe that although $\mathcal{E}$ is a linearly independent set in ℓ^1, it is not a Hamel basis for ℓ^1, because $\mathrm{span}(\mathcal{E}) = c_{00}$ is only a proper subspace of ℓ^1. Can you find a Hamel basis for ℓ^1? Since $\mathcal{E}$ does not span ℓ^1, perhaps we just need to add some "extra vectors" to $\mathcal{E}$ to obtain a larger but still countable set that is still independent yet spans ℓ^1. According to Problem 4.4.10, there is no such countable set. Moreover, we prove next that every Hamel basis for every infinite-dimensional complete normed space must be *uncountable*. Hence, even if we could find a Hamel basis for ℓ^1, it would be too large and unwieldy to be of much use.

Theorem 4.5.3. *If X is an infinite-dimensional Banach space and A is a Hamel basis for X, then A is uncountable.*

Proof. Suppose that X had a countable Hamel basis, say $\{x_n\}_{n \in \mathbb{N}}$. Since $\{x_n\}_{n \in \mathbb{N}}$ is linearly independent, we must have $x_n \neq 0$ for every n. For each $N \in \mathbb{N}$, define

$$F_N = \mathrm{span}\{x_1, \ldots, x_N\}.$$

Since F_N is finite-dimensional, Corollary 3.7.3 implies that it is closed. Given a vector $y \in F_N$ and given $r > 0$, let

$$z = y + \frac{r}{2} \frac{x_{N+1}}{\|x_{N+1}\|}.$$

Then $z \notin F_N$, yet $\|y - z\| = \frac{r}{2} < r$. Therefore $B_r(y)$ is not contained in F_N. Hence F_N contains no open balls, and since F_N is closed, this implies that it is nowhere dense (see Lemma 2.11.2). Consequently, $X = \cup F_N$ is a countable union of nowhere dense sets. Since X is complete, this contradicts the Baire Category Theorem. $\square$

When we work with a Hamel basis, we can only consider finite linear combinations of vectors. However, in an infinite-dimensional Banach space it is much more advantageous to consider "infinite linear combinations." This idea leads us to the following definition of a *Schauder basis* for a Banach space.

Definition 4.5.4 (Schauder Basis). Let X be a Banach space. A countably infinite sequence $\{x_n\}_{n \in \mathbb{N}}$ of elements of X is a *Schauder basis* for X if for each vector $x \in X$ there exist *unique* scalars $c_n(x)$ such that

$$x = \sum_{n=1}^{\infty} c_n(x)\, x_n, \tag{4.17}$$

where this series converges in the norm of X. $\diamond$

Most of our focus will be on infinite-dimensional spaces, but we often need to consider finite-dimensional spaces. A finite-dimensional normed space has a finite Hamel basis, and as a consequence, it cannot contain an *infinite* sequence $\{x_n\}_{n \in \mathbb{N}}$ that satisfies equation (4.17). Taking Definition 4.5.4 literally, it follows that finite-dimensional spaces do not have Schauder bases. However, if X is finite-dimensional, then we often let "Schauder basis" simply mean a Hamel basis for X. Thus for a finite-dimensional space the terms *Schauder basis*, *Hamel basis*, *vector space basis*, and *basis* are all synonymous. A detailed introduction to Schauder bases and related topics and generalizations can be found in [Heil11].

Recall from Definition 4.1.1 that equation (4.17) means that the partial sums of the series converge to x in the norm of X:

$$\lim_{N \to \infty} \left\| x - \sum_{n=1}^{N} c_n(x)\, x_n \right\| = 0.$$

Each partial sum $s_N = \sum_{n=1}^{N} c_n(x)\, x_n$ belongs to $\mathrm{span}\{x_n\}_{n \in \mathbb{N}}$, so this shows that $\mathrm{span}\{x_n\}_{n \in \mathbb{N}}$ is dense in X. Therefore a Schauder basis for X is a complete sequence in the sense of Definition 4.4.4. The reader should check that the uniqueness requirement in the definition of a Schauder basis implies that a Schauder basis is linearly independent. Thus every Schauder basis is both complete and linearly independent. However, we will see in Section 4.6 that there are complete, linearly independent sequences that are not Schauder bases.

Here is an example of a Schauder basis for ℓ^1.

Example 4.5.5. Let $\mathcal{E} = \{\delta_n\}_{n \in \mathbb{N}}$ be the sequence of standard basis vectors. If $x \in \ell^1$, then $x = (x_n)_{n \in \mathbb{N}} = (x_1, x_2, \dots)$, where $\sum |x_n| < \infty$. We claim that

$$x = \sum_{n=1}^{\infty} x_n \delta_n, \tag{4.18}$$

where this series *converges in ℓ^1-norm.* To prove this, let

$$s_N = \sum_{n=1}^{N} x_n \delta_n = (x_1, \dots, x_N, 0, 0, \dots)$$

be the Nth partial sum of the series. The difference between x and s_N is

$$x - s_N = (0, \dots, 0, x_{N+1}, x_{N+2}, \dots),$$

so $\|x - s_N\|_1 = \sum_{n=N+1}^{\infty} |x_n|$. Since $\sum_{n=1}^{\infty} |x_n| < \infty$, Lemma 1.8.4 implies that

$$\lim_{N \to \infty} \|x - s_N\|_1 = \lim_{N \to \infty} \sum_{n=N+1}^{\infty} |x_n| = 0.$$

Thus, the partial sums s_N converge to x in ℓ^1-norm, which proves that equation (4.18) holds. Further, the *only* way to write $x = \sum c_n \delta_n$ is with $c_n = x_n$ for every n (why?), so we conclude that $\mathcal{E}$ is a Schauder basis for ℓ^1. $\Diamond$

In summary, the standard basis $\mathcal{E} = \{\delta_n\}_{n \in \mathbb{N}}$ is a Schauder basis (but not a Hamel basis) for ℓ^1. An entirely similar argument shows that $\mathcal{E}$ is a Schauder basis for ℓ^p for every *finite* value of p (the only difference between ℓ^1 and ℓ^p being the norm with respect to which the series in question must converge). We consider $p = \infty$ next.

Example 4.5.6. Let $x = (1, 1, 1, \dots)$. This is an element of ℓ^∞, but there are *no scalars c_n* such that

$$x = \sum_{n=1}^{\infty} c_n \delta_n \tag{4.19}$$

with convergence of the series *in ℓ^∞-norm* (see Example 4.1.2). Therefore $\mathcal{E}$ is not a Schauder basis for ℓ^∞. According to Problem 4.5.10, the standard basis $\mathcal{E}$ is a Schauder basis for c_0, which is a closed proper subset of ℓ^∞. $\Diamond$

Problem 4.5.9 shows that every Banach X that has a Schauder basis must be separable. We saw in Theorem 2.6.8 that ℓ^∞ is not separable, so not only is the standard basis not a Schauder basis for ℓ^∞, but there is *no sequence* in ℓ^∞ that is a Schauder basis for that space.

Does every separable Banach space have a Schauder basis? This question was a long-standing open mathematical problem, known as the *Basis Problem*. It was shown by Enflo [Enf73] that there exist separable Banach spaces that have no Schauder bases!

If $\{x_n\}_{n \in \mathbb{N}}$ is a Schauder basis for a Banach space X, then for each vector $x \in X$ there exist unique scalars $c_n(x)$ such that

$$x = \sum_{n=1}^{\infty} c_n(x)\, x_n, \tag{4.20}$$

where the series converges with respect to the norm of X. As we discussed in Section 4.3, we say that the series in equation (4.20) *converges unconditionally* if it converges regardless of ordering, i.e., if $\sum_{n=1}^{\infty} c_{\sigma(n)}(x)\, x_{\sigma(n)}$ converges for every bijection $\sigma \colon \mathbb{N} \to \mathbb{N}$. Unconditional convergence is highly desirable, so we give the following name to a Schauder basis for which the series in equation (4.20) converges unconditionally for every x.

Definition 4.5.7 (Unconditional Basis). Let $\{x_n\}_{n \in \mathbb{N}}$ be a Schauder basis for a Banach space X. If the series given in equation (4.20) converges unconditionally for every $x \in X$, then we say that $\{x_n\}_{n \in \mathbb{N}}$ is an *unconditional Schauder basis* for X; otherwise, $\{x_n\}_{n \in \mathbb{N}}$ is a *conditional Schauder basis* for X. $\Diamond$

The standard basis is an unconditional Schauder basis for ℓ^p (when $p < \infty$) and for c_0 (with respect to the sup-norm $\| \cdot \|_\infty$). Not every Schauder basis is unconditional; an example of a conditional basis is given in Problem 4.5.10. We will see more examples of Schauder bases when we consider Hilbert spaces in Chapter 5.

Problems

4.5.8. Suppose that $A = \{x_n\}_{n\in\mathbb{N}}$ is a Schauder basis for a Banach space X. Prove that equality holds in equation (4.16).

4.5.9. Let X be a Banach space, and assume that X has a Schauder basis (either a countably infinite Schauder basis $\{x_n\}_{n\in\mathbb{N}}$ or a finite Hamel basis $\{x_1, \ldots, x_d\}$). Prove that X is separable (it may be helpful to compare Problem 2.6.20).

4.5.10. Prove the following statements (the norm on c_0 is $\|\cdot\|_\infty$).

(a) The standard basis $\mathcal{E} = \{\delta_n\}_{n\in\mathbb{N}}$ is an *unconditional* Schauder basis for c_0.

(b) If $y_n = (1, \ldots, 1, 0, 0, \ldots)$, where the 1 is repeated n times, then the sequence $\{y_n\}_{n\in\mathbb{N}}$ is a *conditional* Schauder basis for c_0.

4.5.11. Let c denote the set of all convergent sequences of scalars, i.e.,

$$c = \left\{ x = (x_k)_{k\in\mathbb{N}} : \lim_{k\to\infty} x_k \text{ exists} \right\}.$$

(a) Prove that c is a closed subspace of ℓ^∞, and therefore is a Banach space with respect to the sup-norm.

(b) Let $\{\delta_n\}_{n\in\mathbb{N}}$ denote the sequence of standard basis vectors. Find a sequence $\delta_0 \in c$ such that $\{\delta_n\}_{n\geq 0}$ forms an unconditional basis for c.

(c) For each $n \in \mathbb{N}$, define $z_n = (0, \ldots, 0, 1, 1, \ldots)$, where the 0 is repeated $n - 1$ times. Show that $\{z_n\}_{n\in\mathbb{N}}$ is a conditional basis for c (this is called the *summing basis* for c).

4.5.12. Let $C_0(\mathbb{R})$ be the space of continuous functions that vanish at infinity introduced in Problem 3.5.19. Assume that $g \in C_0(\mathbb{R})$ is not the zero function. Prove that the following set of four functions is linearly independent:

$$\{g(t), \ g(t-1), \ g(t)\cos t, \ g(t-\sqrt{2})\cos \pi t\}. \tag{4.21}$$

Note: As of the time of writing, this is an *open problem*. That is, while it is known that the set given in equation (4.21) is independent for *some* functions g, it is *not* known whether it is independent for *every* nonzero $g \in C_0(\mathbb{R})$. In other words, it is possible that there might be some function $g \in C_0(\mathbb{R})$ for which the set is dependent, but we know no actual examples of functions g

for which this is the case. For more on this problem, which is a special case of the *HRT Conjecture*, see [Heil06] and [HS15].

4.6 The Weierstrass Approximation Theorem

Issues of *approximation* are rampant in analysis. Quite often we need to know how well one object can be approximated by another (hopefully simpler) object. In this section we will consider how well a continuous function f on a bounded closed interval $[a, b]$ can be approximated by a polynomial. There are many ways to quantify this question. For example, Taylor's Theorem tells us that if f is k-times differentiable at a point t, then we can use *Taylor polynomials* to approximate f in some neighborhood of t. However, we are interested here in a somewhat different question. Specifically, given an arbitrary continuous but not necessarily differentiable function f, we want to know whether there is a polynomial p that approximates f well across the *entire* interval $[a, b]$. The *Weierstrass Approximation Theorem* asserts that we can always do this. There are many different proofs of Weierstrass's theorem; we will give a constructive proof due to Bernstein (i.e., a proof that explicitly tells us which polynomial will uniformly approximate f across the interval to within a given accuracy).

We will be dealing with the space of continuous functions on $[a, b]$, which was first introduced in Example 3.1.5 and is denoted by $C[a, b]$. Since $[a, b]$ is a compact set, every continuous function on $[a, b]$ is bounded, and therefore $C[a, b]$ is a Banach space with respect to the uniform norm

$$\|f\|_{\mathrm{u}} = \sup_{t \in [a,b]} |f(t)|.$$

For simplicity of notation we will state Bernstein's theorem for the interval $[a, b] = [0, 1]$ (but, as we will explain afterwards, by making a change of variables the result carries over to other bounded closed intervals). Given a function $f \in C[0, 1]$, Bernstein constructed explicit polynomials B_n that converge uniformly to f. Specifically, the nth *Bernstein polynomial* for f is

$$B_n(t) = \sum_{k=0}^{n} f(k/n) \binom{n}{k} t^k (1 - t)^{n-k}, \qquad (4.22)$$

where $\binom{n}{k}$, or "n choose k," is the binomial coefficient defined by

$$\binom{n}{k} = \frac{n!}{k! \, (n - k)!}.$$

The *Binomial Theorem* tells us that if x, $y \in \mathbb{R}$, then

$$(x + y)^n = \sum_{k=0}^{n} \binom{n}{k} x^k y^{n-k}.$$

Consequently,

$$1 = 1^n = \left(t + (1 - t)\right)^n = \sum_{k=0}^{n} \binom{n}{k} t^k (1 - t)^{n-k}. \tag{4.23}$$

Note that this equation is very similar to the right-hand side of equation (4.22), except that the factors $f(k/n)$ are omitted in equation (4.23). This will be important in our proof of Bernstein's theorem. Another fact we will need is that

$$\frac{t(1 - t)}{n} = \sum_{k=0}^{n} (t - k/n)^2 \binom{n}{k} t^k (1 - t)^{n-k}. \tag{4.24}$$

This equality can be obtained from equation (4.23) after some manipulation; for one derivation of equation (4.24), see [Bar76, Sec. 24]. Now we state and prove Bernstein's theorem.

Theorem 4.6.1 (Bernstein Approximation Theorem). *If $f \in C[0, 1]$, then $B_n \to f$ uniformly as $n \to \infty$, where B_n is the nth Bernstein polynomial for f defined by equation (4.22).*

Proof. Fix any $\varepsilon > 0$. Since $[0, 1]$ is a compact set, f is uniformly continuous on this domain, so there must exist some $\delta > 0$ such that

$$|x - y| < \delta \implies |f(x) - f(y)| < \varepsilon. \tag{4.25}$$

We will show that $\|f - B_n\|_u < 3\varepsilon$ for all n larger than the number

$$N = \max\left\{ \frac{1}{\delta^4}, \frac{\|f\|_u^2}{\varepsilon^2} \right\}. \tag{4.26}$$

The reason for defining N in this way may be mysterious at the moment, but we will see below why we chose this value.

Suppose that $n > N$, and fix a point $t \in [0, 1]$. We use equation (4.23) to rewrite and estimate $f(t) - B_n(t)$ as follows:

$$|f(t) - B_n(t)|$$
$$= \left| f(t) \sum_{k=0}^{n} \binom{n}{k} t^k (1 - t)^{n-k} - \sum_{k=0}^{n} f(k/n) \binom{n}{k} t^k (1 - t)^{n-k} \right|$$

$$= \left| \sum_{k=0}^{n} \left(f(t) - f(k/n) \right) \binom{n}{k} t^k (1-t)^{n-k} \right|$$

$$\leq \sum_{k=0}^{n} |f(t) - f(k/n)| \binom{n}{k} t^k (1-t)^{n-k}. \tag{4.27}$$

In order to analyze this last sum more carefully, we will break it into two parts, one that contains those terms for which k/n is "close" to t, and the other containing the terms for which k/n is "far" from t. Specifically, let J be the set of those indices k between 0 and n for which k/n is close enough to t that

$$|t - k/n| < n^{-1/4}.$$

Since $n > N \geq \delta^{-4}$, it follows that if $k \in J$ then $|t - k/n| < n^{-1/4} \leq \delta$, and therefore $|f(t) - f(k/n)| < \varepsilon$ by equation (4.25). Consequently,

$$\sum_{k \in J} |f(t) - f(k/n)| \binom{n}{k} t^k (1-t)^{n-k} < \sum_{k \in J} \varepsilon \binom{n}{k} t^k (1-t)^{n-k}$$

$$\leq \varepsilon \sum_{k=0}^{N} \binom{n}{k} t^k (1-t)^{n-k}$$

$$= \varepsilon, \tag{4.28}$$

where at the last step we used equation (4.23).

Now we consider the terms corresponding to $k \notin J$. In this case we have $|t - k/n| \geq n^{-1/4}$, which implies that

$$\frac{1}{\left(t - k/n \right)^2} \leq n^{1/2}.$$

Since we also have

$$|f(t) - f(k/n)| \leq |f(t)| + |f(k/n)| \leq \|f\|_u + \|f\|_u = 2\|f\|_u,$$

it follows that

$$\sum_{k \notin J} |f(t) - f(k/n)| \binom{n}{k} t^k (1-t)^{n-k}$$

$$\leq \sum_{k \notin J} 2\|f\|_u \binom{n}{k} t^k (1-t)^{n-k}$$

$$= 2\|f\|_u \sum_{k \notin J} \frac{(t - k/n)^2}{(t - k/n)^2} \binom{n}{k} t^k (1-t)^{n-k}$$

$$\leq 2\|f\|_u \sum_{k \notin J} n^{1/2} \left(t - k/n\right)^2 \binom{n}{k} t^k \left(1 - t\right)^{n-k}$$

$$\leq 2\|f\|_u \, n^{1/2} \sum_{k=0}^{n} \left(t - k/n\right)^2 \binom{n}{k} t^k \left(1 - t\right)^{n-k}$$

$$= 2\|f\|_u \, n^{1/2} \frac{t(1-t)}{n} \qquad \text{(by equation (4.24))}$$

$$\leq \frac{2\|f\|_u}{n^{1/2}} \qquad \text{(since } 0 \leq t \leq 1)$$

$$< 2\varepsilon \qquad \text{(since } n > N \geq \|f\|_u^2/\varepsilon^2).$$

Since $\sum_{k=0}^{n} = \sum_{k \in J} + \sum_{k \notin J}$, by combining the above estimates with equation (4.27) we see that

$$|f(t) - B_n(t)| \leq \sum_{k=0}^{n} |f(t) - f(k/n)| \binom{n}{k} t^k \left(1 - t\right)^{n-k} < 3\varepsilon.$$

This inequality holds for each $t \in [0, 1]$ and all $n > N$. Taking the supremum over $t \in [0, 1]$, we see that if $n > N$, then

$$\|f - B_n\|_u = \sup_{t \in [0,1]} |f(t) - B_n(t)| \leq 3\varepsilon.$$

This shows that B_n converges uniformly to f as $n \to \infty$. $\square$

Although Theorem 4.6.1 ensures that f will be well approximated by a polynomial on the interval $[0, 1]$, this may not be the case for points outside of that interval. For example, in Figure 4.2 we show the function $f(t) = \cos 2\pi t$ and its eighth Bernstein polynomial $B_8(t)$. The Bernstein polynomial does approximate f across the interval $[0, 1]$, but the two functions quickly diverge outside of $[0, 1]$.

Restricting our attention now to just the interval $[0, 1]$, we know from Theorem 4.6.1 that B_n converges uniformly to f on that interval as n increases. Hence by choosing n large enough we can make $\|f - B_n\|_u$ as small as we like. However, the *rate* of convergence given by Bernstein's Theorem is quite slow in general. For example, in Figure 4.3 we show the Bernstein polynomial B_{50} for $f(t) = \cos 2\pi t$ on $[0, 1]$. Comparing Figures 4.2 and 4.3, we see that B_{50} is a better approximation than B_8, but even with $n = 50$ the accuracy of the approximation is poor (numerically, $\|f - B_{50}\|_u \approx 0.1$). Given even a modest value of ε, the value of N specified by equation (4.26) can be very large (see Problem 4.6.4).

By making a change of variables, Theorem 4.6.1 carries over to other intervals $[a, b]$. For example, if we choose $g \in C[a, b]$, then the function

$$f(t) = g\big((b - a)t + a\big), \qquad t \in [0, 1],$$

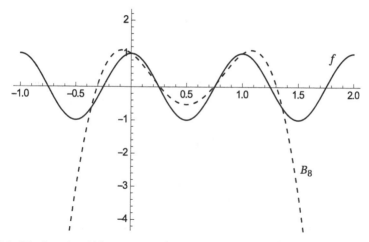

Fig. 4.2 The function $f(t) = \cos 2\pi t$ (solid) and its approximation $B_8(t)$ (dashed). The approximation is close on the interval $[0, 1]$, but quickly diverges outside that interval.

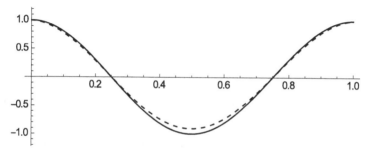

Fig. 4.3 The function $f(t) = \cos 2\pi t$ (solid) and its approximation $B_{50}(t)$ (dashed), shown just on the interval $[0, 1]$. Even with $n = 50$, we can still see the difference between f and B_{50} (in fact, $\|f - B_{50}\|_u$ is about 0.1; see Problem 4.6.4)

belongs to $C[0, 1]$. If we let B_n be the nth Bernstein polynomial for f, then $B_n \to f$ uniformly on the interval $[0, 1]$ by Theorem 4.6.1. The function

$$p_n(x) = B_n\left(\frac{x - a}{b - a}\right), \qquad x \in [a, b],$$

is a polynomial in x with degree at most n, and $B_n(t) = p_n\big((b - a)t + a\big)$ for $t \in [0, 1]$. Therefore

$$\|f - B_n\|_u = \sup_{t \in [0,1]} |f(t) - B_n(t)|$$

$$= \sup_{t \in [0,1]} \big|g\big((b - a)t + a\big) - p_n\big((b - a)t + a\big)\big|$$

$$= \sup_{x \in [a,b]} |g(x) - p_n(x)|$$

$$= \|g - p_n\|_u.$$

Since $\|f - B_n\|_u \to 0$, we conclude that $p_n \to g$ uniformly on $[a, b]$. As a corollary, this proves the Weierstrass Approximation Theorem, which we state next.

Theorem 4.6.2 (Weierstrass Approximation Theorem). *If $f \in C[a, b]$ and $\varepsilon > 0$, then there exists a polynomial $p(t) = \sum_{k=0}^{N} c_k t^k$ such that*

$$\|f - p\|_u = \sup_{t \in [a,b]} |f(t) - p(t)| < \varepsilon. \qquad \diamondsuit$$

In summary, if f is any function in $C[a, b]$, then Bernstein's Theorem tells us how to construct polynomials p_n, with degree at most n, that converge uniformly to f. By the definition of a polynomial, we can write

$$p_n(t) = \sum_{k=0}^{n} c_{kn} t^k$$

for some scalars c_{kn}. Hence

$$f(t) = \lim_{n \to \infty} p_n(t) = \lim_{n \to \infty} \sum_{k=0}^{n} c_{kn} t^k, \qquad (4.29)$$

where the convergence is with respect to the uniform norm. Does this mean that we can write

$$f(t) = \sum_{k=0}^{\infty} c_k t^k = \lim_{n \to \infty} \sum_{k=0}^{n} c_k t^k \qquad (4.30)$$

for some scalars c_k? Not necessarily, since the scalars c_{kn} in equation (4.29) can depend on n, whereas in order for equation (4.30) to hold we must be able to choose the scalars c_k to be *independent of n*.

In fact, an infinite series of the form given in equation (4.30) is called a *power series*. Such series are studied in detail in undergraduate calculus texts, and we review some of their properties in Problem 4.6.6. In particular, that problem shows that if the series in equation (4.30) converges for $t \in [a, b]$, then f is *infinitely differentiable* on (a, b). Not every function in $C[a, b]$ is infinitely differentiable (e.g., consider $f(t) = |t - c|$, where $a < c < b$), so there are functions $f \in C[a, b]$ that *cannot* be written in the form that appears in equation (4.30). We formalize these remarks as the following result.

Theorem 4.6.3. *The sequence of monomials*

$$\mathcal{M} = \{t^k\}_{k=0}^{\infty} = \{1, t, t^2, \dots\}$$

is complete and linearly independent *in* $C[a, b]$, *but* $\mathcal{M}$ *is not a Schauder basis for* $C[a, b]$.

Proof. By Theorem 4.6.2, each $f \in C[a, b]$ is the uniform limit of polynomials p_n. Every polynomial is, by definition, a finite linear combination of monomials t^k, so $p_n \in \text{span}(\mathcal{M})$ for each n. Therefore, by Corollary 2.6.4, the span of $\mathcal{M}$ is dense in $C[a, b]$. Hence $\mathcal{M}$ is a complete sequence in $C[a, b]$ in the sense of Definition 4.4.4, and we have seen before that $\mathcal{M}$ is linearly independent.

In order for $\mathcal{M}$ to be a Schauder basis for $C[a, b]$, it would have to be the case that every $f \in C[a, b]$ could be written as $f(t) = \sum_{k=1}^{\infty} c_k t^k$ for a unique choice of scalars c_k (see Definition 4.5.4). However, we have seen that this is not the case, so $\mathcal{M}$ is not a Schauder basis. □

Problems

4.6.4. Let $f(t) = \cos 2\pi t$.

(a) Prove that $|f(x) - f(y)| \leq 2\pi |x - y|$ for all $x, y \in \mathbb{R}$.

(b) Theorem 4.6.1 tells us that if $\varepsilon > 0$ is given, then there is an N such that $\|f - B_N\|_u < \varepsilon$ for all $n > N$. Use the proof of Theorem 4.6.1, and equation (4.26) in particular, to determine an explicit value for N for which this is sure to hold for the choices $\varepsilon = 0.1$ and $\varepsilon = 0.01$.

(c) Show that $|f(0.5) - B_{50}(0.5)| > 0.09$, and therefore $\|f - B_{50}\|_u > 0.09$. How does this compare to the value for N that you derived in part (b)?

4.6.5. Suppose that $f \in C[0, 1]$ satisfies

$$\int_0^1 f(t) \, t^n \, dt = 0, \qquad n = 0, 1, 2, \ldots.$$

Prove that $\int_0^1 |f(t)|^2 \, dt = 0$, and hence $f = 0$.

4.6.6. Let $(c_k)_{k \geq 0}$ be a fixed sequence of scalars. Suppose that the series $\sum_{k=0}^{\infty} c_k t^k$ converges for some number $t \in \mathbb{R}$, and set $r = |t|$. Prove the following statements.

(a) For each $|t| < r$ we have $\sum_{k=0}^{\infty} |c_k| |t|^k < \infty$, and therefore the series $f(t) = \sum_{k=0}^{\infty} c_k t^k$ converges for all $|t| < r$.

(b) f is infinitely differentiable on the interval $(-r, r)$.

4.7 The Stone–Weierstrass Theorem

We will prove a useful (and elegant) generalization of the Weierstrass Approximation Theorem known as the *Stone–Weierstrass Theorem*. While the original Weierstrass theorem shows only that the set of polynomials is dense in $C[a, b]$, the Stone–Weierstrass Theorem allows us to prove that many different collections of continuous functions are dense in $C(K)$ when K is a compact subset of a metric space.

First we prove an intermediary result for real-valued functions (so $\mathbb{F} = \mathbb{R}$ for this theorem and its proof). In the statement of this result, $h = \max\{f, g\}$ is the function obtained by letting $h(x)$ be the larger of $f(x)$ and $g(x)$, and similarly $k = \min\{f, g\}$ is defined by letting $k(x)$ be the smaller of $f(x)$ and $g(x)$.

Theorem 4.7.1. *Assume that K is a compact subset of a metric space X, and let $\mathcal{F}$ be any collection of continuous functions $f \colon K \to \mathbb{R}$ that satisfies the following two conditions.*

(a) *If f, $g \in \mathcal{F}$, then $\max\{f, g\} \in \mathcal{F}$ and $\min\{f, g\} \in \mathcal{F}$.*

(b) *If a, $b \in \mathbb{R}$ and $x \neq y \in K$ are given, then there exists some $f \in \mathcal{F}$ such that $f(x) = a$ and $f(y) = b$.*

Then $\mathcal{F}$ is dense in $C(K)$. Consequently, given any $g \in C(K)$, there exist functions $f_n \in \mathcal{F}$ such that $f_n \to g$ uniformly on K.

Proof. If K contains only a single point, then the result is trivial, so we can assume that K contains at least two points. Also note that by induction, hypothesis (a) extends to finitely many functions, i.e., if $f_1, \dots, f_N$ belong to $\mathcal{F}$, then $\max\{f_1, \dots, f_N\}$ and $\min\{f_1, \dots, f_N\}$ also belong to $\mathcal{F}$.

Choose any function $f \in C(K)$ and any $\varepsilon > 0$, and let x be a fixed point in K. By hypothesis (b), for each $y \in K$ with $y \neq x$ there exists some function $g_{xy} \in \mathcal{F}$ such that $g_{xy}(x) = f(x)$ and $g_{xy}(y) = f(y)$. For the case $y = x$, choose any point $z \neq x$ in K; then there exists some function $g_{xx} \in \mathcal{F}$ such that $g_{xx}(x) = f(x)$ and $g_{xx}(z) = f(z)$. In any case, for each $y \in K$ (including $y = x$), the function g_{xy} satisfies

$$g_{xy}(x) = f(x) \qquad \text{and} \qquad g_{xy}(y) = f(y). \tag{4.31}$$

Keeping x and y fixed for now, since g_{xy} and f are both continuous on K, there is some $\delta_y > 0$ such that for all $z \in K$ with $d(y, z) < \delta_y$ we have

$$|g_{xy}(z) - g_{xy}(y)| < \varepsilon \qquad \text{and} \qquad |f(z) - f(y)| < \varepsilon.$$

For simplicity of notation, let $U_y = B_{\delta_y}(y)$ be the open ball of radius δ_y centered at the point y. Then for each $z \in K \cap U_y$,

$$|g_{xy}(z) - f(z)| \leq |g_{xy}(z) - g_{xy}(y)| + |g_{xy}(y) - f(z)|$$
$$= |g_{xy}(z) - g_{xy}(y)| + |f(y) - f(z)|$$
$$< \varepsilon + \varepsilon = 2\varepsilon. \tag{4.32}$$

The collection $\{U_y\}_{y \in K}$ of all the open balls U_y with $y \in K$ is an open cover of the compact set K. Hence there must be finitely many points $y_1, \ldots, y_N \in K$ such that

$$K \subseteq U_{y_1} \cup \cdots \cup U_{y_N}.$$

Let $h_x = \max\{g_{xy_1}, \ldots, g_{xy_N}\}$. Hypothesis (a) tells us that $h_x \in \mathcal{F}$, and by equation (4.31) we have

$$h_x(x) = \max\{g_{xy_1}(x), \ldots, g_{xy_N}(x)\} = \max\{f(x), \ldots, f(x)\} = f(x).$$

If we choose any point $z \in K$, then $z \in K \cap U_{y_n}$ for some n, and therefore $|g_{xy_n}(z) - f(z)| < 2\varepsilon$ by equation (4.32). Hence

$$f(z) < g_{xy_n}(z) + 2\varepsilon \leq h_x(z) + 2\varepsilon, \qquad \text{for all } z \in K. \tag{4.33}$$

To summarize our work to this point, given $x \in K$, we defined functions g_{xy} for $y \in K$, and then used these functions and the compactness of K to define a function h_x for which equation (4.33) holds. Next we will use compactness again, this time to reduce our focus from all of these functions h_x to a single function h.

To do this, note that h_x and f are each continuous on K, so there exists some δ_x such that for all $z \in K$ with $d(x, z) < \delta_x$ we have

$$|h_x(z) - h_x(x)| < \varepsilon \qquad \text{and} \qquad |f(z) - f(x)| < \varepsilon.$$

Let $V_x = B_{\delta_x}(x)$. Then for each $z \in K \cap V_x$,

$$|h_x(z) - f(z)| \leq |h_x(z) - h_x(x)| + |h_x(x) - f(z)|$$
$$= |h_x(z) - h_x(x)| + |f(x) - f(z)|$$
$$< \varepsilon + \varepsilon = 2\varepsilon. \tag{4.34}$$

Since $\{V_x\}_{x \in K}$ is an open cover of the compact set K, there must be finitely many points $x_1, \ldots, x_M \in K$ such that

$$K \subseteq V_{x_1} \cup \cdots \cup V_{x_M}.$$

Let

$$h = \min\{h_{x_1}, \ldots, h_{x_M}\}.$$

Hypothesis (a) ensures that h belongs to $\mathcal{F}$.

Now, if $z \in K$, then $z \in V_{x_m}$ for *some* m. Therefore, by equation (4.34),

$$h(z) - 2\varepsilon \leq h_{x_m}(z) - 2\varepsilon < f(z). \tag{4.35}$$

On the other hand, equation (4.33) implies that for *every* m we have

$$f(z) < h_{x_m}(z) + 2\varepsilon.$$

Taking the minimum over $m = 1, \ldots, M$, we conclude that

$$f(z) < \min\{h_{x_1}(z), \ldots, h_{x_M}(z)\} + 2\varepsilon = h(z) + 2\varepsilon. \tag{4.36}$$

Combining equations (4.35) and (4.36), we see that

$$|f(z) - h(z)| \leq 2\varepsilon, \qquad \text{for all } z \in K.$$

Therefore $\|f - h\|_{\mathrm{u}} \leq 2\varepsilon$. Since h belongs to $\mathcal{F}$, this proves that $\mathcal{F}$ is dense in $C(K)$. $\square$

For example, according to Problem 4.7.4, the collection $\mathcal{L}$ of all continuous piecewise linear functions on $[a, b]$ satisfies hypotheses (a) and (b) in Theorem 4.7.1, and therefore $\mathcal{L}$ is dense in $C[a, b]$. Hence every continuous real-valued function on $[a, b]$ can be uniformly approximated as closely as we like by a piecewise linear function (although this is not difficult to prove directly).

Next we will use Theorem 4.7.1 and the Weierstrass Approximation Theorem to prove the Stone–Weierstrass Theorem for real-valued functions (so we take $\mathbb{F} = \mathbb{R}$ in this theorem and its proof).

Theorem 4.7.2 (Stone–Weierstrass Theorem). *Let K be a compact subset of a metric space X, and assume that $\mathcal{A}$ is a collection of continuous functions $f \colon K \to \mathbb{R}$ that satisfies the following four conditions.*

(a) *The constant function 1 belongs to $\mathcal{A}$.*

(b) *$\mathcal{A}$ is a subspace of $C(K)$, i.e., if $f, g \in \mathcal{A}$ and $a, b \in \mathbb{R}$, then $af + bg \in \mathcal{A}$.*

(c) *$\mathcal{A}$ is an algebra, i.e., it is a subspace and if $f, g \in \mathcal{A}$, then $fg \in \mathcal{A}$.*

(d) *$\mathcal{A}$ separates points, i.e., if $x, y \in K$ and $x \neq y$, then there is some $f \in \mathcal{A}$ such that $f(x) \neq f(y)$.*

Then $\mathcal{A}$ is dense in $C(K)$. Consequently, given any $g \in C(K)$, there exist functions $f_n \in \mathcal{A}$ such that $f_n \to g$ uniformly on K.

Proof. Let $\mathcal{F}$ be the closure of $\mathcal{A}$. By Corollary 2.6.4, $\mathcal{F}$ is the collection of all functions that can be uniformly approximated by functions from $\mathcal{A}$:

$$\mathcal{F} = \overline{\mathcal{A}} = \{g \in C(K) : \exists f_n \in \mathcal{A} \text{ such that } f_n \to g \text{ uniformly}\}.$$

Our goal is to prove that $\mathcal{F} = C(K)$.

Step 1. We will show that $\mathcal{F}$ satisfies hypothesis (b) of Theorem 4.7.1. To do this, choose any numbers $a, b \in \mathbb{R}$ and any points $x \neq y \in K$. Since we have assumed that $\mathcal{A}$ separates points, there must exist some function $g \in \mathcal{A}$ such that $g(x) \neq g(y)$. Choose numbers α and β (why can we do this?) such that

$$\alpha g(x) + \beta = a \quad \text{and} \quad \alpha g(y) + \beta = b.$$

Let 1 be the constant function, defined by $1(t) = 1$ for every t. Since $1 \in \mathcal{A}$ and since $\mathcal{A}$ is a subspace, the function

$$f(t) = \alpha g(t) + \beta 1(t) = \alpha g(t) + \beta, \qquad t \in K,$$

belongs to $\mathcal{A}$. Hence $f \in \mathcal{F}$, since $\mathcal{A} \subseteq \mathcal{F}$. Since $f(x) = a$ and $f(y) = b$, this shows that $\mathcal{F}$ satisfies hypothesis (b) of Theorem 4.7.1.

Step 2. Next we will show that if $f \in \mathcal{F}$, then $|f| \in \mathcal{F}$ as well. To simplify the notation, let $F(x) = |f(x)|$ for $x \in K$.

Fix any $\varepsilon > 0$. Since $f \in \mathcal{F}$ and $\mathcal{F}$ is the closure of $\mathcal{A}$, there exists some function $h \in \mathcal{A}$ such that

$$\|f - h\|_{\mathrm{u}} = \sup_{x \in K} |f(x) - h(x)| < \varepsilon.$$

That is, h is within ε of f as measured by the uniform norm. Applying the Reverse Triangle Inequality, it follows that $H = |h|$ is equally close to $F = |f|$, because $|F(x) - H(x)| = \big||f(x)| - |h(x)|\big| \leq |f(x) - h(x)|$ and therefore

$$\begin{aligned}
\|F - H\|_{\mathrm{u}} &= \sup_{x \in K} |F(x) - H(x)| \\
&\leq \sup_{x \in K} |f(x) - h(x)| \\
&= \|f - h\|_{\mathrm{u}} < \varepsilon.
\end{aligned} \tag{4.37}$$

Since h is continuous and K is compact, h must be bounded. Set $R = \|h\|_{\mathrm{u}}$, so we have $|h(x)| \leq R$ for $x \in K$. Let a denote the absolute value function on the interval $[-R, R]$:

$$a(t) = |t|, \qquad -R \leq t \leq R.$$

Since a is a continuous function on the closed, bounded interval $[-R, R]$, the Weierstrass Approximation Theorem implies that there is some polynomial $p(t) = \sum_{k=0}^{n} c_k t^k$ such that

$$\|a - p\|_{\mathrm{u}} = \sup_{|t| \leq R} |a(t) - p(t)| < \varepsilon. \tag{4.38}$$

Let $g = p \circ h$, the composition of p with h. Since h belongs to $\mathcal{A}$ and $\mathcal{A}$ is closed under products, we have $h^n = h \cdot h \cdots h \in \mathcal{A}$ for $n = 1, 2, \ldots$. We also have $h^0 = 1 \in \mathcal{A}$. Since $\mathcal{A}$ is a subspace, it follows that

$$g(x) = (p \circ h)(x) = p(h(x)) = \sum_{k=0}^{n} c_k \, h(x)^k \in \mathcal{A}.$$

Now, $g = p \circ h$ and $H = |h| = a \circ h$. Since a is close to p, we hope that g will be close to H. To prove this, note that if $x \in K$ and we set $t = h(x)$, then $t \in [-R, R]$ and therefore

$$\bigl| H(x) - g(x) \bigr| = \bigl| a(h(x)) - p(h(x)) \bigr| = \bigl| a(t) - p(t) \bigr| < \varepsilon.$$

Taking the supremum over all $x \in K$, we conclude that

$$\|H - g\|_{\mathrm{u}} = \sup_{x \in K} \bigl| H(x) - g(x) \bigr| \leq \varepsilon. \tag{4.39}$$

Finally, applying the Triangle Inequality and using equations (4.37) and (4.39), we obtain

$$\bigl\| |f| - g \bigr\|_{\mathrm{u}} = \|F - g\|_{\mathrm{u}} \leq \|F - H\|_{\mathrm{u}} + \|H - g\|_{\mathrm{u}} < \varepsilon + \varepsilon = 2\varepsilon.$$

Thus $|f|$ is approximated to within a distance of 2ε by the function g, which belongs to $\mathcal{A}$. Therefore $|f|$ belongs to the closure of $\mathcal{A}$, which is $\mathcal{F}$.

Step 3. Now we will show that $\mathcal{F}$ satisfies hypothesis (a) of Theorem 4.7.1. Choose any two functions f and g in $\mathcal{F}$. We know that $\mathcal{A}$ is a subspace of $C(K)$, and Problem 3.3.12 tells us that the closure of a subspace of a normed space is itself a subspace. Therefore $\mathcal{F}$ is a subspace, and hence $f - g$ belongs to $\mathcal{F}$. By Step 2, it follows that $|f - g| \in \mathcal{F}$. Applying again the fact that $\mathcal{F}$ is a subspace, we see that

$$\max\{f, g\} = \frac{f + g + |f - g|}{2} \in \mathcal{F},$$

and similarly

$$\min\{f, g\} = \frac{f + g - |f - g|}{2} \in \mathcal{F}.$$

Therefore hypothesis (a) of Theorem 4.7.1 holds.

Step 4. We have shown that $\mathcal{F}$ satisfies all of the hypotheses of Theorem 4.7.1. Consequently, that theorem implies that $\mathcal{F}$ is dense in $C(K)$. That is, the closure of $\mathcal{F}$ is all of $C(K)$. But $\mathcal{F}$ is closed, so it equals its own closure. Therefore $\mathcal{F} = C(K)$. $\square$

Theorem 4.7.2 can be extended to complex-valued functions if one more hypothesis is added. The exact formulation of the complex version is given in Problem 4.7.10.

To give an application of the Stone–Weierstrass Theorem, we introduce a special set of continuous, real-valued functions on the interval $[-\pi, \pi]$:

$$C_{\mathrm{per}}[-\pi, \pi] = \{ f \in C[-\pi, \pi] : f(-\pi) = f(\pi) \}. \tag{4.40}$$

Each element f can be extended to all of $\mathbb{R}$ by setting

$$f(t + 2\pi n) = f(t), \qquad t \in [-\pi, \pi] \text{ and } n \in \mathbb{Z}.$$

The extended function f is 2π-*periodic*, which means that it repeats after 2π units:

$$f(t + 2\pi) = f(t), \qquad t \in \mathbb{R}.$$

Conversely, if g is a 2π-periodic function on the domain $\mathbb{R}$, then its restriction to the domain $[-\pi, \pi]$ is an element of $C_{\mathrm{per}}[-\pi, \pi]$. Thus there is a one-to-one correspondence between elements of $C_{\mathrm{per}}[-\pi, \pi]$ and 2π-periodic functions on the real line, and so this space is useful whenever we need to deal with periodic phenomena. Problem 4.7.7 shows that $C_{\mathrm{per}}[-\pi, \pi]$ is a closed, proper subspace of $C[-\pi, \pi]$.

The *real trigonometric system* in $C[-\pi, \pi]$ is the following set of sines and cosines together with the constant function 1:

$$\mathcal{T} = \{1\} \cup \{\cos nt\}_{n \in \mathbb{N}} \cup \{\sin nt\}_{n \in \mathbb{N}}.$$

The number n is called the *frequency* of the function $\cos nt$ or $\sin nt$. Since $\cos 0t = 1$, the constant function 1 can be considered to be the cosine of frequency zero.

The trigonometric system plays a fundamental role in the field of *harmonic analysis*, and we will examine the trigonometric system in more detail in Section 5.10. Here we will consider $\mathrm{span}(\mathcal{T})$, the finite linear span of the trigonometric system, as a subset of $C_{\mathrm{per}}[-\pi, \pi]$. A typical element of $\mathrm{span}(\mathcal{T})$ has the form

$$a_0 + \sum_{n=1}^{N} a_n \cos nt + \sum_{n=1}^{N} b_n \sin nt$$

for some integer N and some real scalars a_n and b_n. Each such function belongs to $C_{\mathrm{per}}[-\pi, \pi]$, and we will show that every element of $C_{\mathrm{per}}[-\pi, \pi]$ is the uniform limit of functions from $\mathrm{span}(\mathcal{T})$.

Theorem 4.7.3 (Uniform Approximation by Trigonometric Polynomials). *The finite linear span of the trigonometric system, $\mathrm{span}(\mathcal{T})$, is dense in $C_{\mathrm{per}}[-\pi, \pi]$ with respect to the uniform norm.*

Proof. We would like to apply the Stone–Weierstrass Theorem, but there are two difficulties. First, $C_{\mathrm{per}}[-\pi, \pi]$ is not a space of the form $C(K)$ where K is a compact metric space. Instead, $C_{\mathrm{per}}[-\pi, \pi]$ is a proper subspace of $C[-\pi, \pi]$. Second, although $\mathrm{span}(\mathcal{T})$ satisfies hypotheses (a)–(c) of Theorem 4.7.2, it does not separate points, because every function $f \in \mathrm{span}(\mathcal{T})$ satisfies $f(-\pi) = f(\pi)$.

To circumvent this difficulty we translate our problem to a new, "isomorphic" setting in which these two particular issues do not arise. Let

$$S^1 = \{(x, y) \in \mathbb{R}^2 : x^2 + y^2 = 1\}$$

be the unit circle in $\mathbb{R}^2$, and note that S^1 is compact. Each point in S^1 has the form $(\cos t, \sin t)$ for some $t \in \mathbb{R}$, and values of t differing by an integer multiple of 2π correspond to the same element of S^1. Therefore, given $n \in \mathbb{N}$ we can define a continuous function $c_n \colon S^1 \to \mathbb{R}$ by

$$c_n((\cos t, \sin t)) = \cos nt, \qquad t \in \mathbb{R}.$$

Similarly,

$$s_n((\cos t, \sin t)) = \sin nt, \qquad t \in \mathbb{R},$$

is continuous on S^1. Let 1 denote the constant function on S^1, and set

$$\mathcal{A} = \operatorname{span}\{1, c_n, s_n\}_{n \in \mathbb{N}}. \tag{4.41}$$

According to Problem 4.7.8, $\mathcal{A}$ satisfies all of the hypotheses of the Stone–Weierstrass Theorem when we take $X = K = S^1$. Consequently, $\mathcal{A}$ is dense in $C(S^1)$.

Now we translate these facts from $C(S^1)$ back to $C_{\text{per}}[-\pi, \pi]$. Choose f in $C_{\text{per}}[-\pi, \pi]$ and fix $\varepsilon > 0$. Define $F \colon S^1 \to \mathbb{R}$ by

$$F((\cos t, \sin t)) = f(t), \qquad t \in [-\pi, \pi].$$

Because f is continuous and $f(-\pi) = f(\pi)$, the function F is well-defined and continuous on S^1. Hence $F \in C(S^1)$, so there is some $G \in \mathcal{A}$ such that

$$\|F - G\|_{\mathrm{u}} = \sup_{z \in S^1} |F(z) - G(z)| < \varepsilon.$$

Every element of $\mathcal{A}$ is a finite linear combination of the functions 1, c_n, and s_n. Therefore

$$G = a_0 + \sum_{n=1}^{N} a_n c_n + \sum_{n=1}^{N} b_n s_n$$

for some integer N and real scalars a_n, b_n. Define $g \colon [-\pi, \pi] \to \mathbb{R}$ by

$$g(t) = G((\cos t, \sin t)) = a_0 + \sum_{n=1}^{N} a_n \cos nt + \sum_{n=1}^{N} b_n \sin nt,$$

and observe that $g \in \operatorname{span}(\mathcal{T})$. Since

$$
\begin{aligned}
\|f - g\|_{\mathrm{u}} &= \sup_{t \in [-\pi, \pi]} |f(t) - g(t)| \\
&= \sup_{t \in [-\pi, \pi]} |F((\cos t, \sin t)) - G((\cos t, \sin t))| \\
&= \sup_{z \in S^1} |F(z) - G(z)| \\
&< \varepsilon,
\end{aligned}
$$

we see that span($\mathcal{T}$) is dense in $C_{\text{per}}[-\pi, \pi]$. $\square$

Using the terminology of Definition 4.4.4, an equivalent wording of Theorem 4.7.3 is that the trigonometric system $\mathcal{T}$ is a *complete sequence* in $C_{\text{per}}[-\pi, \pi]$.

Problems

4.7.4. For this problem we take $\mathbb{F} = \mathbb{R}$.

(a) A function f on $[a, b]$ is *piecewise linear* if there exist finitely many points $a = x_0 < x_1 < \cdots < x_{n-1} < x_n = b$ such that the graph of f is linear on each interval $[x_{j-1}, x_j]$. Let $\mathcal{F}$ be the set of continuous, real-valued, piecewise linear functions on $[a, b]$, and prove that $\mathcal{F}$ satisfies the hypotheses of Theorem 4.7.1. Conclude that $\mathcal{F}$ is dense in $C[a, b]$.

(b) Let $\mathcal{P}$ be the set of all real-valued polynomials on $[a, b]$. Show that $\mathcal{P}$ does not satisfy the hypotheses of Theorem 4.7.1. Does this contradict the Weierstrass Approximation Theorem?

4.7.5. For this problem we take $\mathbb{F} = \mathbb{R}$.

(a) Prove that $\mathcal{A} = \text{span}\{e^{nt}\}_{n \geq 0}$ is dense in $C[0, 1]$.

(b) Suppose that a function $f \in C[0, 1]$ satisfies $\int_0^1 f(t)\, e^{nt}\, dt = 0$ for every $n \geq 0$. Prove that $f = 0$.

4.7.6. For this problem we take $\mathbb{F} = \mathbb{R}$.

(a) Given a real-valued function $f \in C[0, \pi]$, use the Stone–Weierstrass Theorem directly (i.e., without appealing to Theorem 4.7.3) to prove that there exist "polynomials in $\cos t$" that converge uniformly to f. That is, show that each f in $C[0, \pi]$ is the uniform limit of functions of the form $\sum_{k=0}^{n} c_k \cos^k t$.

(b) Use trigonometric identities to prove that for each $n \in \mathbb{N}$ there exist real numbers $c_0, c_1, \ldots, c_n$ such that

$$\cos^n t = \sum_{k=0}^{n} c_k \cos kt.$$

(c) Use parts (a) and (b) to prove that $\text{span}\{\cos nt\}_{n \geq 0}$ is dense in $C[0, \pi]$, i.e., each f in $C[0, \pi]$ is the uniform limit of functions of the form $\sum_{k=0}^{n} c_k \cos kt$.

Note: It is *not* true that every $f \in C[0, \pi]$ can be written as $f(t) = \sum_{k=0}^{\infty} c_k \cos kt$ for some choice of scalars c_k, with uniform (or even just pointwise) convergence of the series. Why does this not contradict statement (c)?

4.7.7. We take $\mathbb{F} = \mathbb{R}$ in this problem. Prove the following statements about the space $C_{\mathrm{per}}[-\pi, \pi]$ defined in equation (4.40).

(a) $C_{\mathrm{per}}[-\pi, \pi]$ is a closed, proper subspace of $C[-\pi, \pi]$.

(b) There exists a function $w \in C[-\pi, \pi]$ with $w(-\pi) \neq w(\pi)$ such that every $f \in C[-\pi, \pi]$ can be written as $f = g + cw$ for some $g \in C_{\mathrm{per}}[-\pi, \pi]$ and $c \in \mathbb{R}$.

4.7.8. For this problem we take $\mathbb{F} = \mathbb{R}$. Prove that the set $\mathcal{A}$ defined in equation (4.41) satisfies hypotheses (a)–(d) of the Stone–Weierstrass Theorem (using $X = K = S^1$).

Hint: Trigonometric identities, e.g., $2 \cos x \cos y = \cos(x - y) + \cos(x + y)$.

4.7.9. For this problem we take $\mathbb{F} = \mathbb{R}$.

(a) Suppose that $\mathcal{A}$ is a subset of $C[a, b]$ that satisfies hypotheses (a)–(d) of the Stone–Weierstrass Theorem. Prove that every function $F(x, y)$ in $C([a, b]^2)$ can be written as the uniform limit of functions of the form

$$f_1(x)\, g_1(y) + \cdots + f_n(x)\, g_n(y),$$

where f_k, g_k belong to $\mathcal{A}$.

(b) Can every $F \in C([a, b]^2)$ be written as the uniform limit of functions of the form $f(x)\, g(y)$, where $f, g \in \mathcal{A}$?

4.7.10. We take $\mathbb{F} = \mathbb{C}$ in this problem. Let K be a compact subset of a metric space X, and assume that $\mathcal{A}$ is a collection of continuous functions $f : K \to \mathbb{C}$ that satisfies the following five conditions.

(a) The constant function 1 belongs to $\mathcal{A}$.

(b) $\mathcal{A}$ is a *subspace* of $C(K)$, i.e., if $f, g \in \mathcal{A}$ and $a, b \in \mathbb{C}$, then $af + bg \in \mathcal{A}$.

(c) $\mathcal{A}$ is an *algebra*, i.e., it is a subspace *and* if $f, g \in \mathcal{A}$, then $fg \in \mathcal{A}$.

(d) $\mathcal{A}$ *separates points*, i.e., if $x, y \in K$ and $x \neq y$, then there is some $f \in \mathcal{A}$ such that $f(x) \neq f(y)$.

(e) $\mathcal{A}$ is *self-adjoint*, i.e., if $f \in \mathcal{A}$, then its complex conjugate $\overline{f}$, defined by $\overline{f}(x) = \overline{f(x)}$ for $x \in K$, belongs to $\mathcal{A}$.

Let $C(K)$ be the set of all continuous, complex-valued functions $f : K \to \mathbb{C}$, and prove that $\mathcal{A}$ is dense in $C(K)$.

Hint: Let $\mathcal{A}_{\mathbb{R}}$ be the set of real-valued functions that belong $\mathcal{A}$. Use Theorem 4.7.2 to prove that every real-valued continuous function $g : K \to \mathbb{R}$ is the uniform limit of functions $g_n \in \mathcal{A}_{\mathbb{R}}$. Given a complex-valued continuous function $f : K \to \mathbb{C}$, write $f = g + ih$, where g and h are real-valued, and note that $f + \overline{f} = 2g$.

4.7.11. We take $\mathbb{F} = \mathbb{C}$ in this problem. Let $\mathbb{T} = \{z \in \mathbb{C} : |z| = 1\}$ be the unit circle in the complex plane, and observe that $\mathbb{T}$ is a compact subset of $\mathbb{C}$. Prove the following statements.

(a) The function z^n is continuous on $\mathbb{T}$ for each integer $n \in \mathbb{Z}$.

(b) The collection

$$\mathcal{A} = \text{span}\{z^n\}_{n \in \mathbb{Z}} = \left\{ \sum_{n=-N}^{N} c_n z^n : N \in \mathbb{N}, c_n \in \mathbb{C} \right\}$$

satisfies hypotheses (a)–(e) of Problem 4.7.10.

(c) $\mathcal{A}$ is dense in $C(\mathbb{T})$.

4.8 The Tietze Extension Theorem

In this section we will prove the Tietze Extension Theorem, which is a generalization of Urysohn's Lemma (Theorem 2.10.2). Tietze's Theorem states that if F is a closed set and $f \colon F \to \mathbb{R}$ is continuous, then there exists a continuous function $g \colon X \to \mathbb{R}$ that equals f on the set F. Thus every real-valued function that is continuous on a closed subset of X can be extended to a function that is continuous on all of X (and by applying the theorem separately to the real and imaginary parts of a function, the theorem also holds for complex-valued functions).

The first step in the proof is given in the following lemma, which constructs a continuous function on X that is related to f on F in a certain way.

Lemma 4.8.1. *Let F be a closed subset of a metric space X. If $f \colon F \to \mathbb{R}$ is continuous and $|f(x)| \le M$ for all $x \in F$, then there exists a continuous function $g \colon X \to \mathbb{R}$ such that*

(a) $|f(x) - g(x)| \le \frac{2}{3}M$ *for $x \in F$,*

(b) $|g(x)| \le \frac{1}{3}M$ *for $x \in F$, and*

(c) $|g(x)| < \frac{1}{3}M$ *for $x \notin F$.*

Proof. Since f is real-valued and continuous on the closed set F, the sets

$$A = \left\{ x \in F : f(x) \le -\tfrac{1}{3}M \right\} \quad \text{and} \quad B = \left\{ x \in F : f(x) \ge \tfrac{1}{3}M \right\}$$

are closed subsets of X (why?). Further, they are disjoint by construction. The reader should now show (this is Problem 4.8.3) that in each of the following cases the given function g has the required properties.

Case 1: $A = B = \varnothing$, $g = 0$.

$$Case\ 2\colon A \neq \varnothing,\ B \neq \varnothing,\ g(x)\ =\ \frac{M}{3}\,\frac{\operatorname{dist}(x, A) - \operatorname{dist}(x, B)}{\operatorname{dist}(x, A) + \operatorname{dist}(x, B)}.$$

$$Case\ 3\colon A = \varnothing,\ B \neq \varnothing,\ g(x)\ =\ \frac{M}{3}\,\frac{\operatorname{dist}(x, B)}{1 + \operatorname{dist}(x, B)}.$$

$$Case\ 4\colon A \neq \varnothing,\ B = \varnothing,\ g(x)\ =\ \frac{M}{3}\,\frac{\operatorname{dist}(x, A)}{1 + \operatorname{dist}(x, A)}.\quad \square$$

Now we prove the Tietze Extension Theorem.

Theorem 4.8.2 (Tietze Extension Theorem). *If F is a closed subset of a metric space X and $f\colon F \to \mathbb{R}$ is continuous, then the following statements hold.*

(a) *There exists a continuous function $g\colon X \to \mathbb{R}$ such that $g = f$ on F.*

(b) *If $|f(x)| \leq M$ for all $x \in F$, then we can choose the function g in statement (a) so that $|g(x)| \leq M$ for all x and $|g(x)| < M$ for $x \notin F$.*

Proof. (b) We will prove this statement first, i.e., we assume that f is continuous on F and $|f(x)| \leq M$ on F. For integers $n = 0, 1, 2, \ldots$ we will inductively define a function g_n that is continuous on X and satisfies

$$\sup_{x \in F} \left| f(x) - \sum_{k=0}^{n} g_k(x) \right| \leq \left(\tfrac{2}{3}\right)^n M. \tag{4.42}$$

For $n = 0$, equation (4.42) holds if we simply take $g_0 = 0$. Assume that $m \geq 0$ and $g_0, \ldots, g_m$ are continuous functions on X that satisfy equation (4.42) for $n = 0, \ldots, m$. Set

$$h_m\ =\ f\ -\ \sum_{k=0}^{m} g_k.$$

This function is continuous on F, and $|h_m(x)| \leq (2/3)^m M$ for all $x \in F$. Applying Lemma 4.8.1 to h_m, we obtain a continuous function g_{m+1} that satisfies

$$|h_m(x) - g_{m+1}(x)|\ \leq\ \tfrac{2}{3}\left(\tfrac{2}{3}\right)^m M, \qquad x \in F, \tag{4.43}$$

$$|g_{m+1}(x)|\ \leq\ \tfrac{1}{3}\left(\tfrac{2}{3}\right)^m M, \qquad x \in F, \tag{4.44}$$

$$|g_{m+1}(x)|\ <\ \tfrac{1}{3}\left(\tfrac{2}{3}\right)^m M, \qquad x \notin F. \tag{4.45}$$

In particular, equation (4.43) implies that

$$\sup_{x \in F} \left| f(x) - \sum_{k=0}^{m+1} g_k(x) \right|\ =\ \sup_{x \in F} |h_m(x) - g_{m+1}(x)|\ \leq\ \left(\tfrac{2}{3}\right)^{m+1} M.$$

This completes the inductive step in the definition of $g_0, g_1, \ldots$.

Each function $g_0, g_1, \ldots$ belongs to $C_b(X)$, and equations (4.44) and (4.45) imply that

$$\sum_{k=0}^{\infty} \|g_k\|_u = \sum_{k=1}^{\infty} \sup_{x \in X} |g_k(x)| \le \sum_{k=1}^{\infty} \frac{1}{3} \left(\frac{2}{3}\right)^{k-1} M < \infty.$$

Hence the series

$$g = \sum_{k=0}^{\infty} g_k.$$

converges absolutely in $C_b(X)$ with respect to the uniform norm. In a Banach space, every absolutely convergent series is actually a convergent series (see Theorem 4.2.6). Therefore the series defining g converges uniformly and g belongs to $C_b(X)$. In fact, using equation (4.44) we compute that if $x \in F$, then

$$|g(x)| \le \sum_{k=0}^{\infty} |g_k(x)| \le \sum_{k=1}^{\infty} \frac{1}{3} \left(\frac{2}{3}\right)^{k-1} M = M.$$

If $x \notin F$, then a similar calculation based on equation (4.45) shows that $|g(x)| < M$. Finally, taking the limit as $n \to \infty$ in equation (4.42), we obtain $g(x) = f(x)$ for all $x \in F$.

(a) Now let f be an arbitrary continuous function on F. Set $h(x) = \arctan x$ and consider the composition $h \circ f$. This function is bounded and continuous on F, so by applying statement (b) to $h \circ f$ we conclude that there is a continuous function G that coincides with $h \circ f$ on F. But then the function $g(x) = (h^{-1} \circ G)(x) = \tan(G(x))$ is continuous on X and coincides with f on F. $\square$

Problems

4.8.3. Complete the proof of Lemma 4.8.1.

4.8.4. Show by example that the assumption in the Tietze Extension Theorem that F is closed is necessary.

4.9 The Arzelà–Ascoli Theorem

If K is a compact metric space, then every continuous scalar-valued function on K is bounded, and the space $C(K)$ of all continuous functions $f \colon K \to \mathbb{F}$ is a Banach space with respect to the uniform norm

$$\|f\|_u = \sup_{x \in K} |f(x)|.$$

The *Arzelà–Ascoli Theorem* characterizes the subsets of $C(K)$ that have compact closures (assuming that K is compact). We will prove this theorem in this section. First, we introduce some terminology.

Definition 4.9.1 (Equicontinuity and Pointwise Boundedness). Let X be a metric space, and let $\mathcal{F}$ be a subset of $C(X)$.

(a) We say that the family $\mathcal{F}$ is *equicontinuous at a point* $x \in X$ if for every $\varepsilon > 0$ there is a $\delta > 0$ such that

$$d(x, y) < \delta \quad \Longrightarrow \quad |f(x) - f(y)| < \varepsilon \text{ for every } f \in \mathcal{F}.$$

If $\mathcal{F}$ is equicontinuous at every point $x \in X$, then we simply say that $\mathcal{F}$ *is equicontinuous.*

(b) The family $\mathcal{F}$ is *pointwise bounded* if for each individual point $x \in X$ we have

$$\sup_{f \in \mathcal{F}} |f(x)| < \infty. \qquad \diamond \tag{4.46}$$

Equicontinuity requires a kind of uniformity in the behavior of the functions in the family $\mathcal{F}$. For a single function f to be continuous at a point x, for each $\varepsilon > 0$ there has to be a $\delta > 0$ such that $|f(x) - f(y)| < \varepsilon$ whenever $d(x, y) < \delta$. However, the number δ can depend on f (and also on ε and x). In order for $\mathcal{F}$ to be equicontinuous at x, there must be a single $\delta > 0$ such that if $d(x, y) < \delta$, then we simultaneously have $|f(x) - f(y)| < \varepsilon$ for every function $f \in \mathcal{F}$.

By definition, $\mathcal{F}$ *is a bounded subset of* $C(X)$ if

$$\sup_{f \in \mathcal{F}} \|f\|_{\mathrm{u}} = \sup_{f \in \mathcal{F}} \sup_{x \in X} |f(x)| < \infty. \tag{4.47}$$

To emphasize that we are dealing with the uniform norm, we usually say that $\mathcal{F}$ *is uniformly bounded* if equation (4.47) holds. If $\mathcal{F}$ is uniformly bounded, then equation (4.46) holds for each $x \in X$, so every uniformly bounded set is pointwise bounded. However, as we will demonstrate, not every pointwise bounded set is uniformly bounded.

Example 4.9.2. Set $X = [0, 1]$ and let $\mathcal{F} = \{f_n\}_{n \in \mathbb{N}}$ be the sequence of "Shrinking Triangles" from Example 3.5.2. Each f_n satisfies $\|f_n\|_{\mathrm{u}} = 1$, so $\mathcal{F}$ is a uniformly bounded subset of $C[0, 1]$. If we set $h_n(x) = n f_n(x)$, then $\mathcal{H} = \{h_n\}_{n \in \mathbb{N}}$ is not uniformly bounded. In fact, while the function h_n is identically zero outside of the interval $[0, \frac{1}{n}]$, its maximum value on $[0, \frac{1}{n}]$ is n, so $\|h_n\|_{\mathrm{u}} = n$. Even so, for any *particular* $x \in [0, 1]$ we have $h_n(x) \neq 0$ for at most *finitely many* n, and therefore

$$\sup_{n \in \mathbb{N}} |h_n(x)| < \infty.$$

Hence $\mathcal{H}$ is pointwise bounded, even though it is not uniformly bounded. $\diamond$

Now we prove the Arzelà–Ascoli Theorem (with some steps in the proof assigned as exercises).

Theorem 4.9.3 (Arzelà–Ascoli). *Let K be a compact subset of a metric space X. Given a collection $\mathcal{F}$ of functions in $C(K)$, the following two statements are equivalent.*

(a) *$\mathcal{F}$ is equicontinuous and pointwise bounded.*

(b) *$\overline{\mathcal{F}}$ is a compact subset of $C(K)$.*

Proof. (a) $\Rightarrow$ (b). Suppose that $\mathcal{F} \subseteq C(K)$ is both equicontinuous and pointwise bounded. In order to prove that $\mathcal{F}$ has compact closure, we will apply Problem 2.8.20, which tells us that it is sufficient to prove that $\mathcal{F}$ is totally bounded. So, fix any $\varepsilon > 0$. We must show that $\mathcal{F}$ can be covered by finitely many open balls of radius ε. Note that these are open balls in the space $C(K)$, and the norm on $C(K)$ is the uniform norm.

Choose any point $x \in K$. Since $\mathcal{F}$ is equicontinuous at x, there exists some $\delta_x > 0$ such that for every $f \in \mathcal{F}$ and $y \in K$ we have

$$d(x, y) < \delta_x \implies |f(x) - f(y)| < \frac{\varepsilon}{4}. \tag{4.48}$$

Since $\{B_{\delta_x}(x)\}_{x \in K}$ is an open cover of the compact set K, there must exist finitely many points $x_1, \ldots, x_N$ such that

$$K \subseteq \bigcup_{k=1}^{N} B_{\delta_{x_k}}(x_k).$$

For simplicity of notation, set $\delta_k = \delta_{x_k}$. Then it follows from equation (4.48) that for each $f \in \mathcal{F}$, each $k = 1, \ldots, N$, and every $y \in K$ we have

$$d(x_k, y) < \delta_k \implies |f(x_k) - f(y)| < \frac{\varepsilon}{4}. \tag{4.49}$$

The reader should show (this is Problem 4.9.6) that this implies that $\mathcal{F}$ is uniformly bounded, i.e., it is a bounded subset of $C(K)$ with respect to the uniform norm, and therefore

$$M = \sup_{f \in \mathcal{F}} \|f\|_{\mathrm{u}} = \sup_{f \in \mathcal{F}} \sup_{x \in K} |f(x)| < \infty.$$

Let $D = \{z \in \mathbb{C} : |z| \leq M\}$, i.e., D is the closed disk in the complex plane with radius M (if $\mathbb{F} = \mathbb{R}$ then it suffices to just let D be the closed interval $[-M, M]$). This is a compact subset of $\mathbb{C}$, so there exist finitely many points $z_1, \ldots, z_m \in \mathbb{C}$ such that each point in D lies less than a distance $\varepsilon/4$ from some z_j. Let $Z = \{z_1, \ldots, z_m\}$, and let Γ be the set of all ordered N-tuples of the points z_j, i.e.,

$$\Gamma = Z^N = \{z = (w_1, \ldots, w_N) \in \mathbb{C}^N : w_k \in Z, \ k = 1, \ldots, N\}.$$

Note that Γ is a finite set. For each $z \in \Gamma$, let $\mathcal{F}_z$ consist of all those functions $f \in \mathcal{F}$ such that for each $k = 1, \ldots, m$ and every $y \in K$ we have

$$d(x_k, y) < \delta_k \implies |z_k - f(y)| < \frac{\varepsilon}{2}.$$

Let

$$\Gamma_0 = \{z \in \Gamma : \mathcal{F}_z \neq \varnothing\}.$$

Suppose that $f \in \mathcal{F}$, and fix $1 \leq k \leq N$. Since $|f(x_k)| \leq M$, there is some point $z_{j_k} \in Z$ such that $|f(x_k) - z_{j_k}| < \varepsilon/4$. Combining this with equation (4.49), we see that if $y \in K$ satisfies $d(x_k, y) < \delta_k$, then

$$|z_{j_k} - f(y)| \leq |z_{j_k} - f(x_k)| + |f(x_k) - f(y)| < \frac{\varepsilon}{4} + \frac{\varepsilon}{4} = \frac{\varepsilon}{2}.$$

This is true for each $k = 1, \ldots, N$, so if we set $z = (z_{j_1}, \ldots, z_{j_N})$, then $f \in \mathcal{F}_z$. This shows that

$$\bigcup_{z \in \Gamma_0} \mathcal{F}_z = \mathcal{F}. \tag{4.50}$$

Now choose one function $f_z \in \mathcal{F}_z$ for each point $z \in \Gamma_0$. The reader should show (this is Problem 4.9.6) that

$$\mathcal{F}_z \subseteq B_\varepsilon(f_z) \qquad \text{for each } z \in \Gamma_0. \tag{4.51}$$

Combining equations (4.50) and (4.51), we see that $\mathcal{F}$ is totally bounded, and therefore its closure is compact by Problem 2.8.20.

(b) $\Rightarrow$ (a). Assume that $\mathcal{F}$ has compact closure. Then $\mathcal{F}$ must be a bounded subset of $C(K)$, which means that $\sup_{f \in \mathcal{F}} \|f\|_u < \infty$. Thus $\mathcal{F}$ is uniformly bounded, and consequently it is pointwise bounded.

Suppose that $\mathcal{F}$ is not equicontinuous at some point x. Then there exists some $\varepsilon > 0$ such that for each integer $n > 0$ there exists some function $f_n \in \mathcal{F}$ and some point $x_n \in K$ such that $d(x, x_n) < \frac{1}{n}$ but

$$|f_n(x) - f_n(x_n)| \geq \varepsilon. \tag{4.52}$$

Since $\overline{\mathcal{F}}$ is a compact subset of the Banach space $C(K)$, it is sequentially compact. Consequently, $\{f_n\}_{n \in \mathbb{N}}$ contains a uniformly convergent subsequence $\{f_{n_k}\}_{k \in \mathbb{N}}$. Let $f = \lim f_{n_k}$, where the convergence is with respect to the uniform norm. Since f is continuous and $x_{n_k} \to x$ in K, we have $f(x_{n_k}) \to f(x)$. Hence

$$|f_{n_k}(x) - f_{n_k}(x_{n_k})|$$
$$\leq |f_{n_k}(x) - f(x)| + |f(x) - f(x_{n_k})| + |f(x_{n_k}) - f_{n_k}(x_{n_k})|$$
$$\leq \|f_{n_k} - f\|_u + |f(x) - f(x_{n_k})| + \|f - f_{n_k}\|_u$$
$$\to 0 \quad \text{as } k \to \infty,$$

which contradicts equation (4.52). $\quad\square$

The following corollary gives one of the typical forms in which the Arzelà–Ascoli Theorem is applied in practice.

Corollary 4.9.4. *Let K be a compact subset of a metric space X. If $\mathcal{F}$ is a subset of $C(K)$ that is both equicontinuous and pointwise bounded, then every sequence $\{f_n\}_{n\in\mathbb{N}}$ of elements of $\mathcal{F}$ contains a uniformly convergent subsequence.*

Proof. By the Arzelà–Ascoli Theorem, we know that $\overline{\mathcal{F}}$ is a compact subset of $C(K)$. Since $C(K)$ is a Banach space, every compact subset of $C(K)$ is sequentially compact. Therefore any sequence $\{f_n\}_{n\in\mathbb{N}}$ that is contained in $\overline{\mathcal{F}}$ must have a subsequence that converges with respect to the norm of $C(K)$, i.e., it has a uniformly convergent subsequence. (The limit of this subsequence need not belong to $\mathcal{F}$, although it will belong to $\overline{\mathcal{F}}$.) $\quad\square$

Problems

4.9.5. Let K be a compact subset of a metric space X. Assume that f_n is continuous on K, and $f_n \to f$ uniformly on K. Prove that $\{f_n\}_{n\in\mathbb{N}}$ is both equicontinuous and pointwise bounded on K.

Remark: Problem 4.9.12 shows that there exist sequences that are both equicontinuous and pointwise bounded but do not converge uniformly.

4.9.6. Prove the two "reader should show" statements made in the proof of Theorem 4.9.3. Specifically, prove that:

(a) the family $\mathcal{F}$ is uniformly bounded, and

(b) if $z \in \Gamma_0$, then $\mathcal{F}_z \subseteq B_\varepsilon(f_z)$.

4.9.7. Assume that K is a compact subset of a metric space X, and $\{f_n\}_{n\in\mathbb{N}}$ is a sequence of functions that are equicontinuous on K and converge pointwise on K, i.e., $f(x) = \lim_{n\to\infty} f_n(x)$ exists for each $x \in K$. Prove that f_n converges uniformly to f on K.

4.9.8. Let I be an interval in $\mathbb{R}$. Suppose that $\mathcal{F} \subseteq C_b(I)$ is such that every function in $\mathcal{F}$ is Lipschitz continuous with the same Lipschitz constant, i.e., there is a single constant $K \geq 0$ such that

$$\forall f \in \mathcal{F}, \quad \forall x, y \in I, \quad |f(x) - f(y)| \leq K\,|x - y|.$$

Prove that $\mathcal{F}$ is equicontinuous. Must $\mathcal{F}$ be pointwise bounded?

4.9.9. Fix $K \geq 0$, and let $\mathcal{F}$ be the set of all functions on $[a, b]$ that are Lipschitz continuous with constant K and satisfy $f(a) = 0$:

$$\mathcal{F} = \{f \in C[a, b] : f(a) = 0 \text{ and } |f(x) - f(y)| \leq K\,|x - y|, \text{ all } x, y \in [a, b]\}.$$

Prove that $\mathcal{F}$ is a compact subset of $C[a, b]$.

4.9.10. Suppose that $\{f_n\}_{n\in\mathbb{N}}$ is a uniformly bounded sequence in $C[a,b]$. Set

$$F_n(x) = \int_a^x f_n(t)\,dt, \qquad x \in [a,b],$$

and prove that there exists a subsequence $\{F_{n_k}\}_{k\in\mathbb{N}}$ that converges uniformly on $[a,b]$.

4.9.11. Let X be a compact metric space, and suppose that $\mathcal{F} \subseteq C(X)$ is equicontinuous. Prove that $\mathcal{F}$ is *uniformly equicontinuous*, i.e., for each $\varepsilon > 0$ there exists a $\delta > 0$ such that if x and y are any two points of X such that $d(x,y) < \delta$, then $|f(x) - f(y)| < \varepsilon$ for every $f \in \mathcal{F}$.

4.9.12. This problem will show that the hypothesis in the Arzelà–Ascoli Theorem that X is compact is necessary.

Let $f_n(x) = \sin(x/n)$ for $x \in [0, \infty)$, and let $\mathcal{F} = \{f_n\}_{n\in\mathbb{N}}$. Prove the following statements.

(a) $\mathcal{F}$ is an equicontinuous and pointwise bounded subset of $C_b[0,\infty)$; in fact, $\mathcal{F}$ is uniformly bounded.

(b) $f_n \to 0$ pointwise on $[0,\infty)$.

(c) There is no subsequence of $\mathcal{F}$ that converges uniformly.

(d) $\mathcal{F}$ is not contained in any compact subset of $C_b[0,\infty)$.

4.9.13. Repeat Problem 4.9.12 using $f_n(x) = \sin\sqrt{x + 4\pi^2 n^2}$ for $x \in [0,\infty)$.

4.9.14. Recall from Problem 4.2.15 that $\mathrm{Lip}[0,1]$, the space of functions that are Lipschitz continuous on $[0,1]$, is a Banach space with respect to the norm

$$\|f\|_{\mathrm{Lip}} = \|f\|_u + \sup_{x\neq y} \frac{|f(x) - f(y)|}{|x - y|}.$$

Let A and B be the closed unit balls in $C[0,1]$ and $\mathrm{Lip}[0,1]$, respectively. That is,

$$A = \{f \in C[0,1] : \|f\|_u \leq 1\},$$

$$B = \{f \in \mathrm{Lip}[0,1] : \|f\|_{\mathrm{Lip}} \leq 1\}.$$

Prove the following statements.

(a) A is a closed and bounded subset of $C[0,1]$, but it is not compact (with respect to the norm $\|\cdot\|_u$).

(b) B is a closed and bounded subset of $\mathrm{Lip}[0,1]$, but it is not compact (with respect to the norm $\|\cdot\|_{\mathrm{Lip}}$).

(c) B is a compact subset of $C[0,1]$ (with respect to the norm $\|\cdot\|_u$).

Chapter 5
Inner Products and Hilbert Spaces

In a normed vector space each vector has an assigned *length*, and from this we obtain the *distance* from x to y as the length of the vector $x - y$. For vectors in $\mathbb{R}^d$ or $\mathbb{C}^d$ we also know how to measure the *angle* between vectors. In particular, we know how to determine when two vectors are *perpendicular*, or *orthogonal* (this happens precisely when their dot product is zero). In this chapter we will explore vector spaces that have an *inner product*. An inner product is a generalization of the dot product, and it provides us with a generalization of the notion of orthogonality to a much wider range of spaces than just $\mathbb{R}^d$ and $\mathbb{C}^d$.

In the first half of this chapter we define inner products, explore their properties, and give examples of inner products on various spaces. In particular, we see that each inner product induces a norm, so every vector space that has an inner product is a normed space. If the space is complete with respect to this norm, then we call it a *Hilbert space*. Thus a Hilbert space is a special type of Banach space, one whose norm is induced from an inner product. Not all norms are induced from inner products. For example, out of all the ℓ^p spaces, only ℓ^2 is a Hilbert space.

One of the most important advantages of an inner product is that we have a notion of orthogonality of vectors. We explore orthogonality in the second half of the chapter. Because of the existence of the inner product, we see that every subset of an inner product space has an *orthogonal complement*. Moreover, when our space is complete we can solve an important optimization problem for convex sets; specifically, we prove that if K is a closed, convex subset of a Hilbert space H, then for every point $x \in H$ there is a *unique closest point* to x in the set K. Specializing to the case that K is a closed subspace leads us to the idea of *orthogonal projections*. This allows us to construct *orthonormal bases* for separable Hilbert spaces. Important examples of orthonormal bases include the standard basis for ℓ^2 and the trigonometric system in the *Lebesgue space* $L^2[0, 1]$.

© Springer International Publishing AG, part of Springer Nature 2018
C. Heil, *Metrics, Norms, Inner Products, and Operator Theory*, Applied and Numerical Harmonic Analysis, https://doi.org/10.1007/978-3-319-65322-8_5

5.1 The Definition of an Inner Product

The inspiration for the definition of an inner product is the dot product on $\mathbb{R}^d$ and $\mathbb{C}^d$. Following our usual convention of letting $\mathbb{F}$ denote our choice of scalar field (either $\mathbb{R}$ or $\mathbb{C}$), we can write the dot product of two vectors $x, y \in \mathbb{F}^d$ as

$$x \cdot y = x_1 \overline{y_1} + \cdots + x_d \overline{y_d}. \tag{5.1}$$

If the scalar field is $\mathbb{F} = \mathbb{R}$, then the complex conjugate in the definition of the dot product is superfluous, since x_k and y_k are real in that case. However, by writing out the complex conjugates, equation (5.1) represents the dot product on $\mathbb{F}^d$ for both of the choices $\mathbb{F} = \mathbb{R}$ and $\mathbb{F} = \mathbb{C}$.

Here are four important properties of the dot product. These hold for all vectors $x, y, z \in \mathbb{F}^d$ and all scalars $a, b \in \mathbb{F}$.

- $x \cdot x \geq 0$.
- $x \cdot x = 0$ if and only if $x = 0$.
- $x \cdot y = \overline{y \cdot x}$.
- $(ax + by) \cdot z = a(x \cdot z) + b(y \cdot z)$.

An *inner product* is a function of two variables on a vector space that satisfies similar properties.

Definition 5.1.1 (Semi-inner Product, Inner Product). Let H be a vector space. A *semi-inner product* on H is a scalar-valued function $\langle \cdot, \cdot \rangle$ on $H \times H$ such that for all vectors $x, y, z \in H$ and all scalars $a, b \in \mathbb{F}$ we have:

(a) Nonnegativity: $\langle x, x \rangle \geq 0$,

(b) Conjugate Symmetry: $\langle x, y \rangle = \overline{\langle y, x \rangle}$, and

(c) Linearity in the First Variable: $\langle ax + by, z \rangle = a\langle x, z \rangle + b\langle y, z \rangle$.

If a semi-inner product $\langle \cdot, \cdot \rangle$ also satisfies:

(d) Uniqueness: $\langle x, x \rangle = 0$ if and only if $x = 0$,

then it is called an *inner product* on H. In this case, H is called an *inner product space* or a *pre-Hilbert space*. $\Diamond$

Thus, an inner product is a semi-inner product that also satisfies the uniqueness requirement. We will mostly be interested in inner products, but semi-inner products arise in certain situations of interest.

Here are some basic properties of a semi-inner product (the proof is assigned as Problem 5.1.3).

Lemma 5.1.2. *If $\langle \cdot, \cdot \rangle$ is a semi-inner product on a vector space H, then the following statements hold for all vectors $x, y, z \in H$ and all scalars $a, b \in \mathbb{F}$.*

(a) $\langle x, y \rangle + \langle y, x \rangle = 2\operatorname{Re}\langle x, y \rangle = 2\operatorname{Re}\langle y, x \rangle$.

(b) $\langle x, 0 \rangle = 0 = \langle 0, x \rangle$. $\Diamond$

If $\langle \cdot, \cdot \rangle$ is a semi-inner product on a vector space H, then we define

$$\|x\| = \langle x, x \rangle^{1/2}, \qquad x \in H.$$

We will prove in Lemma 5.2.4 that $\| \cdot \|$ is a seminorm on H, and therefore we refer to $\| \cdot \|$ as the *seminorm induced by* $\langle \cdot, \cdot \rangle$. Likewise, if $\langle \cdot, \cdot \rangle$ is an inner product, then we will prove that $\| \cdot \|$ is a norm on H, and in this case we refer to $\| \cdot \|$ as the *norm induced by* $\langle \cdot, \cdot \rangle$. It may be possible to place other norms on H, but unless we explicitly state otherwise, we assume that all norm-related statements on an inner product space are taken with respect to the induced norm.

Problems

5.1.3. Prove Lemma 5.1.2.

5.2 Properties of an Inner Product

The next lemma gives some properties of semi-inner products.

Lemma 5.2.1. *If $\langle \cdot, \cdot \rangle$ is a semi-inner product on a vector space H, then the following statements hold for all vectors x, y, $z \in H$ and all scalars $a, b \in \mathbb{F}$.*

(a) Antilinearity in the Second Variable: $\langle x, ay + bz \rangle = \overline{a} \langle x, y \rangle + \overline{b} \langle x, z \rangle$.

(b) Polar Identity: $\|x + y\|^2 = \|x\|^2 + 2 \operatorname{Re}\langle x, y \rangle + \|y\|^2$.

(c) Pythagorean Theorem: *If* $\langle x, y \rangle = 0$, *then* $\|x \pm y\|^2 = \|x\|^2 + \|y\|^2$.

(d) Parallelogram Law: $\|x + y\|^2 + \|x - y\|^2 = 2\big(\|x\|^2 + \|y\|^2\big)$.

Proof. (a) Given x, y, $z \in H$ and $a, b \in \mathbb{F}$, by applying conjugate symmetry and linearity in the first variable we see that

$$\langle x, ay + bz \rangle = \overline{\langle ay + bz, x \rangle} = \overline{a \langle y, x \rangle + b \langle z, x \rangle} = \overline{a} \langle x, y \rangle + \overline{b} \langle x, z \rangle.$$

(b) If x, $y \in H$, then by applying linearity in the first variable, antilinearity in the second variable, and conjugate symmetry, we obtain

$$\begin{aligned}
\|x + y\|^2 &= \langle x + y, x + y \rangle \\
&= \langle x, x \rangle + \langle x, y \rangle + \langle y, x \rangle + \langle y, y \rangle \\
&= \|x\|^2 + \langle x, y \rangle + \overline{\langle x, y \rangle} + \|y\|^2 \\
&= \|x\|^2 + 2 \operatorname{Re}\langle x, y \rangle + \|y\|^2.
\end{aligned}$$

(c) This follows immediately from part (b).

(d) Given $x, y \in H$, we compute that

$$\|x + y\|^2 + \|x - y\|^2$$
$$= \|x\|^2 + \langle x, y \rangle + \langle y, x \rangle + \|y\|^2 + \|x\|^2 - \langle x, y \rangle - \langle y, x \rangle + \|y\|^2$$
$$= 2 \|x\|^2 + 2 \|y\|^2. \quad \square$$

If the scalar field is $\mathbb{F} = \mathbb{R}$, then antilinearity in the second variable reduces to $\langle x, ay + bz \rangle = a \langle x, y \rangle + b \langle x, z \rangle$ (linearity), and the Polar Identity becomes $\|x + y\|^2 = \|x\|^2 + 2 \langle x, y \rangle + \|y\|^2$.

Remark 5.2.2. A function of two variables that is linear in the first variable and antilinear in the second variable is called a *sesquilinear form* (the prefix "sesqui-" means "one and a half"). Hence a semi-inner product $\langle \cdot, \cdot \rangle$ is an example of a sesquilinear form. Sometimes an inner product is required to be antilinear in the first variable and linear in the second (this is common in the physics literature). There are many different standard notations for semi-inner products. While our preferred notation is $\langle x, y \rangle$, the notations $[x, y]$, (x, y), and $\langle x | y \rangle$ are also common. $\diamond$

The inequality that we prove next is known as the *Schwarz Inequality*, the *Cauchy–Schwarz Inequality*, the *Cauchy–Bunyakovski–Schwarz Inequality*, or simply the *CBS Inequality*.

Theorem 5.2.3 (Cauchy–Bunyakovski–Schwarz Inequality). *If $\langle \cdot, \cdot \rangle$ is a semi-inner product on a vector space H, then*

$$|\langle x, y \rangle| \leq \|x\| \, \|y\|, \qquad \text{for all } x, y, \in H.$$

Proof. If $x = 0$ or $y = 0$, then there is nothing to prove, so suppose that x and y are both nonzero.

Given any scalar $z \in \mathbb{F}$, there is a scalar α with modulus $|\alpha| = 1$ such that $\alpha z = |z|$ (if z is real, then we can take α to be ± 1, and if $z = re^{i\theta}$ is complex, then we can take $\alpha = e^{-i\theta}$; if $z \neq 0$, then α is unique). In particular, if we set $z = \langle x, y \rangle$, then there is a scalar with $|\alpha| = 1$ such that

$$\langle x, y \rangle = \alpha \, |\langle x, y \rangle|.$$

Multiplying both sides by $\overline{\alpha}$, we see that we also have $\overline{\alpha} \langle x, y \rangle = |\langle x, y \rangle|$.

For each $t \in \mathbb{R}$, using the Polar Identity and antilinearity in the second variable, we compute that

$$0 \leq \|x - \alpha t y\|^2 = \|x\|^2 - 2 \operatorname{Re}(\langle x, \alpha t y \rangle) + t^2 \|y\|^2$$
$$= \|x\|^2 - 2t \operatorname{Re}(\overline{\alpha} \langle x, y \rangle) + t^2 \|y\|^2$$

$$= \|x\|^2 - 2t \,|\langle x, y\rangle| + t^2 \,\|y\|^2$$
$$= at^2 + bt + c,$$

where $a = \|y\|^2$, $b = -2|\langle x, y\rangle|$, and $c = \|x\|^2$. This is a real-valued quadratic polynomial in the variable t. Since this polynomial is nonnegative, it can have at most one real root. This implies that the discriminant $b^2 - 4ac$ cannot be strictly positive. Hence

$$b^2 - 4ac = \left(-2\,|\langle x, y\rangle|\right)^2 - 4\,\|x\|^2\,\|y\|^2 \leq 0,$$

and the desired result follows by rearranging this inequality. □

By combining the Polar Identity with the Cauchy–Bunyakovski–Schwarz Inequality, we can now prove that the induced seminorm $\|\cdot\|$ satisfies the Triangle Inequality.

Lemma 5.2.4. *Let H be a vector space. If $\langle\cdot,\cdot\rangle$ is a semi-inner product on H then $\|\cdot\|$ is a seminorm on H, and if $\langle\cdot,\cdot\rangle$ is an inner product on H then $\|\cdot\|$ is a norm on H.*

Proof. The only property that is not obvious is the Triangle Inequality. To prove this, we compute that

$$
\begin{aligned}
\|x + y\|^2 &= \|x\|^2 + 2\operatorname{Re}\langle x, y\rangle + \|y\|^2 \quad &\text{(Polar Identity)} \\
&\leq \|x\|^2 + 2\,|\langle x, y\rangle| + \|y\|^2 \quad &(\operatorname{Re} z \leq |\operatorname{Re} z| \leq |z| \text{ for scalars } z) \\
&\leq \|x\|^2 + 2\,\|x\|\,\|y\| + \|y\|^2 \quad &\text{(CBS Inequality)} \\
&= \left(\|x\| + \|y\|\right)^2.
\end{aligned}
$$

The Triangle Inequality follows by taking square roots. □

In particular, every inner product space is a normed space, so all of the definitions and results that we obtained for metric spaces in Chapter 2 and for normed spaces in Chapter 3 apply to inner product spaces, using the induced norm $\|x\| = \langle x, x\rangle^{1/2}$ and the induced metric $d(x, y) = \|x - y\|$. It may be possible to place a norm on H other than the induced norm, but unless we explicitly state otherwise, all norm-related statements on an inner product space are taken with respect to the induced norm. For example, we formalize the definition of convergence in an inner product space as follows.

Definition 5.2.5. Let H be an inner product space. Given vectors x_n, $x \in H$, we say that x_n *converges to x in H*, and write $x_n \to x$, if

$$\lim_{n\to\infty} \|x - x_n\| = 0,$$

where $\|x\| = \langle x, x\rangle^{1/2}$ is the induced norm on H. ◇

If $x_1, \ldots, x_N$ are finitely many elements of an inner product space H and $y \in H$, then linearity in the first variable combined with an argument by induction shows that

$$\left\langle \sum_{n=1}^{N} x_n, y \right\rangle = \sum_{n=1}^{N} \langle x_n, y \rangle.$$

However, it is not immediately obvious whether an analogous result will hold for infinite series. This is because an infinite series is a *limit* of the partial sums of the series. Therefore we need to know how an inner product behaves under limits, and these facts are given in the following lemma (whose proof is Problem 5.2.9).

Lemma 5.2.6. *If H is an inner product space, then the following statements hold.*

(a) Continuity of the inner product: *If $x_n \to x$ and $y_n \to y$ in H, then $\langle x_n, y_n \rangle \to \langle x, y \rangle$.*

(b) *If the series $\sum_{n=1}^{\infty} x_n$ converges in H, then*

$$\left\langle \sum_{n=1}^{\infty} x_n, y \right\rangle = \sum_{n=1}^{\infty} \langle x_n, y \rangle, \qquad y \in H. \qquad \Diamond$$

Just as in a metric or normed space, it is important to know whether Cauchy sequences in an inner product space converge. We give the following name to those inner product spaces where this is the case.

Definition 5.2.7 (Hilbert Space). An inner product space H is called a *Hilbert space* if it is complete with respect to the induced norm. $\quad \Diamond$

Thus, an inner product space is a Hilbert space if and only if every Cauchy sequence in H converges to an element of H. Equivalently, a Hilbert space is an inner product space that is a Banach space with respect to the induced norm. For example, $\mathbb{F}^d$ is a Hilbert space with respect to the usual dot product given in equation (5.1). Other inner products on $\mathbb{F}^d$ are given in Problems 5.2.13 and 5.2.14. In the next section we will construct inner products on ℓ^2 and $C[a, b]$ and determine whether they are Hilbert spaces with respect to those inner products.

Problems

5.2.8. Given an inner product space H, prove the following statements.

(a) If $x, y \in H$ and $\langle x, z \rangle = \langle y, z \rangle$ for every $z \in H$, then $x = y$.

(b) If $x \in H$, then $\|x\| = \sup_{\|y\|=1} |\langle x, y \rangle|$.

5.2.9. Prove Lemma 5.2.6.

5.2.10. Let $\langle \cdot, \cdot \rangle$ be a semi-inner product on a vector space H. Show that equality holds in the Cauchy–Bunyakovski–Schwarz Inequality if and only if there exist scalars a, b, not both zero, such that $\|ax + by\| = 0$. In particular, if $\langle \cdot, \cdot \rangle$ is an inner product, then either $x = cy$ or $y = cx$, where c is a scalar.

5.2.11. Let H, K be Hilbert spaces. Show that the Cartesian product $H \times K$ is a Hilbert space with respect to the inner product

$$\langle (h_1, k_1), (h_2, k_2) \rangle = \langle h_1, h_2 \rangle_H + \langle k_1, k_2 \rangle_K, \qquad h_1, h_2 \in H, \ k_1, k_2 \in K.$$

5.2.12. Let H be a Hilbert space. Recall that the Pythagorean Theorem states that if $\langle x, y \rangle = 0$, then $\|x + y\|^2 = \|x\|^2 + \|y\|^2$.

(a) Prove that if H is a real Hilbert space ($\mathbb{F} = \mathbb{R}$), then the converse to the Pythagorean Theorem holds, i.e., if $\|x + y\|^2 = \|x\|^2 + \|y\|^2$ then $\langle x, y \rangle \neq 0$.

(b) Show by example that if H is a complex Hilbert space ($\mathbb{F} = \mathbb{C}$), then there can exist vectors $x, y \in H$ such that $\|x + y\|^2 = \|x\|^2 + \|y\|^2$ but $\langle x, y \rangle \neq 0$.

5.2.13. For this problem we take $\mathbb{F} = \mathbb{R}$. A $d \times d$ matrix A with real entries is *positive definite* if it is symmetric ($A = A^{\mathrm{T}}$) and $Ax \cdot x > 0$ for every nonzero vector $x \in \mathbb{R}^d$, where $x \cdot y$ denotes the dot product of vectors in $\mathbb{R}^d$. Prove the following statements.

(a) If S is an invertible $d \times d$ matrix and Λ is a diagonal matrix whose diagonal entries are all positive, then $A = S\Lambda S^{\mathrm{T}}$ is positive definite.

(b) If A is a positive definite $d \times d$ matrix, then

$$\langle x, y \rangle_A = Ax \cdot y, \qquad x, y \in \mathbb{R}^d,$$

defines an inner product on $\mathbb{R}^d$.

(c) If $\langle \cdot, \cdot \rangle$ is an inner product on $\mathbb{R}^d$, then there exists some positive definite $d \times d$ matrix A such that $\langle \cdot, \cdot \rangle = \langle \cdot, \cdot \rangle_A$.

5.2.14. A $d \times d$ matrix A with complex entries is *positive definite* if A is *Hermitian* (i.e., $A = A^{\mathrm{H}}$, where $A^{\mathrm{H}} = \overline{A^{\mathrm{T}}}$ is the *conjugate transpose* of A) and $Ax \cdot x > 0$ for all nonzero $x \in \mathbb{C}^d$. Adapt Problem 5.2.13 to the case $\mathbb{F} = \mathbb{C}$ (replace transposes with conjugate transposes).

Remark: Theorem 7.3.7 will show that if a $d \times d$ matrix satisfies $Ax \cdot x \in \mathbb{R}$ for every $x \in \mathbb{C}^d$, then we automatically have $A = A^{\mathrm{H}}$.

5.3 Examples

Recall from Section 3.2 that ℓ^2 denotes the space of all square-summable sequences of scalars. According to Theorem 3.4.3, ℓ^2 is a Banach space with respect to the norm

$$\|x\|_2 = \left(\sum_{k=1}^{\infty} |x_k|^2 \right)^{1/2}.$$

We will define an inner product on ℓ^2 and prove that it is a Hilbert space with respect to that inner product.

The dual index to $p = 2$ is $p' = 2$, so Hölder's Inequality for the case $p = 2$ tells us that if $x = (x_k)_{k \in \mathbb{N}}$ and $y = (y_k)_{k \in \mathbb{N}}$ belong to ℓ^2, then

$$\sum_{k=1}^{\infty} |x_k y_k| \leq \left(\sum_{k=1}^{\infty} |x_k|^2 \right)^{1/2} \left(\sum_{k=1}^{\infty} |y_k|^2 \right)^{1/2} = \|x\|_2 \|y\|_2 < \infty. \quad (5.2)$$

Consequently, we can define

$$\langle x, y \rangle = \sum_{k=1}^{\infty} x_k \overline{y_k}, \qquad x, y \in \ell^2, \quad (5.3)$$

because this series of scalars converges absolutely by equation (5.2). In particular, if $y = x$, then

$$\langle x, x \rangle = \sum_{k=1}^{\infty} x_k \overline{x_k} = \sum_{k=1}^{\infty} |x_k|^2 = \|x\|_2^2. \quad (5.4)$$

This is nonnegative and finite for each $x \in \ell^2$. Further, $\langle x, x \rangle = 0$ if and only if $\|x\|_2 = 0$, which happens if and only if $x = 0$. This establishes the nonnegativity and uniqueness requirements of an inner product. The conjugate symmetry and linearity in the first variable requirements likewise follow directly. Hence $\langle \cdot, \cdot \rangle$ is an inner product on ℓ^2.

By equation (5.4), the norm induced from the inner product on ℓ^2 defined in equation (5.3) is exactly the ℓ^2-norm $\| \cdot \|_2$ (which is commonly referred to as the *Euclidean norm*). We know that ℓ^2 is complete with respect to this norm (see Problem 3.4.11), so ℓ^2 is a Hilbert space with respect to this inner product.

In contrast, we show next that if $p \neq 2$, then there is no inner product on ℓ^p whose induced norm equals the ℓ^p-norm $\| \cdot \|_p$.

Example 5.3.1. Fix $1 \leq p < \infty$ with $p \neq 2$, and suppose that there were an inner product $\langle \cdot, \cdot \rangle$ on ℓ^p whose induced norm equals $\| \cdot \|_p$. That is, suppose that there were an inner product such that $\langle x, x \rangle = \|x\|_p^2$ for all $x \in \ell^p$. Then, by Lemma 5.2.1, the ℓ^p-norm would satisfy the Parallelogram Law. Yet if we let δ_1 and δ_2 be the first two standard basis vectors, then we have

$$\|\delta_1 + \delta_2\|_p^2 + \|\delta_1 - \delta_2\|_p^2 = (1+1)^{2/p} + (1+1)^{2/p} = 2 \cdot 2^{2/p} = 2^{1+(2/p)},$$

while

$$2 \left(\|\delta_1\|_p^2 + \|\delta_2\|_p^2 \right) = 2 \left(1^2 + 1^2 \right) = 4.$$

These are not equal, because $p \neq 2$. Thus the Parallelogram Law fails, which is a contradiction. Hence no such inner product can exist. The reader should check that a similar argument applies when $p = \infty$. □

There do exist inner products on ℓ^p when $p \neq 2$, but Example 5.3.1 shows that the norm induced from such an inner product can never be the "standard norm" $\| \cdot \|_p$.

Next we will construct an inner product on $C[a, b]$, where $[a, b]$ is a bounded closed interval. We know from Theorem 3.5.10 that $C[a, b]$ is a Banach space with respect to the uniform norm. However, the uniform norm does not satisfy the Parallelogram Law, so, just as in Example 5.3.1, it follows that there is no inner product on $C[a, b]$ whose induced norm is $\| \cdot \|_u$. Consequently, the norm induced from the inner product that we will define will not be the uniform norm, but something else.

All functions in $C[a, b]$ are bounded and Riemann integrable. If we choose $f, g \in C[a, b]$, then the product $f(t) \overline{g(t)}$ is continuous and Riemann integrable, so we can define

$$\langle f, g \rangle = \int_a^b f(t) \overline{g(t)} \, dt, \qquad f, g \in C[a, b]. \tag{5.5}$$

We call $\langle f, g \rangle$ the L^2-*inner product* of f and g. We can easily see that $\langle \cdot, \cdot \rangle$ satisfies the nonnegativity, conjugate symmetry, and linearity in the first variable requirements stated in Definition 5.1.1, and hence is at least a semi-inner product on $C[a, b]$. According to Problem 5.3.4, the uniqueness requirement also holds (continuity is important here!), so $\langle \cdot, \cdot \rangle$ is an inner product on $C[a, b]$. The norm induced from this inner product is

$$\|f\|_2 = \langle f, f \rangle^{1/2} = \left(\int_a^b |f(t)|^2 \, dt \right)^{1/2}, \qquad f \in C[a, b],$$

which is precisely the L^2-*norm* on $C[a, b]$ that we defined in Theorem 3.2.6. We proved in Lemma 3.4.7 that $C[a, b]$ is not complete with respect to this norm. Therefore $C[a, b]$ is an inner product space but not a Hilbert space with respect to the inner product defined in equation (5.5).

Remark 5.3.2. As we discussed just before Definition 3.4.8, if we use *Lebesgue integrals* instead of Riemann integrals, then we can define a larger *Lebesgue space* $L^2[a, b]$ that contains $C[a, b]$ and whose norm is an extension of the L^2-norm on $C[a, b]$. Using Lebesgue integrals,

$$\langle f, g \rangle = \int_a^b f(t) \overline{g(t)} \, dt, \qquad f, g \in L^2[a, b],$$

defines an inner product on $L^2[a, b]$, and $L^2[a, b]$ is a Hilbert space with respect to this inner product.

We refer to texts such as [Fol99], [SS05], [WZ77], and [Heil19] for precise definitions and discussion of the Lebesgue integral, the Hilbert space $L^2[a,b]$, and more general Lebesgue spaces $L^p(E)$.   ◇

Problems

5.3.3. (a) Given vectors x_n, x in an inner product space H, prove that $x_n \to x$ if and only if $\|x_n\| \to \|x\|$ and $\langle x_n, y \rangle \to \langle x, y \rangle$ for every $y \in H$.

(b) Show by example that part (a) can fail if we do not assume that $\|x_n\| \to \|x\|$.

5.3.4. (a) Prove that the function $\langle \cdot, \cdot \rangle$ defined in equation (5.5) is an inner product on $C[a,b]$. Where does continuity play a role in your proof?

(b) Exhibit a piecewise continuous but discontinuous function f that satisfies $\int_a^b |f(t)|^2 \, dt = 0$.

5.3.5. Prove the following statements.

(a) The uniform norm on $C[a,b]$ does not satisfy the Parallelogram Law, and there is no inner product on $C[a,b]$ whose induced norm equals $\| \cdot \|_u$.

(b) If $1 \le p \le \infty$ with $p \ne 2$, then the L^p-norm $\| \cdot \|_p$ does not satisfy the Parallelogram Law on $C[a,b]$, and there is no inner product on $C[a,b]$ whose induced norm equals $\| \cdot \|_p$.

5.3.6. Prove that the norms $\| \cdot \|_u$ and $\| \cdot \|_2$ on $C[a,b]$ are not *equivalent* in the sense of Definition 3.6.1. Even so, prove that uniform convergence implies L^2-norm convergence, i.e.,

$$\lim_{n \to \infty} \|f - f_n\|_u = 0 \quad \Longrightarrow \quad \lim_{n \to \infty} \|f - f_n\|_2 = 0.$$

5.3.7. Let ℓ_w^2 be the weighted ℓ^2 space defined in Problem 3.4.14. Define an inner product on ℓ_w^2, and show that ℓ_w^2 is a Hilbert space with respect to this inner product. What happens if we allow $w(k) = 0$ for some k?

5.3.8. Let $I = (a,b)$ be a finite open interval, and let $C_b^1(I)$ be the space introduced in Definition 3.5.11. Prove that

$$\langle f, g \rangle = \int_a^b f(t) \overline{g(t)} \, dt + \int_a^b f'(t) \overline{g'(t)} \, dt$$

is an inner product on $C_b^1(I)$. Is the norm induced from this inner product the same as the norm $\| \cdot \|_{C_b^1}$ defined in equation (3.23)?

5.4 Orthogonal and Orthonormal Sets

The existence of a notion of orthogonality gives inner product spaces a much more tractable structure than that of generic Banach spaces, and leads to many beautiful results that have natural, constructive proofs. We will derive some of the properties of inner product spaces related to orthogonality in this section.

We declare that two vectors are orthogonal if their inner product is zero (consequently, the zero vector is orthogonal to every other vector). We say that a collection of vectors is orthogonal if every pair of vectors from the collection is orthogonal. Often it is convenient to work with orthogonal unit vectors; we refer to these as orthonormal vectors. Here are the precise definitions (the "Kronecker delta" referred to in part (c) is defined by $\delta_{ij} = 1$ if $i = j$, and $\delta_{ij} = 0$ if $i \neq j$).

Definition 5.4.1 (Orthogonality). Let H be an inner product space, and let I be an arbitrary index set.

(a) Two vectors $x, y \in H$ are *orthogonal*, denoted by $x \perp y$, if $\langle x, y \rangle = 0$.

(b) A set of vectors $\{x_i\}_{i \in I}$ is *orthogonal* if $\langle x_i, x_j \rangle = 0$ whenever $i \neq j$.

(c) A set of vectors $\{x_i\}_{i \in I}$ is *orthonormal* if it is orthogonal and each vector x_i is a unit vector. Using the Kronecker delta notation, $\{x_i\}_{i \in I}$ is an orthonormal set if for all $i, j \in I$ we have

$$\langle x_i, x_j \rangle = \delta_{ij} = \begin{cases} 1, & i = j, \\ 0, & i \neq j. \end{cases} \qquad \diamondsuit$$

The zero vector may be an element of a set of orthogonal vectors. Any orthogonal set $\{x_i\}_{i \in I}$ of nonzero vectors can be rescaled to form an orthonormal set. Specifically, if $\{x_i\}_{i \in I}$ is orthogonal, $x_i \neq 0$ for every i, and we set

$$y_i = \frac{x_i}{\|x_i\|},$$

then $\{y_i\}_{i \in I}$ is an orthonormal set.

Most of the orthogonal or orthonormal sets that we will encounter in this chapter will be finite or countable collections of vectors. We refer to a countable orthogonal set $\{x_n\}_{n \in \mathbb{N}}$ as an *orthogonal sequence*, and if each x_n is also a unit vector, then we call it an *orthonormal sequence*.

Example 5.4.2. The sequence of standard basis vectors $\{\delta_n\}_{n \in \mathbb{N}}$ is an orthonormal sequence in ℓ^2. In fact, it is a *complete* orthonormal sequence, so, using the terminology we will introduce in Definition 5.8.2, we call it an *orthonormal basis* for ℓ^2. The sequence $\{\delta_{2n}\}_{n \in \mathbb{N}}$ is also an orthonormal sequence, but it is not complete in ℓ^2. $\diamondsuit$

Example 5.4.3. Consider the sequence of monomials $\mathcal{M} = \{t^k\}_{k=0}^{\infty}$ in the space $C[-1,1]$. The inner product of t^{2j} with t^{2k+1} is

$$\langle t^{2j}, t^{2k+1} \rangle = \int_{-1}^{1} t^{2j} \overline{t^{2k+1}} \, dt = \int_{-1}^{1} t^{2j+2k+1} \, dt,$$

and this is zero, since $t^{2j+2k+1}$ is an odd function on $[-1,1]$. Thus every even-degree monomial is orthogonal to every odd-degree monomial. However, $\mathcal{M}$ is not an orthogonal sequence. For example,

$$\langle t^{2j}, t^{2k} \rangle = \int_{-1}^{1} t^{2j} \overline{t^{2k}} \, dt = \int_{-1}^{1} t^{2j+2k} \, dt = \frac{2}{2j + 2k + 1} \neq 0. \qquad \Diamond$$

Although the sequence of monomials is not orthogonal, we did prove in Section 1.11 that it is linearly independent. Later we will show how the *Gram–Schmidt orthonormalization procedure* can be used to take any independent sequence and create a related sequence that is orthogonal and has the same span (see Section 5.9).

Problems

5.4.4. Extend the Pythagorean Theorem to finite orthogonal sets of vectors. That is, prove that if $x_1, \ldots, x_N \in H$ are orthogonal vectors in an inner product space H, then

$$\left\| \sum_{n=1}^{N} x_n \right\|^2 = \sum_{n=1}^{N} \|x_n\|^2.$$

Does the result still hold if we assume only that $\langle \cdot, \cdot \rangle$ is a semi-inner product?

5.4.5. Prove that any orthogonal sequence of nonzero vectors $\{x_n\}_{n \in \mathbb{N}}$ in an inner product space is finitely linearly independent (i.e., every finite subset is linearly independent).

5.4.6. Given a Hilbert space H, prove the following statements.

(a) The distance between two orthonormal vectors x, y is $\|x - y\| = \sqrt{2}$.

(b) If H is separable, then every orthogonal set in H is countable.

5.4.7. Let $\ell^2(\mathbb{R})$ consist of all sequences of scalars $x = (x_t)_{t \in \mathbb{R}}$ indexed by the real line such that *at most countably many of the x_t are nonzero and* $\sum_{t \in \mathbb{R}} |x_t|^2 < \infty$.

(a) Prove that $\ell^2(\mathbb{R})$ is a Hilbert space with respect to the inner product $\langle x, y \rangle = \sum_{t \in \mathbb{R}} x_t \overline{y_t}$.

(b) For each $t \in \mathbb{R}$, define $\delta_t = (\delta_{tu})_{u \in \mathbb{R}}$, where δ_{tu} is the Kronecker delta (1 if $t = u$, 0 otherwise). Show that $\{\delta_t\}_{t \in \mathbb{R}}$ is an (uncountable) orthonormal sequence in $\ell^2(\mathbb{R})$. Is $\ell^2(\mathbb{R})$ separable?

5.5 Orthogonal Complements

We defined what it means for vectors to be orthogonal, but sometimes we need to consider subsets or subspaces that are orthogonal. For example, if we consider $\mathbb{R}^3$, then we often say that the z-axis is orthogonal to the x-y plane in $\mathbb{R}^3$. What we mean by this statement is that every vector on the z-axis is orthogonal to every vector in the x-y plane. Here is a definition that extends this idea to arbitrary subsets of an inner product space.

Definition 5.5.1 (Orthogonal Subsets). Let H be an inner product space, and let A and B be subsets of H.

(a) We say that a vector $x \in H$ is *orthogonal to the set* A, and write $x \perp A$, if $x \perp y$ for every vector $y \in A$.

(b) We say that A and B are *orthogonal sets*, and write $A \perp B$, if $x \perp y$ for every $x \in A$ and $y \in B$. $\diamondsuit$

The *largest possible* set that is orthogonal to a given set A is called the *orthogonal complement* of A, defined precisely as follows.

Definition 5.5.2 (Orthogonal Complement). Let A be a subset of an inner product space H. The *orthogonal complement* of A is

$$A^\perp = \{x \in H : x \perp A\} = \{x \in H : \langle x, y \rangle = 0 \text{ for all } y \in A\}. \quad \diamondsuit$$

For example, although the x-axis in $\mathbb{R}^3$ is orthogonal to the z-axis, it is not the largest set that is orthogonal to the z-axis. The largest set that is orthogonal to the z-axis is the x-y plane, and this plane is the orthogonal complement of the z-axis in $\mathbb{R}^3$. The orthogonal complement $A^\perp$ contains *all* (not just *some*) vectors x in H that are orthogonal to all elements of A.

Here is another example.

Example 5.5.3. Let E be the set of all those vectors in ℓ^2 that have the form $x = (0, x_2, 0, x_4, \dots)$, and let O be the set of all vectors in ℓ^2 of the form $x = (x_1, 0, x_3, 0, \dots)$. The reader should check that E and O are each closed subspaces of ℓ^2. If $x \in E$, then $x \perp y$ for every $y \in O$. This shows that $E \subseteq O^\perp$. On the other hand, if $x \in O^\perp$, then x is orthogonal to every vector in O, including $\delta_1, \delta_3, \dots$. This implies that every odd component of x is zero, and so $x \in E$. Hence $O^\perp \subseteq E$, and therefore $O^\perp = E$. A similar argument shows that $E^\perp = O$. $\square$

Often the set A in Definition 5.5.2 is a subspace of H (as in the preceding example), but it does not have to be. The next lemma gives some basic properties of orthogonal complements (the proof is assigned as Problem 5.5.6). In particular, note that the orthogonal complement of any subset A is a *closed subspace* of H, even if A is not.

Lemma 5.5.4. *If A is a subset of an inner product space H, then the following statements hold.*

(a) $A^\perp$ *is a closed subspace of H.*

(b) $H^\perp = \{0\}$ *and* $\{0\}^\perp = H$.

(c) *If $A \subseteq B$, then $B^\perp \subseteq A^\perp$.*

(d) $A \subseteq (A^\perp)^\perp$. $\Diamond$

Later we will prove that *if M is a closed subspace of a Hilbert space, then* $(M^\perp)^\perp = M$ (see Lemma 5.6.5).

Problems

5.5.5. Find the orthogonal complements of the following subsets of ℓ^2.

(a) $\{0\}$.   (b) ℓ^2.   (c) $\{\delta_k\}_{k \in J}$, where $J \subseteq \mathbb{N}$.   (d) c_{00}.   (e) $\{\delta_k + \delta_{k+1}\}_{k \in \mathbb{N}}$.

5.5.6. Prove Lemma 5.5.4.

5.5.7. Suppose that $\{x_n\}_{n \in \mathbb{N}}$ is a sequence in a Hilbert space H, and $y \in H$ is orthogonal to x_n for every n. Prove the following statements.

(a) y is orthogonal to every vector in $\operatorname{span}\{x_n\}_{n \in \mathbb{N}}$.

(b) y is orthogonal to every vector in $\overline{\operatorname{span}}\{x_n\}_{n \in \mathbb{N}}$.

(c) $\left(\{x_n\}_{n \in \mathbb{N}}\right)^\perp = \left(\overline{\operatorname{span}}\{x_n\}_{n \in \mathbb{N}}\right)^\perp$.

5.6 Orthogonal Projections and the Closest Point Theorem

Given a point x and a set S, there need not be a point in S that is *closest* to x. For example, if in the real line we take $S = (0,1)$ and $x = 2$, then there is no point in the interval S that is closer to x than every other element of S. No matter which element of S we consider, there is always another element of S that is closer to x. Still, every point in S is at least 1 unit away from x, and there are elements of S whose distance from x is as close to 1 as we like. Hence we say that "the distance from x to S is 1." We extend this idea to arbitrary normed spaces as follows.

Definition 5.6.1 (Distance from a Point to a Set). Let X be a normed space. The *distance* from $x \in X$ to a subset $S \subseteq X$ is the infimum of all distances between x and points $y \in S$:

$$\text{dist}(x, S) = \inf_{y \in S} \|x - y\|.$$

If this infimum is achieved by some vector $y \in S$, then we say that y *is a point in S that is closest to x*. That is, $y \in S$ is closest to x if and only if $\|x - y\| \leq \|x - z\|$ for every $z \in S$. $\diamondsuit$

Finding a point that is closest to a given set is a type of *optimization problem* that arises in a wide variety of circumstances. Unfortunately, in general it can be difficult to compute the exact distance from a point x to a set S, or to determine whether there is a vector in S that is closest to x. Even if a closest point exists, it need not be unique. To illustrate this, consider $X = \mathbb{R}^2$, $S = [-1, 1]^2$, and $x = (2, 0)$. The distance from x to S depends on what norm we choose to place on $\mathbb{R}^2$. For example, if we place the ℓ^∞ norm on $\mathbb{R}^2$, then $\|x - y\|_\infty \geq 1$ for every $y \in S$, and each vector $z_t = (1, t)$ with $-1 \leq t \leq 1$ lying on the right-hand side of the square S satisfies $\|x - z_t\|_\infty = 1$ (see the illustration in Figure 5.1). Consequently, $\text{dist}(x, S) = 1$ and *every* point z_t on the right edge of S is a *closest* point to x. On the other hand, if we place the Euclidean norm on $\mathbb{R}^2$, then we can see from inspection that $y = (1, 0)$ is the unique point in the square S that is closest to x.

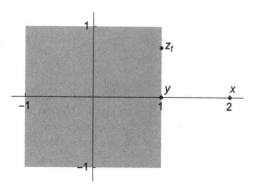

Fig. 5.1 With respect to the sup-norm, every point $z_t = (1, t)$ with $-1 \leq t \leq 1$ is a point in the square $S = [-1, 1]^2$ that is closest to $x = (2, 0)$. With respect to the Euclidean norm, the point $y = (1, 0)$ is the unique point in the square that is closest to $x = (2, 0)$.

We will prove that if S is a *closed and convex subset of a Hilbert space H*, then for each vector $x \in H$ there exists a *unique* vector $y \in S$ that is closest to x. Looking at the proof, we can see why we must assume in this result that H is a Hilbert space and not just an inner product space—we need to know that every Cauchy sequence in H will converge to an element of H.

Theorem 5.6.2 (Closest Point Theorem). *Let H be a Hilbert space, and let S be a nonempty closed, convex subset of H. Given any $x \in H$ there exists a unique vector $y \in S$ that is closest to x. That is, there is a unique vector $y \in S$ that satisfies*

$$\|x - y\| = \text{dist}(x, S) = \inf_{z \in S} \|x - z\|.$$

Proof. Set $d = \text{dist}(x, S)$. Then, by the definition of an infimum, there exist vectors $y_n \in S$ such that $\|x - y_n\| \to d$ as $n \to \infty$, and for each of these vectors we have $\|x - y_n\| \geq d$. Therefore, if we fix $\varepsilon > 0$, then we can find an integer $N > 0$ such that

$$d^2 \leq \|x - y_n\|^2 \leq d^2 + \varepsilon^2 \qquad \text{for all } n > N.$$

Set

$$p = \frac{y_m + y_n}{2}.$$

Since S is convex and p is the midpoint of the line segment joining y_m to y_n, we have $p \in S$, and therefore

$$\|x - p\| \geq \text{dist}(x, S) = d.$$

Using the Parallelogram Law, it follows that if $m, n > N$, then

$$
\begin{aligned}
\|y_n - y_m\|^2 + 4d^2 &\leq \|y_n - y_m\|^2 + 4\|x - p\|^2 \\
&= \|(x - y_n) - (x - y_m)\|^2 + \|(x - y_n) + (x - y_m)\|^2 \\
&= 2\left(\|x - y_n\|^2 + \|x - y_m\|^2\right) \qquad \text{(Parallelogram Law)} \\
&\leq 4\left(d^2 + \varepsilon^2\right).
\end{aligned}
$$

Rearranging, we see that $\|y_m - y_n\| \leq 2\varepsilon$ for all $m, n > N$. Therefore $\{y_n\}_{n \in \mathbb{N}}$ is a Cauchy sequence in H. Since H is complete, this sequence must converge, say to y. Since S is closed and $y_n \in S$ for every n, the vector y must belong to S. Also, since $x - y_n \to x - y$, it follows from the continuity of the norm (Lemma 5.2.6) that

$$\|x - y\| = \lim_{n \to \infty} \|x - y_n\| = d.$$

Hence y is a point in S that is closest to x.

It remains only to show that y is the *unique* point in S that is closest to x. If $z \in S$ is also a closest point, then $\|x - y\| = d = \|x - z\|$. The midpoint $p = (y + z)/2$ belongs to S, so $\|x - p\| \geq d$. Applying the Parallelogram Law again, we see that

$$
\begin{aligned}
4d^2 &= 2\left(\|x - y\|^2 + \|x - z\|^2\right) \\
&= \|(x - y) - (x - z)\|^2 + \|(x - y) + (x - z)\|^2
\end{aligned}
$$

$$= \|y - z\|^2 + 4\|x - p\|^2$$
$$\geq \|y - z\|^2 + 4d^2.$$

Consequently $\|y - z\| \leq 0$, which implies that $y = z$. $\square$

In particular, every closed subspace M of H is nonempty, closed, and convex. For this setting we introduce a special name for the point p in M that is closest to a given vector x. We also use the same name to denote the function that maps x to the point p in M that is closest to x.

Definition 5.6.3 (Orthogonal Projection). Let M be a closed subspace of a Hilbert space H.

(a) Given $x \in H$, the unique vector $p \in M$ that is closest to x is called the *orthogonal projection of x onto M.*

(b) The function $P: H \to H$ defined by $Px = p$, where p is the orthogonal projection of x onto M, is called the *orthogonal projection of H onto M.* ◇

Since the orthogonal projection p is the vector in M that is closest to x, we can think of p as being the best approximation to x by vectors from M. The difference vector $e = x - p$ is the error in this approximation (see Figure 5.2).

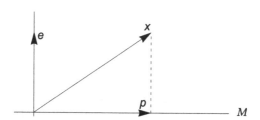

Fig. 5.2 The orthogonal projection of a vector x onto a subspace M.

The following lemma gives us several equivalent reformulations of the orthogonal projection. In particular, it states that the orthogonal projection of x onto M is the unique vector $p \in H$ such that the error vector $e = x - p$ is orthogonal to M.

Lemma 5.6.4. *Let M be a closed subspace of a Hilbert space H. Given vectors x and p in H, the following four statements are equivalent.*

(a) *p is the orthogonal projection of x onto M, i.e., p is the unique point in M that is closest to x.*

(b) *$p \in M$ and $x - p \perp M$.*

(c) *$x = p + e$, where $p \in M$ and $e \in M^\perp$.*

(d) *$e = x - p$ is the orthogonal projection of x onto $M^\perp$.*

Proof. We will prove one implication, and assign the task of proving the remaining (easier) implications as Problem 5.6.7.

(a) $\Rightarrow$ (b). Let p be the (unique) point in M closest to x, and let $e = x - p$. Choose any vector $y \in M$. We must show that $\langle y, e \rangle = 0$. Since M is a subspace, we have $p + \lambda y \in M$ for every scalar λ. But p is closer to x than $p + \lambda y$, so we compute that

$$
\begin{aligned}
\|e\|^2 \; = \; \|x - p\|^2 \; &\leq \; \|x - (p + \lambda y)\|^2 \\
&= \; \|e - \lambda y\|^2 \\
&= \; \|e\|^2 - 2\operatorname{Re}\langle \lambda y, e \rangle + |\lambda|^2 \, \|y\|^2.
\end{aligned}
$$

Rearranging, we obtain

$$
\forall \lambda \in \mathbb{F}, \quad 2\operatorname{Re}(\lambda \langle y, e \rangle) \; \leq \; |\lambda|^2 \, \|y\|^2. \tag{5.6}
$$

If we consider $\lambda = t > 0$ in equation (5.6), then we can divide through by t to get
$$
\forall t > 0, \quad 2\operatorname{Re}\langle y, e \rangle \; \leq \; t \, \|y\|^2.
$$

Letting $t \to 0^+$, we conclude that $\operatorname{Re}\langle y, e \rangle \leq 0$. If we similarly take $\lambda = t < 0$ and let $t \to 0^-$, we obtain $\operatorname{Re}\langle y, e \rangle \geq 0$, so $\operatorname{Re}\langle y, e \rangle = 0$. If $\mathbb{F} = \mathbb{R}$, then this implies that $\langle y, e \rangle = 0$, and so we are done. If $\mathbb{F} = \mathbb{C}$, then by taking $\lambda = it$ with $t > 0$ and then $\lambda = it$ with $t < 0$, we can show that $\operatorname{Im}\langle y, e \rangle = 0$, and it follows from this that $\langle y, e \rangle = 0$. $\quad\square$

As an application, we will use Lemma 5.6.4 to compute the orthogonal complement of the orthogonal complement of a set. Recall from Section 4.4 that the closed span of a set $A \subseteq H$ is the closure of $\operatorname{span}(A)$ in H. Equivalently, $\overline{\operatorname{span}}(A)$ is the set of all limits of elements of $\operatorname{span}(A)$, and by Problem 4.4.5 it is also the *smallest closed subspace* that contains the set A.

Lemma 5.6.5. *Let H be a Hilbert space.*
(a) *If M is a closed subspace of H, then $(M^\perp)^\perp = M$.*
(b) *If A is any subset of H, then*

$$
A^\perp \; = \; \operatorname{span}(A)^\perp \; = \; \overline{\operatorname{span}}(A)^\perp \quad \textit{and} \quad (A^\perp)^\perp \; = \; \overline{\operatorname{span}}(A).
$$

Proof. (a) We are given a closed subspace M. If $x \in M$, then $\langle x, y \rangle = 0$ for every $y \in M^\perp$, so $x \in (M^\perp)^\perp$. Hence $M \subseteq (M^\perp)^\perp$.

Conversely, suppose that $x \in (M^\perp)^\perp$, and let p be the orthogonal projection of x onto M. Since M is a closed subspace, we can write $x = p + e$, where $p \in M$ and $e \in M^\perp$. Note that $p \in M \subseteq (M^\perp)^\perp$. Since x also belongs to the subspace $(M^\perp)^\perp$, it follows that $e = x - p \in (M^\perp)^\perp$. However, we also know from Theorem 5.6.4 that the error vector belongs to $M^\perp$. Thus e belongs to both $M^\perp$ and $(M^\perp)^\perp$. Every vector in $M^\perp$ is orthogonal to every vector in $(M^\perp)^\perp$, so this implies that e is orthogonal to itself. Therefore

$\|e\|^2 = \langle e, e \rangle = 0$, which implies that $e = 0$. Hence $x = p + 0 \in M$, so we conclude that $(M^\perp)^\perp \subseteq M$.

(b) Now we are given an arbitrary subset A of H. Let $M = \overline{\text{span}}(A)$. We must show that $A^\perp = M^\perp$. Since $A \subseteq M$, the reader can check that the inclusion $M^\perp \subseteq A^\perp$ follows directly.

Suppose that $x \in A^\perp$. Then $x \perp A$, i.e., x is orthogonal to every vector in A. By forming linear combinations, it follows that $x \perp \text{span}(A)$, and then by taking limits, it follows from this that $x \perp \overline{\text{span}}(A) = M$ (the details of this argument are part of Problem 5.5.7). Therefore $x \in M^\perp$, which proves that $A^\perp \subseteq M^\perp$.

Thus, we have shown that $A^\perp = M^\perp$. Applying part (a), we conclude that $(A^\perp)^\perp = (M^\perp)^\perp = M$. $\square$

Specializing to the case that A is a sequence gives us the following useful result. Recall from Definition 4.4.4 that a sequence $\{x_n\}_{n \in \mathbb{N}}$ is *complete* if its closed span is all of H.

Corollary 5.6.6. *Given a sequence $\{x_n\}_{n \in \mathbb{N}}$ in a Hilbert space H, the following two statements are equivalent.*

(a) *$\{x_n\}_{n \in \mathbb{N}}$ is a complete sequence, i.e., $\text{span}\{x_n\}_{n \in \mathbb{N}}$ is dense in H.*

(b) *The only vector in H that is orthogonal to every x_n is the zero vector, i.e.,*

$$x \in H \text{ and } \langle x, x_n \rangle = 0 \text{ for every } n \quad \Longrightarrow \quad x = 0. \qquad (5.7)$$

Proof. Let $M = \overline{\text{span}}\{x_n\}_{n \in \mathbb{N}}$. By Lemma 5.6.5 we have $M^\perp = \left(\{x_n\}_{n \in \mathbb{N}}\right)^\perp$. Since $H^\perp = \{0\}$, it follows that

$$\begin{aligned}
\{x_n\}_{n \in \mathbb{N}} \text{ is complete} &\iff M = H \\
&\iff M^\perp = H^\perp \\
&\iff \left(\{x_n\}_{n \in \mathbb{N}}\right)^\perp = \{0\} \\
&\iff \text{equation (5.7) holds.} \qquad \square
\end{aligned}$$

The use of orthogonal complement notation makes our proof of Corollary 5.6.6 very concise, but it is instructive for the reader to write out a detailed direct proof.

Problems

5.6.7. Prove the remaining implications in Lemma 5.6.4.

5.6.8. For this problem we take $\mathbb{F} = \mathbb{R}$. Let S be the "closed first quadrant" in ℓ^2, i.e., S is the set of all sequences $y = (y_k)_{k \in \mathbb{N}}$ in ℓ^2 such that $y_k \geq 0$ for every k.

(a) Show that S is a closed, convex subset of ℓ^2.

(b) Given $x \in \ell^2$, find the sequence $y \in S$ that is closest to x.

5.6.9. Let M be a closed subspace of a Hilbert space H, and let P be the orthogonal projection of H onto M. Show that $I - P$ is the orthogonal projection of H onto $M^\perp$.

5.6.10. Let M and N be orthogonal closed subspaces of a Hilbert space H. We define the *orthogonal direct sum* of M and N to be

$$M \oplus N = \{x + y : x \in M, y \in N\}.$$

Prove that $M \oplus N$ is a closed subspace of H, and $M \oplus M^\perp = H$.

5.6.11. Let H be a Hilbert space, and let U, V be any two subspaces of H (not necessarily closed or orthogonal). The *sum* of U and V is

$$U + V = \{x + y : x \in U, y \in V\}.$$

Prove the following statements.

(a) $(U + V)^\perp = U^\perp \cap V^\perp$.

(b) $\overline{U^\perp + V^\perp} \subseteq (U \cap V)^\perp$.

(c) If U and V are closed subspaces, then $\overline{U^\perp + V^\perp} = (U \cap V)^\perp$.

5.7 Orthogonal and Orthonormal Sequences

Now we take a closer look at orthogonal sets, and especially countable orthogonal sequences. Any set of nonzero orthogonal vectors can be turned into an orthonormal set simply by dividing each vector in the set by its length. Therefore, for simplicity of presentation we will state our results for orthonormal sets (the reader consider what modifications are needed for sets that are orthogonal but not necessarily orthonormal).

If $\{e_n\}_{n \in \mathbb{N}}$ is a sequence of orthonormal vectors in a Hilbert space, then its closed span $M = \overline{\text{span}}\{e_n\}_{n \in \mathbb{N}}$ is a closed subspace of H. The next theorem gives an explicit formula for the orthogonal projection of a vector onto M, along with other useful facts. Entirely similar results hold for a finite orthonormal sequence $\{e_1, \ldots, e_d\}$ by replacing $n \in \mathbb{N}$ by $n = 1, \ldots, d$ in Theorem 5.7.1. In fact, the proof is easier for finite sequences, since in that case there are no infinite series and hence no issues about convergence (we assign this as Problem 5.7.2). In the statement of the following theorem we implicitly assume that c_n denotes a scalar.

Theorem 5.7.1. *Let H be a Hilbert space, let $\mathcal{E} = \{e_n\}_{n \in \mathbb{N}}$ be an orthonormal sequence in H, and set*

$$M = \overline{\operatorname{span}}(\mathcal{E}) = \overline{\operatorname{span}}\{e_n\}_{n\in\mathbb{N}}. \tag{5.8}$$

Then the following statements hold.

(a) Bessel's Inequality:

$$\sum_{n=1}^{\infty} |\langle x, e_n\rangle|^2 \le \|x\|^2, \qquad \text{for all } x \in H.$$

(b) *If the series* $x = \sum_{n=1}^{\infty} c_n e_n$ *converges (where the* c_n *are scalars), then* $c_n = \langle x, e_n\rangle$ *for each* $n \in \mathbb{N}$.

(c) *The following equivalence holds:*

$$\sum_{n=1}^{\infty} c_n e_n \text{ converges} \quad \Longleftrightarrow \quad \sum_{n=1}^{\infty} |c_n|^2 < \infty. \tag{5.9}$$

Furthermore, in this case the series $\sum_{n=1}^{\infty} c_n e_n$ *converges unconditionally, i.e., it converges regardless of the ordering of the index set.*

(d) *If* $x \in H$, *then*

$$p = \sum_{n=1}^{\infty} \langle x, e_n\rangle\, e_n$$

is the orthogonal projection of x *onto* M, *and* $\|p\|^2 = \sum_{n=1}^{\infty} |\langle x, e_n\rangle|^2$.

(e) *If* $x \in H$, *then*

$$x \in M \quad \Longleftrightarrow \quad x = \sum_{n=1}^{\infty} \langle x, e_n\rangle\, e_n \quad \Longleftrightarrow \quad \|x\|^2 = \sum_{n=1}^{\infty} |\langle x, e_n\rangle|^2. \tag{5.10}$$

Proof. (a) Choose $x \in H$. For each $N \in \mathbb{N}$ define

$$p_N = \sum_{n=1}^{N} \langle x, e_n\rangle\, e_n \qquad \text{and} \qquad q_N = x - p_N.$$

Since the e_n are orthonormal, the Pythagorean Theorem implies that

$$\|p_N\|^2 = \sum_{n=1}^{N} \|\langle x, e_n\rangle\, e_n\|^2 = \sum_{n=1}^{N} |\langle x, e_n\rangle|^2.$$

The vectors p_N and q_N are orthogonal, because

$$\langle p_N, q_N\rangle = \langle p_N, x\rangle - \langle p_N, p_N\rangle = \sum_{n=1}^{N} \langle x, e_n\rangle\, \langle e_n, x\rangle - \|p_N\|^2 = 0.$$

Consequently,

$$\|x\|^2 = \|p_N + q_N\|^2 = \|p_N\|^2 + \|q_N\|^2 \quad \text{(Pythagorean Theorem)}$$

$$\geq \|p_N\|^2 = \sum_{n=1}^{N} |\langle x, e_n \rangle|^2.$$

Letting $N \to \infty$, we obtain Bessel's Inequality.

(b) If $x = \sum c_n e_n$ converges, then for each fixed m we have

$$\langle x, e_m \rangle = \left\langle \sum_{n=1}^{\infty} c_n e_n, e_m \right\rangle = \sum_{n=1}^{\infty} c_n \langle e_n, e_m \rangle = \sum_{n=1}^{\infty} c_n \delta_{mn} = c_m,$$

where we have used Lemma 5.2.6(b) to move the infinite series outside the inner product.

(c) If $x = \sum c_n e_n$ converges, then $c_n = \langle x, e_n \rangle$ by part (b), and therefore $\sum |c_n|^2 < \infty$ by Bessel's Inequality.

For the converse direction, suppose that $\sum |c_n|^2 < \infty$. Set

$$s_N = \sum_{n=1}^{N} c_n e_n \quad \text{and} \quad t_N = \sum_{n=1}^{N} |c_n|^2.$$

We know that $\{t_N\}_{N \in \mathbb{N}}$ is a convergent (hence Cauchy) sequence of scalars, and we must show that $\{s_N\}_{N \in \mathbb{N}}$ is a convergent sequence of vectors. If $N > M$, then

$$\|s_N - s_M\|^2 = \left\| \sum_{n=M+1}^{N} c_n e_n \right\|^2$$

$$= \sum_{n=M+1}^{N} \|c_n e_n\|^2 \quad \text{(Pythagorean Theorem)}$$

$$= \sum_{n=M+1}^{N} |c_n|^2 = |t_N - t_M|.$$

Since $\{t_N\}_{N \in \mathbb{N}}$ is a Cauchy sequence of scalars, it follows that $\{s_N\}_{N \in \mathbb{N}}$ is a Cauchy sequence in H (why?). But H is complete (since it is a Hilbert space), so the Cauchy sequence $\{s_N\}_{N \in \mathbb{N}}$ must converge. Therefore, by the definition of an infinite series, $\sum c_n e_n$ converges.

We have shown that equation (5.9) holds. Now we must prove that if $\sum c_n e_n$ converges, then it converges unconditionally. Choose any bijection $\sigma : \mathbb{N} \to \mathbb{N}$. Theorem 4.3.3 tells us that every convergent series of nonnegative scalars converges unconditionally, and by combining this with the equivalence in equation (5.9) we see that

$$\sum_{n=1}^{\infty} c_n e_n \text{ converges} \iff \sum_{n=1}^{\infty} |c_n|^2 < \infty \qquad \text{(by equation (5.9))}$$

$$\iff \sum_{n=1}^{\infty} |c_\sigma(n)|^2 < \infty \qquad \text{(by Theorem 4.3.3)}$$

$$\iff \sum_{n=1}^{\infty} c_{\sigma(n)} e_{\sigma(n)} \text{ converges} \qquad \text{(by equation (5.9))}.$$

Thus, if $\sum c_n e_n$ converges, then so does any reordering of the series.

(d) By Bessel's Inequality and part (c), we know that the series

$$p = \sum_{n=1}^{\infty} \langle x, e_n \rangle e_n$$

converges. Our task is to show that this vector p is the orthogonal projection of x onto M. If we fix k, then, since $\langle e_n, e_k \rangle = 0$ for $n \neq k$ and $\langle e_k, e_k \rangle = 1$, we have

$$\langle x - p, e_k \rangle = \langle x, e_k \rangle - \left\langle \sum_{n=1}^{\infty} \langle x, e_n \rangle e_n, e_k \right\rangle$$

$$= \langle x, e_k \rangle - \sum_{n=1}^{\infty} \langle x, e_n \rangle \langle e_n, e_k \rangle \qquad \text{(by Lemma 5.2.6)}$$

$$= \langle x, e_k \rangle - \langle x, e_k \rangle \qquad \text{(since } \langle e_n, e_k \rangle = \delta_{nk})$$

$$= 0.$$

Thus $x - p$ is orthogonal to each vector e_k. By forming linear combinations and then limits, it follows that $x - p$ is orthogonal to every vector in M (see Problem 5.5.7). Therefore we have $p \in M$ and $x - p \in M^{\perp}$, so Lemma 5.6.4 implies that p is the orthogonal projection of x onto M.

(e) Statement (d) tells us that $p = \sum \langle x, e_n \rangle e_n$ is the orthogonal projection of x onto M, and

$$\|p\|^2 = \langle p, p \rangle = \sum_{n=1}^{\infty} |\langle x, e_n \rangle|^2.$$

Let i, ii, iii denote the three statements that appear in equation (5.10). We must prove that i, ii, and iii are equivalent.

i $\Rightarrow$ ii. If $x \in M$, then the orthogonal projection of x onto M is x itself (why?), so $x = p = \sum \langle x, e_n \rangle e_n$ in this case.

ii $\Rightarrow$ iii. If $x = p$, then $\|x\|^2 = \|p\|^2 = \sum |\langle x, e_n \rangle|^2$.

iii $\Rightarrow$ i. Suppose $\|x\|^2 = \sum |\langle x, e_n \rangle|^2$. Then, since $x - p \perp p$,

$$
\begin{aligned}
\|x\|^2 &= \|(x - p) + p\|^2 \\
&= \|x - p\|^2 + \|p\|^2 && \text{(Pythagorean Theorem)} \\
&= \|x - p\|^2 + \sum_{n=1}^{\infty} |\langle x, e_n \rangle|^2 && \text{(by part(d))} \\
&= \|x - p\|^2 + \|x\|^2 && \text{(by statement (iii))}.
\end{aligned}
$$

Hence $\|x - p\| = 0$, so $x = p \in M$. $\square$

Problems

5.7.2. Formulate and prove an analogue of Theorem 5.7.1 for a finite ortho-normal sequence $\{e_1, \ldots, e_d\}$.

5.7.3. Let $\{e_n\}_{n \in \mathbb{N}}$ be an infinite orthonormal sequence in a Hilbert space H. Prove that the series $\sum_{n=1}^{\infty} \frac{1}{n} e_n$ converges *unconditionally but not absolutely*.

5.7.4. Suppose that $\{e_n\}_{n \in \mathbb{N}}$ is an infinite orthonormal sequence in a Hilbert space H. Prove the following statements.

(a) $\{e_n\}_{n \in \mathbb{N}}$ contains no convergent subsequences.

(b) If $x \in H$, then $\langle x, e_n \rangle \to 0$ as $n \to \infty$.

5.7.5. A sequence $\{y_n\}_{n \in \mathbb{N}}$ in a Hilbert space H is a *Bessel sequence* if there exists a constant $B < \infty$ such that

$$
\sum_{n=1}^{\infty} |\langle x, y_n \rangle|^2 \leq B \|x\|^2, \qquad \text{for all } x \in H.
$$

Prove the following statements.

(a) Every orthonormal sequence is a Bessel sequence, but not every Bessel sequence is orthonormal.

(b) If $\{y_n\}_{n \in \mathbb{N}}$ is a Bessel sequence and $\sum |c_n|^2 < \infty$, then the series

$$
\sum_{n=1}^{\infty} c_n y_n
$$

converges in H (in fact, it converges unconditionally).

Hint: Let $s_N = \sum_{n=1}^{N} c_n x_n$ be the Nth partial sum of the series, write $\|s_N - s_M\|^2 = \sup_{\|z\|=1} |\langle s_N - s_M, z \rangle|^2$, expand, and apply the CBS inequality in ℓ^2.

5.7.6. We say that a sequence $\{x_n\}_{n\in\mathbb{N}}$ in a Hilbert space H *converges weakly* to $x \in H$, denoted by $x_n \xrightarrow{w} x$, if

$$\lim_{n\to\infty} \langle x_n, y \rangle = \langle x, y \rangle \quad \text{for every } y \in H.$$

Prove the following statements.

(a) If $x_n \to x$ (i.e., $\|x - x_n\| \to 0$), then $x_n \xrightarrow{w} x$. Hence convergence in norm implies weak convergence.

(b) If $\{e_n\}_{n\in\mathbb{N}}$ is an orthonormal sequence in H, then $e_n \xrightarrow{w} 0$ but $e_n \not\to 0$. Hence weak convergence does not imply norm convergence in general.

(c) $x_n \to x$ if and only if $x_n \xrightarrow{w} x$ and $\|x_n\| \to \|x\|$.

(d) If S is a dense subset of H and $\sup \|x_n\| < \infty$, then $x_n \xrightarrow{w} x$ if and only if $\langle x_n, y \rangle \to \langle x, y \rangle$ for all $y \in S$.

(e) The conclusion of part (d) can fail if $\sup \|x_n\| = \infty$.

5.8 Orthonormal Bases

Recall from Definition 4.4.4 that if the closed span of a sequence $\{e_n\}_{n\in\mathbb{N}}$ is all of H, then we say that the sequence $\{e_n\}_{n\in\mathbb{N}}$ is *complete, total,* or *fundamental* in H. Part (d) of Theorem 5.7.1 implies that if $\{e_n\}_{n\in\mathbb{N}}$ is *both* orthonormal and complete, then every vector $x \in H$ can be written as $x = \sum \langle x, e_n \rangle e_n$. The following theorem states that this property *characterizes* completeness (assuming that our sequence $\{e_n\}_{n\in\mathbb{N}}$ is orthonormal), and gives several other useful characterizations of complete orthonormal sequences. A completely analogous theorem holds for a finite orthonormal sequence $\{e_1, \ldots, e_d\}$ (see Problem 5.8.6).

Theorem 5.8.1. *If H is a Hilbert space and $\{e_n\}_{n\in\mathbb{N}}$ is an orthonormal sequence in H, then the following five statements are equivalent.*

(a) $\{e_n\}_{n\in\mathbb{N}}$ *is complete, i.e.,* $\overline{\text{span}}\{e_n\}_{n\in\mathbb{N}} = H$.

(b) $\{e_n\}_{n\in\mathbb{N}}$ *is a Schauder basis for H, i.e., for each $x \in H$ there exists a unique sequence of scalars $(c_n)_{n\in\mathbb{N}}$ such that $x = \sum c_n e_n$.*

(c) *If $x \in H$, then*

$$x = \sum_{n=1}^{\infty} \langle x, e_n \rangle e_n, \tag{5.11}$$

where this series converges in the norm of H.

(d) Plancherel's Equality:

$$\|x\|^2 = \sum_{n=1}^{\infty} |\langle x, e_n \rangle|^2, \qquad \text{for all } x \in H.$$

(e) Parseval's Equality:

$$\langle x, y \rangle = \sum_{n=1}^{\infty} \langle x, e_n \rangle \langle e_n, y \rangle, \qquad \text{for all } x, y \in H.$$

Proof. We will prove the implication (d) $\Rightarrow$ (c), and assign the proof of the remaining implications as Problem 5.8.5.

(d) $\Rightarrow$ (c). Suppose that $\|x\|^2 = \sum_{n=1}^{\infty} |\langle x, e_n \rangle|^2$ for every $x \in H$. Fix $x \in H$, and define $s_N = \sum_{n=1}^{N} \langle x, e_n \rangle e_n$. Then, by direct calculation,

$$\|x - s_N\|^2 = \|x\|^2 - \langle x, s_N \rangle - \langle s_N, x \rangle + \|s_N\|^2$$

$$= \|x\|^2 - \sum_{n=1}^{N} |\langle x, e_n \rangle|^2 - \sum_{n=1}^{N} |\langle x, e_n \rangle|^2 + \sum_{n=1}^{N} |\langle x, e_n \rangle|^2$$

$$= \|x\|^2 - \sum_{n=1}^{N} |\langle x, e_n \rangle|^2$$

$$\to 0 \qquad \text{as } N \to \infty.$$

Hence $x = \sum_{n=1}^{\infty} \langle x, e_n \rangle e_n$. In fact, part (c) of Theorem 5.7.1 implies that this series converges *unconditionally*, i.e., regardless of the ordering of the index set. $\square$

Since the Plancherel and Parseval Equalities are equivalent, these terms are often used interchangeably.

As formalized in the next definition, we refer to a sequence that satisfies the equivalent conditions in Theorem 5.8.1 as an *orthonormal basis*.

Definition 5.8.2 (Orthonormal Basis). Let H be a Hilbert space. A countably infinite orthonormal sequence $\{e_n\}_{n \in \mathbb{N}}$ that is complete in H is called an *orthonormal basis* for H. $\Diamond$

In particular, if $\{e_n\}_{n \in \mathbb{N}}$ is an orthonormal basis for H, then every $x \in H$ can be written uniquely as

$$x = \sum_{n=1}^{\infty} \langle x, e_n \rangle e_n,$$

and by Theorem 5.7.1 this series actually converges unconditionally, i.e., it converges regardless of the ordering of the series. Hence, not only is an orthonormal basis a Schauder basis for H, but it is an *unconditional basis* in the sense of Definition 4.5.7.

Example 5.8.3. The sequence of standard basis vectors $\{\delta_k\}_{k\in\mathbb{N}}$ is both complete and orthonormal in ℓ^2, so it is an orthonormal basis for ℓ^2. Given a sequence $x = (x_k)_{k\in\mathbb{N}}$ in ℓ^2 we have $\langle x, \delta_k \rangle = x_k$, so the representation of x with respect to the standard basis is simply

$$x = \sum_{k=1}^{\infty} \langle x, \delta_k \rangle \, \delta_k = \sum_{k=1}^{\infty} x_k \, \delta_k.$$

The sequence $\{\delta_{2k}\}_{k\in\mathbb{N}}$ is also orthonormal, but it is not complete in ℓ^2, so it is not an orthonormal basis for ℓ^2. Instead, it is an orthonormal basis for the subspace $E = \overline{\text{span}}\{\delta_{2k}\}_{k\in\mathbb{N}}$, which is the set of all sequences in ℓ^2 whose odd components are all zero. $\diamondsuit$

We have discussed separable infinite-dimensional Hilbert spaces, but entirely similar results hold for finite-dimensional Hilbert spaces. If $\{e_1, \ldots, e_d\}$ is a complete orthonormal sequence in a finite-dimensional Hilbert space H, then the same type of arguments as used in the proof of Theorem 5.8.1 show that $\{e_1, \ldots, e_d\}$ is a basis for H in the usual vector space sense, i.e., it is a Hamel basis for H (see Problem 5.8.6). Specifically, for each $x \in H$ we have

$$x = \sum_{k=1}^{d} \langle x, e_n \rangle \, e_n.$$

We therefore extend the terminology of Definition 5.8.2 to cover this case, and refer to $\{e_1, \ldots, e_d\}$ as an *orthonormal basis* for H.

Theorem 5.8.1 gives us a remarkable set of equivalences that hold for orthonormal sequences. The situation for sequences that are not orthogonal is far more delicate. For example, Theorem 5.8.1 tells us that a sequence that is both complete and orthonormal is a Schauder basis, and we know that every Schauder basis is both complete and linearly independent, but it is *not true* that every complete, linearly independent sequence is a Schauder basis. The problems for this section will explore these issues in more detail, using the following terminology.

Definition 5.8.4. Let $\{x_n\}_{n\in\mathbb{N}}$ be a sequence in a Hilbert space H.

(a) We say that $\{x_n\}_{n\in\mathbb{N}}$ is *minimal* if for each $m \in \mathbb{N}$ we have

$$x_m \notin \overline{\text{span}}\{x_n\}_{n\neq m}.$$

(b) If there exists a sequence $\{y_n\}_{n\in\mathbb{N}}$ such that $\langle x_m, y_n \rangle = \delta_{mn}$ for all $m, n \in \mathbb{N}$ then we say that $\{x_n\}_{n\in\mathbb{N}}$ and $\{y_n\}_{n\in\mathbb{N}}$ are *biorthogonal sequences*.

(c) We say that $\{x_n\}_{n\in\mathbb{N}}$ is ω-*dependent* if there exist scalars c_n, not all zero, such that $\sum_{n=1}^{\infty} c_n x_n = 0$, where the series converges in the norm of H. A sequence is ω-*independent* if it is not ω-dependent. $\diamond$

Problem 5.8.9 shows that a sequence is minimal if and only if it has a biorthogonal sequence, and Problem 5.8.10 establishes the following further implications:

orthonormal basis

$\implies$ Schauder basis

$\implies$ minimal

$\implies$ ω-independent

$\implies$ finitely independent.   (5.12)

That is, every orthonormal basis is a Schauder basis, every Schauder basis is minimal, and so forth. However, Problems 5.8.11–5.8.14 show that none of the implications above is reversible, even if we also require the sequence to be complete. That is, there exists a complete, finitely independent sequence that is not ω-independent, and so forth. Even so, Theorem 5.8.1 tells us that every complete *orthonormal* sequence is actually an *orthonormal basis*!

Problems

5.8.5. Prove the remaining implications in Theorem 5.8.1.

5.8.6. Formulate and prove an analogue of Theorem 5.8.1 for a finite orthonormal sequence $\{e_1, \ldots, e_d\}$.

5.8.7. Let $u_1, \ldots, u_d$ be the columns of a $d \times d$ matrix U. Prove that $\{u_1, \ldots, u_d\}$ is an orthonormal basis for $\mathbb{F}^d$ if and only if $U^H U = I$, where $U^H = \overline{U^T}$ is the Hermitian, or conjugate transpose, of U (if $\mathbb{F} = \mathbb{R}$, then $U^H = U^T$, the ordinary transpose of U).

5.8.8. A sequence $\{x_n\}_{n\in\mathbb{N}}$ is called a *Parseval frame* for a separable Hilbert space H if the Plancherel Equality is satisfied, i.e., if $\sum_{n=1}^{\infty} |\langle x, x_n \rangle|^2 = \|x\|^2$ for every vector $x \in H$.

(a) By Theorem 5.8.1, every orthonormal basis is a Parseval frame. Show by example that there exist Parseval frames that are not orthonormal bases.

(b) Make an analogous definition of Parseval frames for finite-dimensional Hilbert spaces. Do there exist three points $x_1, x_2, x_3 \in \mathbb{R}^2$ such that $\|x_1\| = \|x_2\| = \|x_3\|$ and $\{x_1, x_2, x_2\}$ is a Parseval frame for $\mathbb{R}^2$?

5.8.9. Given a sequence $\{x_n\}_{n\in\mathbb{N}}$ in a Hilbert space H, prove that the following two statements are equivalent.

(a) $\{x_n\}_{n \in \mathbb{N}}$ is minimal.

(b) There exists a sequence $\{y_n\}_{n \in \mathbb{N}}$ in H such that $\{x_n\}_{n \in \mathbb{N}}$ and $\{y_n\}_{n \in \mathbb{N}}$ are biorthogonal.

Show further that in case statements (a) and (b) hold, the sequence $\{y_n\}_{n \in \mathbb{N}}$ is unique if and only if $\{x_n\}_{n \in \mathbb{N}}$ is complete.

5.8.10. Let H be a Hilbert space. This problem will establish the implications stated in equation (5.12) (note that the fact that every orthonormal basis is a Schauder basis was proved in Theorem 5.8.1).

(a) If $\{x_n\}_{n \in \mathbb{N}}$ is a Schauder basis for H, then it can be shown that there exist vectors $y_n \in H$ such that for each $x \in H$ we have $x = \sum_{n=1}^{\infty} \langle x, y_n \rangle x_n$, where the series converges in the norm of H (for one proof of this fact, see [Heil11, Thm. 4.13]). Use this to prove that every Schauder basis is minimal.

(b) Show that if $\{x_n\}_{n \in \mathbb{N}}$ is minimal, then it is ω-independent.

(c) Show that if $\{x_n\}_{n \in \mathbb{N}}$ is ω-independent, then it is finitely linearly independent.

5.8.11. Let $\{\delta_n\}_{n \in \mathbb{N}}$ be the sequence of standard basis vectors. Let $y_1 = \delta_1$ and for each integer $n \geq 2$ let y_n be the sequence $y_n = 2^{-n}\delta_1 + \delta_n = (2^{-n}, 0, \ldots, 0, 1, 0, \ldots)$.

(a) Prove that $\{y_n\}_{n \in \mathbb{N}}$ is a Schauder basis for ℓ^2, but it is not an orthonormal basis, and in fact $\langle y_m, y_n \rangle \neq 0$ for every $m \neq n$.

(b) Since $\{y_n\}_{n \in \mathbb{N}}$ is a Schauder basis it is minimal, and therefore it has a biorthogonal sequence $\{w_n\}_{n \in \mathbb{N}}$. What is this biorthogonal sequence?

5.8.12. Let $\{\delta_n\}_{n \in \mathbb{N}}$ be the sequence of standard basis vectors. For each integer $n \geq 2$ let z_n be the sequence $z_n = \delta_1 + \delta_n = (1, 0, \ldots, 0, 1, 0, \ldots)$.

(a) Prove that $\{z_n\}_{n \geq 2}$ and $\{\delta_n\}_{n \geq 2}$ are biorthogonal sequences, and therefore $\{z_n\}_{n \geq 2}$ is minimal by Problem 5.8.9.

(b) Prove that $\{z_n\}_{n \geq 2}$ is complete in ℓ^2.

(c) Prove that the series $\sum_{n=2}^{\infty} (z_n/n)$ does not converge in ℓ^2. Hint: What are the ℓ^2-norms of the partial sums of this series?

(d) Let $x = \left(1, \frac{1}{2}, \frac{1}{3}, \ldots\right)$, and note that $x \in \ell^2$. Suppose that there were scalars c_n such that $x = \sum_{n=2}^{\infty} c_n z_n$ with convergence of the series in ℓ^2-norm. Prove that $c_n = 1/n$ for each $n \geq 2$, and obtain a contradiction.

(e) Show that although $\{z_n\}_{n \geq 2}$ is both minimal and complete, it is not a Schauder basis for ℓ^2.

5.8.13. Let $\{z_n\}_{n \geq 2}$ be the sequence of vectors in ℓ^2 defined in Problem 5.8.12, and let $z_1 = \left(1, \frac{1}{2}, \frac{1}{3}, \ldots\right)$.

(a) Prove that $\{z_n\}_{n \in \mathbb{N}} = \{z_n\}_{n \geq 1}$ is complete, yet $z_1 \in \overline{\text{span}}\{z_n\}_{n \geq 2}$, so $\{z_n\}_{n \in \mathbb{N}}$ is not minimal.

(b) Prove that $\{z_n\}_{n\in\mathbb{N}}$ is ω-independent. Hint: Part (c) of Problem 5.8.12 is helpful.

5.8.14. Let α and β be fixed nonzero scalars such that $|\alpha| > |\beta|$. Let $\{\delta_n\}_{n\in\mathbb{N}}$ be the sequence of standard basis vectors, and define

$$x_0 = \delta_1 \quad \text{and} \quad x_n = \alpha\delta_n + \beta\delta_{n+1}, \ n \in \mathbb{N}.$$

Prove that $\{x_n\}_{n\geq 0}$ is a complete and finitely linearly independent sequence in ℓ^2, but it is not ω-independent (and therefore is not minimal, nor is it a Schauder basis for ℓ^2).

5.9 Existence of an Orthonormal Basis

A Hilbert space is *separable* if it contains a countable dense subset (see the definitions given in Section 2.3). All finite-dimensional Hilbert spaces are separable, and Problem 3.4.11 showed that the infinite-dimensional space ℓ^2 is separable. Not every Hilbert space is separable; an example is given in Problem 5.4.7.

We will show that every separable Hilbert space contains an orthonormal basis. We begin with the *Gram–Schmidt orthonormalization procedure* for finitely many vectors.

Theorem 5.9.1 (Gram–Schmidt). *If $x_1, \ldots, x_d$ are linearly independent vectors in a Hilbert space H, then there exist orthonormal vectors $e_1, \ldots, e_d$ such that*

$$\operatorname{span}\{e_1, \ldots, e_k\} = \operatorname{span}\{x_1, \ldots, x_k\}, \quad k = 1, \ldots, d. \tag{5.13}$$

Proof. For $k = 1$ we set $q_1 = x_1$, and note that $x_1 \neq 0$, since $x_1, \ldots, x_d$ are linearly independent and therefore nonzero. We normalize by setting $e_1 = q_1/\|q_1\|$, and observe that $\operatorname{span}\{e_1\} = \operatorname{span}\{x_1\}$. If $d = 1$, then we stop the process here.

Otherwise, we go on to consider $k = 2$. The vector x_2 does not belong to $M_1 = \operatorname{span}\{x_1\}$, because x_1, x_2 are linearly independent. Let p_2 be the orthogonal projection of x_2 onto M_1. Then $q_2 = x_2 - p_2$ is orthogonal to x_1, and $q_2 \neq 0$, since $x_2 \notin M_1$. We set $e_2 = q_2/\|q_2\|$, and observe that

$$\operatorname{span}\{e_1, e_2\} = \operatorname{span}\{x_1, x_2\},$$

because e_1, e_2 are linear combinations of x_1, x_2 and vice versa. If $d = 2$, then we stop here.

As long as $k \leq d$ we continue to take another step. At step k, we take the orthonormal vectors $e_1, \ldots, e_{k-1}$ that we have constructed so far, and let p_k be the orthogonal projection of x_k onto

$$M_k = \text{span}\{x_1, \ldots, x_{k-1}\} = \text{span}\{e_1, \ldots, e_{k-1}\}.$$

Since $\{e_1, \ldots, e_{k-1}\}$ is an orthonormal basis for M_k, the finite analogue of Theorem 5.7.1(d) gives us a constructive formula for p_k as a linear combination of $e_1, \ldots, e_{k-1}$, namely,

$$p_k = \sum_{j=1}^{k-1} \langle x_k, e_j \rangle e_j. \tag{5.14}$$

We let $q_k = x_k - p_k$ and set $e_k = q_k / \|q_k\|$. Then $e_1, \ldots, e_k$ are orthonormal and satisfy

$$\text{span}\{e_1, \ldots, e_k\} = \text{span}\{x_1, \ldots, x_k\}.$$

Once we reach $k = d$ we stop the process, giving us orthonormal vectors $e_1, \ldots, e_d$ that satisfy equation (5.13). $\square$

To summarize the proof of Theorem 5.9.1, the Gram–Schmidt orthonormalization procedure is the following recursive algorithm.

Algorithm 5.9.2 (Gram–Schmidt). Assume that $x_1, \ldots, x_d$ are linearly independent vectors in a Hilbert space H.

- Base step: Set $e_1 = x_1 / \|x_1\|$.

- Inductive step: Once $e_1, \ldots, e_{k-1}$ have been constructed, compute

$$q_k = x_k - p_k = x_k - \sum_{j=1}^{k-1} \langle x_k, e_j \rangle e_j,$$

and set $e_k = q_k / \|q_k\|$.

Continue this process until we reach $k = d$. Then the vectors $e_1, \ldots, e_d$ will be orthonormal and will satisfy equation (5.13). $\diamond$

To illustrate, we apply the Gram–Schmidt procedure to the monomials x^k in $C[-1, 1]$, using the inner product defined in equation (5.5).

Example 5.9.3 (Legendre Polynomials). Fix any integer $d \in \mathbb{N}$. The set of monomials $\{1, t, t^2, \ldots, t^d\}$ is linearly independent in $C[-1, 1]$, but it is not orthogonal. We will write out the Gram–Schmidt procedure in detail for the first few vectors in this sequence. For convenience of notation, let $m_k(t) = t^k$ for $k = 0, 1, \ldots, d$, and note that we have numbered our sequence starting at $k = 0$ rather than $k = 1$.

For $k = 0$, let e_0 be the vector obtained by normalizing m_0. Since the length of m_0 in $C[-1, 1]$ is

$$\|m_0\|_2 = \left(\int_{-1}^{1} |m_0(t)|^2 \, dt \right)^{1/2} = \left(\int_{-1}^{1} 1^2 \, dt \right)^{1/2} = 2^{1/2},$$

we see that e_0 is the constant polynomial

$$e_0(t) = \frac{m_0(t)}{\|m_0\|_2} = 2^{-1/2} = \frac{1}{\sqrt{2}}.$$

Now consider $k = 1$. We let p_1 be the projection of m_1 onto the line through e_0, which by equation (5.14) is

$$
\begin{aligned}
p_1(t) &= \sum_{j=0}^{0} \langle m_1, e_j \rangle \, e_j(t) \\
&= \langle m_1, e_0 \rangle \, e_0(t) \\
&= \left(\int_{-1}^{1} t \, 2^{-1/2} \, dt \right) e_0(t) = 0.
\end{aligned}
$$

We then set $q_1 = m_1 - p_1 = m_1$. The length of q_1 is

$$\|q_1\|_2 = \left(\int_{-1}^{1} |m_1(t)|^2 \, dt \right)^{1/2} = \left(\int_{-1}^{1} t^2 \, dt \right)^{1/2} = \sqrt{\frac{2}{3}},$$

so the orthonormal vector e_1 that we seek is

$$e_1(t) = \frac{q_1(t)}{\|q_1\|_2} = \frac{m_1(t)}{(2/3)^{1/2}} = \sqrt{\frac{3}{2}} \, t.$$

For $k = 2$, equation (5.14) tells us that the orthogonal projection of m_2 onto $\mathrm{span}\{e_1, e_2\}$ is

$$
\begin{aligned}
p_2(t) &= \sum_{j=0}^{1} \langle m_1, e_j \rangle \, e_j(t) \\
&= \langle m_2, e_0 \rangle \, e_0(t) + \langle m_2, e_1 \rangle \, e_1(t) \\
&= \frac{\sqrt{2}}{3} \, e_0(t) + 0 \, e_1(t) = \frac{1}{3}.
\end{aligned}
$$

Therefore the vector

$$q_2(t) = m_2(t) - p_2(t) = t^2 - \frac{1}{3}$$

is orthogonal to e_0 and e_1, and we normalize to obtain

$$e_2(t) = \frac{q_2(t)}{\|q_2\|_2} = \frac{t^2 - \frac{1}{3}}{(8/45)^{1/2}} = \frac{5}{2\sqrt{2}} \, (3t^2 - 1).$$

Continuing this process, we obtain orthonormal polynomials $e_0, e_1, \ldots, e_d$ in $C[-1,1]$, with d as large as we like. Each polynomial e_k has degree k and

$$\text{span}\{e_0, e_1, \ldots, e_k\} = \text{span}\{1, t, \ldots, t^k\} = \mathcal{P}_k,$$

the set of all polynomials of degree at most k. Hence $\{e_0, e_1, \ldots, e_k\}$ is an orthonormal basis for the subspace $\mathcal{P}_k$ in $C[-1,1]$.

Except for their normalization, the polynomials e_k that we have constructed are known as the *Legendre polynomials*. Traditionally, these polynomials are not normalized so that their L^2-norm is 1, but rather are scaled so that $\|P_k\|_2^2 = \frac{2}{2k+1}$. Using this normalization, the first few Legendre polynomials P_0, P_1, P_2, P_3 are

$$P_0(t) = 1, \quad P_1(t) = t, \quad P_2(t) = \tfrac{1}{2}(3t^2 - 1), \quad P_3(t) = \tfrac{1}{2}(5t^3 - 3t).$$

The Legendre polynomials arise naturally in a variety of applications. For example, they are solutions to *Legendre's differential equation*

$$\frac{d}{dt}\left((1 - t^2)\frac{d}{dt}P_n(t)\right) + n(n+1)P_n(t) = 0.$$

There are many other types of orthogonal polynomials, with applications in approximation theory and other areas. We refer to texts such as [Ask75] and [Sze75] for more details. ◇

If H is a finite-dimensional Hilbert space, then it has a finite Hamel basis, and so we can apply the Gram–Schmidt procedure to construct an orthonormal basis for the space.

Theorem 5.9.4. *If H is a finite-dimensional Hilbert space, then H contains an orthonormal basis $\{e_1, \ldots, e_d\}$, where $d = \dim(H)$ is the dimension of H.*

Proof. By definition, if H has dimension d, then it has a Hamel basis $\mathcal{B}$ consisting of d vectors, say $\mathcal{B} = \{x_1, \ldots, x_d\}$. Such a set is linearly independent and spans H. Applying Gram–Schmidt to $\mathcal{B}$, we obtain orthonormal vectors $e_1, \ldots, e_d$ such that $\text{span}\{e_1, \ldots, e_d\} = \text{span}\{x_1, \ldots, x_d\} = H$. Therefore $\{e_1, \ldots, e_d\}$ is an orthonormal basis for H. □

For the infinite-dimensional case, we prove next that we can use Gram–Schmidt to construct an orthonormal basis if the space is *separable*.

Theorem 5.9.5. *If H is an infinite-dimensional, separable Hilbert space, then H contains an orthonormal basis of the form $\{e_n\}_{n\in\mathbb{N}}$.*

Proof. Since H is separable, it contains a countable dense subset. This subset must be infinite, so let us say that it is $\mathcal{Z} = \{z_n\}_{n\in\mathbb{N}}$. However, $\mathcal{Z}$ need not be linearly independent, so we will extract a linearly independent subsequence via the following procedure.

Let k_1 be the first index such that $z_{k_1} \neq 0$, and set $x_1 = z_{k_1}$. Then let k_2 be the first index larger than k_1 such that $z_{k_2} \notin \text{span}\{x_1\}$, and set $x_2 = z_{k_2}$. Then let k_3 be the first index larger than k_2 such that $z_{k_3} \notin \text{span}\{x_1, x_2\}$, and so forth. In this way we obtain vectors $x_1, x_2, \ldots$ such that $x_1 \neq 0$ and for each $n > 1$ we have both

$$x_n \notin \text{span}\{x_1, \ldots, x_{n-1}\} \tag{5.15}$$

and

$$\text{span}\{x_1, \ldots, x_n\} = \text{span}\{z_1, \ldots, z_{k_n}\}. \tag{5.16}$$

Equation (5.15) ensures that $\{x_n\}_{n \in \mathbb{N}}$ is linearly independent, and equation (5.16) implies that $\text{span}\{x_n\}_{n \in \mathbb{N}} = \text{span}\{z_n\}_{n \in \mathbb{N}}$, which is dense in H.

Now we apply the Gram–Schmidt procedure to the vectors $x_1, x_2, \ldots$, without ever stopping. This gives us orthonormal vectors $e_1, e_2, \ldots$ such that for every n we have

$$\text{span}\{e_1, \ldots, e_n\} = \text{span}\{x_1, \ldots, x_n\}.$$

Consequently, $\text{span}\{e_n\}_{n \in \mathbb{N}} = \text{span}\{x_n\}_{n \in \mathbb{N}}$, which is dense in H. Therefore $\{e_n\}_{n \in \mathbb{N}}$ is a countable complete orthonormal sequence. Hence $\{e_n\}_{n \in \mathbb{N}}$ is, by definition, an orthonormal basis for H. $\square$

Example 5.9.6 (Legendre Polynomials). We saw in Example 5.9.3 that the set of Legendre polynomials $\{P_0, P_1, \ldots\}$ is an infinite orthogonal sequence in $C[-1, 1]$. Although $C[-1, 1]$ is not a Hilbert space, it is a dense subspace of the *Lebesgue space* $L^2[-1, 1]$, which is a Hilbert space (see Remark 5.3.2). Since each polynomial P_n has degree n, the finite span of all of the Legendre polynomials is the set of polynomials $\mathcal{P}$. Thus $\{P_n\}_{n \geq 0}$ is a countable sequence of orthogonal functions whose span is $\mathcal{P}$, which is dense in $L^2[-1, 1]$. Therefore the set of Legendre polynomials $\{P_n\}_{n \geq 0}$ is an orthogonal basis for $L^2[-1, 1]$ (and we can obtain an orthonormal basis by dividing each polynomial P_n by its length). $\diamond$

Theorems 5.9.4 and 5.9.5 show that every separable Hilbert space has an orthonormal basis. It is a finite basis if H is finite-dimensional, and countably infinite if H is an infinite-dimensional separable space. Conversely, every Hilbert space that has a countable orthonormal basis must be separable (see Problem 5.9.8).

Nonseparable Hilbert spaces do exist (one example is given in Problem 5.4.7), and they play important roles in certain contexts. An argument based on the Axiom of Choice in the form of Zorn's Lemma shows that every Hilbert space contains a complete orthonormal set (for one proof, see [Heil11, Thm. 1.56]). If H is nonseparable, then such a complete orthonormal set will be uncountable. An uncountable complete orthonormal set does have certain basis-like properties (see Problems 5.9.10 and 5.9.11), and for this reason some authors refer to a complete orthonormal set of any cardinality as an orthonormal basis. In keeping with the majority of the Banach space literature, we prefer to reserve the word "basis" for use in conjunction with countable sequences only.

Problems

5.9.7. (a) Show that every infinite-dimensional vector space X contains an infinite, linearly independent sequence $\{x_n\}_{n \in \mathbb{N}}$.

(b) Suppose that H is an infinite-dimensional Hilbert space. Prove that the closed unit disk $D = \{x \in H : \|x\| \leq 1\}$ is a closed and bounded subset of H that is not compact.

5.9.8. Prove that a Hilbert space H contains a countable orthonormal basis if and only if H is separable.

5.9.9. Assume that $\{e_n\}_{n \in \mathbb{N}}$ is an orthonormal basis for a Hilbert space H.

(a) Suppose that vectors $y_n \in H$ satisfy $\sum \|e_n - y_n\|^2 < 1$. Prove that $\{y_n\}_{n \in \mathbb{N}}$ is a complete sequence in H.

(b) Show that part (a) can fail if we only have $\sum \|x_n - y_n\|^2 = 1$.

5.9.10. Let H be a (possibly nonseparable) Hilbert space and let I be an index set (possibly uncountable). Show that if $\{x_i\}_{i \in I}$ is an orthonormal set in H, then the following statements hold.

(a) If $x \in H$, then $\langle x, x_i \rangle \neq 0$ for at most countably many $i \in I$.

(b) For each $x \in H$ we have $\sum_{i \in I} |\langle x, x_i \rangle|^2 \leq \|x\|^2$.

(c) For each $x \in H$, the series $p = \sum_{i \in I} \langle x, x_i \rangle x_i$, which contains only countably many nonzero terms, converges unconditionally and p is the orthogonal projection of x onto $\overline{\text{span}}\{x_i\}_{i \in I}$.

5.9.11. Let H be a (possibly nonseparable) Hilbert space and let I be an index set (possibly uncountable). Given an orthonormal set $\{x_i\}_{i \in I}$ in H, prove that the following statements are equivalent.

(a) $\{x_i\}_{i \in I}$ is complete.

(b) For each $x \in H$ we have $x = \sum_{i \in I} \langle x, x_i \rangle x_i$, where only countably many terms of this series are nonzero.

(c) For each $x \in H$ we have $\|x\|^2 = \sum_{i \in I} |\langle x, x_i \rangle|^2$.

5.10 The Real Trigonometric System

We saw in Example 5.9.3 that the Legendre polynomials $\{P_n\}_{n \geq 0}$ form an infinite orthogonal sequence in $C[-1, 1]$ with respect to the L^2-norm (and it is an orthogonal *basis* for the larger Lebesgue space $L^2[-1, 1]$). We will consider a different orthonormal sequence in this section. For convenience of notation, we will construct this sequence in $C[-\pi, \pi]$, but by making a change of variables the same ideas can be applied in $C[a, b]$ for any compact interval $[a, b]$.

The system that we will construct, which is called the *trigonometric system*, is central to the field of mathematics known as *harmonic analysis*. There are two versions of the trigonometric system, one real-valued and one complex-valued. If $\mathbb{F} = \mathbb{R}$, then the trigonometric system is a set of sines and cosines, while if $\mathbb{F} = \mathbb{C}$, then it is a set of *complex exponential functions*. In some ways the complex system is the easier one to work with, but we will begin with the real system in this section and turn to the complex system in Section 5.11. Therefore:

Throughout this section we take $\mathbb{F} = \mathbb{R}$.

The *real trigonometric system* in $C[-\pi, \pi]$ is the following set of sines and cosines together with the constant function 1:

$$\mathcal{T} = \{1\} \cup \{\cos nt\}_{n \in \mathbb{N}} \cup \{\sin nt\}_{n \in \mathbb{N}}. \tag{5.17}$$

The number n is called the *frequency* of $\cos nt$ or $\sin nt$. Since $\cos 0t = 1$, the constant function 1 can be considered the cosine of frequency zero.

We claim that the trigonometric system is an orthogonal sequence in $C[-\pi, \pi]$. For example, the constant function is orthogonal to every other element of the trigonometric system, because

$$\langle 1, \sin nt \rangle = \int_{-\pi}^{\pi} \sin nt \, dt = 0, \quad \langle 1, \cos nt \rangle = \int_{-\pi}^{\pi} \cos nt \, dt = 0.$$

Also, $\sin t$ is orthogonal to $\cos t$, because

$$\langle \sin t, \cos t \rangle = \int_{-\pi}^{\pi} \sin t \, \cos t \, dt = \left. \frac{\sin^2 t}{2} \right|_{-\pi}^{\pi} = \frac{\sin^2 \pi - \sin^2(-\pi)}{2} = 0.$$

And $\sin t$ is orthogonal to $\sin 2t$, because

$$\langle \sin t, \sin 2t \rangle = \int_{-\pi}^{\pi} \sin t \, (2 \sin t \, \cos t) \, dt = 2 \left. \frac{\sin^3 t}{3} \right|_{-\pi}^{\pi} = 0.$$

With more involved calculations based on trigonometric identities we can prove that the inner product of any two distinct elements of $\mathcal{T}$ is zero (this is Problem 5.10.6). Hence $\mathcal{T}$ is an orthogonal (but not orthonormal) system in $C[-\pi, \pi]$.

To create an orthonormal system, we normalize each vector by dividing by its length. We have

$$\|1\|_2^2 = \int_{-\pi}^{\pi} 1^2 \, dt = 2\pi,$$

$$\|\sin nt\|_2^2 = \int_{-\pi}^{\pi} \cos^2 nt \, dt = \pi,$$

$$\|\cos nt\|_2^2 = \int_{-\pi}^{\pi} \sin^2 nt \, dt = \pi,$$

so the *normalized trigonometric system* in $C[-\pi, \pi]$ is

$$\mathcal{T}_{\text{nor}} = \left\{\frac{1}{\sqrt{2\pi}}\right\} \cup \left\{\frac{\cos nt}{\sqrt{\pi}}\right\}_{n \in \mathbb{N}} \cup \left\{\frac{\sin nt}{\sqrt{\pi}}\right\}_{n \in \mathbb{N}}. \qquad (5.18)$$

Unfortunately, $C[-\pi, \pi]$ is not a Hilbert space, since it is not complete with respect to the L^2-norm (see Lemma 3.4.7). However, it is a dense subspace of the larger Lebesgue space $L^2[-\pi, \pi]$, and we will use that fact to prove that the trigonometric system is an orthogonal basis for the Hilbert space $L^2[-\pi, \pi]$.

Theorem 5.10.1. *The trigonometric system $\mathcal{T}$ given in equation (5.17) is an orthogonal basis for $L^2[-\pi, \pi]$, and the normalized trigonometric system $\mathcal{T}_{\text{nor}}$ given in equation (5.18) is an orthonormal basis for $L^2[-\pi, \pi]$.*

Proof. We have seen that the trigonometric system is orthogonal, and therefore the normalized trigonometric system is orthonormal. By Theorem 5.8.1, to prove that we have an orthonormal basis we just have to prove that $\mathcal{T}$ is complete in $L^2[-\pi, \pi]$. That is, we have to prove that the finite linear span of $\mathcal{T}$ is dense in $L^2[-\pi, \pi]$. To do this we will make use of Theorem 4.7.3, which showed that the trigonometric system is complete in the subspace

$$C_{\text{per}}[-\pi, \pi] = \{f \in C[-\pi, \pi] : f(-\pi) = f(\pi)\}$$

with respect to the uniform norm.

Choose any function $f \in L^2[-\pi, \pi]$ and fix $\varepsilon > 0$. Our goal is to show that there is a function $p \in \text{span}(\mathcal{T})$ such that $\|f - p\|_2 < 3\varepsilon$.

Step 1. Since $C[-\pi, \pi]$ is dense in $L^2[-\pi, \pi]$, there is a continuous function g such that

$$\|f - g\|_2^2 = \int_{-\pi}^{\pi} |f(t) - g(t)|^2 \, dt < \varepsilon^2.$$

If g is the zero function, we can take $p = 0$ and we are done with the proof of the theorem. Therefore, we will focus on the case that g is not identically zero.

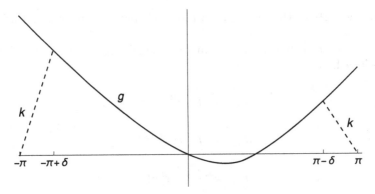

Fig. 5.3 The function k (dashed) is close in L^2-norm to g (solid), and also satisfies $k(-\pi) = k(\pi)$.

Step 2. Although g is continuous, $g(-\pi)$ need not equal $g(\pi)$, and therefore g need not belong to $C_{\text{per}}[-\pi, \pi]$. So; we will create a continuous function k that is in $C_{\text{per}}[-\pi, \pi]$ and is very close to g in L^2-norm. Let

$$\delta = \frac{\varepsilon^2}{8\|g\|_u^2},$$

and let k be the continuous function given by

$$k(t) = \begin{cases} 0, & t = -\pi, \\ \text{linear}, & -\pi < t < -\pi + \delta, \\ g(t), & -\pi + \delta \leq t \leq \pi - \delta, \\ \text{linear}, & \pi - \delta < t < \pi, \\ 0, & t = \pi. \end{cases} \tag{5.19}$$

As illustrated in Figure 5.3, for $t \in [-\pi + \delta, \pi - \delta]$ we have $g(t) - k(t) = 0$, and for any other t we have at worst

$$|g(t) - k(t)| \leq |g(t)| + |k(t)| \leq \|g\|_u + \|g\|_u = 2\|g\|_u.$$

Therefore

$$\|g - k\|_2^2 = \int_{-\pi}^{-\pi+\delta} |g(t) - k(t)|^2 \, dt + \int_{-\pi+\delta}^{\pi-\delta} |g(t) - k(t)|^2 \, dt$$
$$+ \int_{\pi-\delta}^{\pi} |g(t) - k(t)|^2 \, dt$$

$$\leq \int_{-\pi}^{-\pi+\delta} 4\,\|g\|_u^2\,dt \;+\; 0 \;+\; \int_{\pi-\delta}^{\pi} 4\,\|g\|_u^2\,dt$$

$$= 8\delta\,\|g\|_u^2 \;=\; \varepsilon^2.$$

Step 3. Theorem 4.7.3 showed that span($\mathcal{A}$) is dense in $C_{\mathrm{per}}[-\pi,\pi]$ with respect to the uniform norm. Since $k \in C_{\mathrm{per}}[-\pi,\pi]$, it follows that there is some $p \in \mathrm{span}(\mathcal{T})$ such that

$$\|k - p\|_u \;<\; (2\pi)^{-1/2}\varepsilon.$$

However, we need to measure distance in the L^2-norm rather than the uniform norm, so we compute that

$$\|k - p\|_2^2 \;=\; \int_{-\pi}^{\pi} |k(t) - p(t)|^2\,dt$$

$$\leq \int_{-\pi}^{\pi} \|k - p\|_u^2\,dt$$

$$= 2\pi\,\|k - p\|_u^2 \;<\; 2\pi\frac{\varepsilon^2}{2\pi} \;=\; \varepsilon^2.$$

Step 4. Combining the above estimates and using the Triangle Inequality, we see that

$$\|f - p\|_2 \;\leq\; \|f - g\|_2 + \|g - k\|_2 + \|k - p\|_2 \;<\; \varepsilon + \varepsilon + \varepsilon \;=\; 3\varepsilon.$$

Thus, each element f of $L^2[-\pi,\pi]$ can be approximated as closely as we like in L^2-norm by a function p in span($\mathcal{T}$). This shows that span($\mathcal{T}$) is dense in $L^2[-\pi,\pi]$ (see Corollary 2.6.4 or Problem 2.6.12). Hence $\mathcal{T}$ is a complete sequence in the sense of Definition 4.4.4, and therefore the normalized trigonometric system is an orthonormal basis for $L^2[-\pi,\pi]$ by Theorem 5.8.1. $\square$

Remark 5.10.2. A technical point in the proof of Theorem 5.10.1 is that in order to integrate general functions in $L^2[0,1]$ we must use Lebesgue integrals, not Riemann integrals. All functions that are Riemann integrable, such as piecewise continuous functions, are also Lebesgue integrable, and for these functions the Lebesgue integral and the Riemann integral are equal. In particular, the functions g, k, and p that appear in the proof of Theorem 5.10.1 are Riemann integrable. However, those integrals in the proof that involve the function f are generally Lebesgue integrals, not Riemann integrals. $\diamond$

Since we have shown that the normalized trigonometric system is an orthonormal basis for $L^2[-\pi,\pi]$, if f is any function in $L^2[-\pi,\pi]$, then part (c) of Theorem 5.8.1 tells us that

$$f = \left\langle f, \frac{1}{\sqrt{2\pi}} \right\rangle \frac{1}{\sqrt{2\pi}} + \sum_{n=1}^{\infty} \left\langle f, \frac{\cos nt}{\sqrt{\pi}} \right\rangle \frac{\cos nt}{\sqrt{\pi}} + \sum_{n=1}^{\infty} \left\langle f, \frac{\sin nt}{\sqrt{\pi}} \right\rangle \frac{\sin nt}{\sqrt{\pi}}.$$

While this looks unpleasant, we can simplify by collecting factors. It is traditional to let a_n and b_n denote the following quantities, which we call the *Fourier coefficients* of f:

$$a_0 = \frac{1}{\pi} \langle f, 1 \rangle = \frac{1}{\pi} \int_{-\pi}^{\pi} f(t) \, dt,$$

$$a_n = \frac{1}{\pi} \langle f, \cos nt \rangle = \frac{1}{\pi} \int_{-\pi}^{\pi} f(t) \, \cos nt \, dt,$$

$$b_n = \frac{1}{\pi} \langle f, \sin nt \rangle = \frac{1}{\pi} \int_{-\pi}^{\pi} f(t) \, \sin nt \, dt.$$

Using this notation, the series expression for f given above simplifies to

$$f = \frac{a_0}{2} + \sum_{n=1}^{\infty} a_n \cos nt + \sum_{n=1}^{\infty} b_n \sin nt. \tag{5.20}$$

In summary, every square-integrable function f on $[-\pi, \pi]$ can be written as the "infinite linear combination" of sines and cosines that is shown in equation (5.20).

Equation (5.20) is called the *Fourier series expansion* of f. It is important to note that Theorem 5.8.1 tells us only that this series converges *in L^2-norm*. We do not know whether the series will converge pointwise, uniformly, or in any other manner. We know only that the *partial sums*

$$s_N(t) = \frac{a_0}{2} + \sum_{n=1}^{N} a_n \cos nt + \sum_{n=1}^{N} b_n \sin nt$$

converge to f in L^2-norm as $N \to \infty$. That is, we know that

$$\lim_{N \to \infty} \|f - s_N\|_2^2 = \lim_{N \to \infty} \int_{-\pi}^{\pi} |f(t) - s_N(t)|^2 \, dt = 0,$$

but we do not have any assurance that $s_N(t)$ will converge to $f(t)$ for any particular t. Indeed, establishing the convergence of Fourier series in senses other than L^2-norm can be difficult, and there even exist continuous functions whose Fourier series do not converge at any particular point t. These issues are addressed in more detail in courses on harmonic analysis; see texts such as [DM72], [Ben97], [Kat04], and [Heil11, Ch. 14].

We expand a bit on the meaning of equation (5.27). The graph of $\cos nt$ for one value of n is pictured in Figure 5.4. This is a *pure tone*, and $a_n \cos nt$ is a pure tone that has been rescaled in amplitude. This function could represent the displacement of the center of an ideal string vibrating at the frequency n with amplitude a_n. It could also represent the displacement of the center of an ideal stereo speaker from its rest position at time t. If you were listening to this ideal speaker, you would hear a "pure tone." Real strings and speakers are of course quite complicated and do not vibrate as pure tones—there are overtones and other complications. Still, the functions $\cos nt$ and $\sin nt$ represent pure tones, and the idea of *Fourier series* is that we can use these pure tones as elementary building blocks for the construction of other, much more complicated, signals.

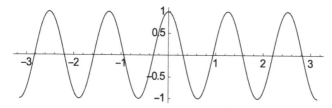

Fig. 5.4 Graph of $\varphi(t) = \cos(3t)$.

A function like $\varphi(t) = a_m \cos mt + b_n \sin nt$ (or any other linear combination of two elements of the trigonometric system) is a superposition of two pure tones. One such superposition is shown in Figure 5.5. A superposition of sines and cosines with frequencies $n = 1, \ldots, 50$ and randomly chosen amplitudes is depicted in Figure 5.6.

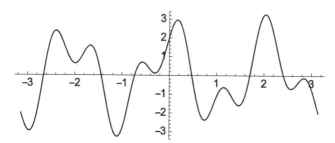

Fig. 5.5 Graph of $\varphi(t) = 2\cos(3t) + 1.3\sin(7t)$.

Equation (5.27) says that any function $f \in L^2[-\pi, \pi]$ can be represented as a sum of pure tones $a_n \cos nt$ and $b_n \sin nt$ over all possible frequencies. By superimposing all the pure tones with the correct amplitudes, we create any square-integrable sound that we like. The pure tones are our simple "building blocks," and by combining them we can create any sound (or signal, or function). Of course, the "superposition" is an infinite sum and the convergence is in the L^2-norm sense, but still the point is that by combining our simple sines and cosines we create very complicated functions f.

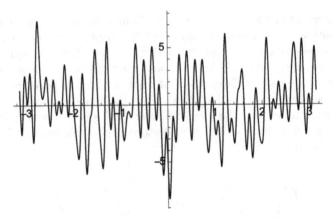

Fig. 5.6 Graph of 50 superimposed pure tones.

Example 5.10.3 (The Square Wave). Let

$$f(t) = \begin{cases} -1, & -\pi \le t < 0, \\ 1, & 0 \le t < \pi. \end{cases}$$

We call f the *square wave function.* Since f is piecewise continuous, it is Riemann integrable. Theorem 5.11.2 tells us that the Fourier series given in equation (5.20) converges to f in L^2-norm. That is, the partial sums

$$s_N = \frac{a_0}{2} + \sum_{n=1}^{N} a_n \cos nt + \sum_{n=1}^{N} b_n \sin nt$$

converge to f in L^2-norm.

We can explicitly compute the coefficients a_n and b_n. Since the square wave is odd and $\cos nt$ is even, we have $a_n = 0$ for every n, including $n = 0$. Problem 5.10.11 asks the reader to verify that if $n > 0$, then

$$
\begin{aligned}
b_n &= \frac{1}{\pi} \int_{-\pi}^{\pi} f(t) \sin nt \, dt \\
&= -\frac{1}{\pi} \int_{-\pi}^{0} \sin nt \, dt + \frac{1}{\pi} \int_{0}^{\pi} \sin nt \, dt = \begin{cases} 0, & n \text{ even}, \\ \frac{4}{\pi n}, & n \text{ odd.} \end{cases}
\end{aligned}
\tag{5.21}
$$

Hence if $N = 2K - 1$ is odd, then the partial sum s_{2K-1} simplifies to

$$s_{2K-1} = \sum_{n=1}^{2K-1} b_n \sin nt = \frac{4}{\pi} \sum_{k=1}^{K} \frac{\sin(2k-1)t}{2k-1}.$$

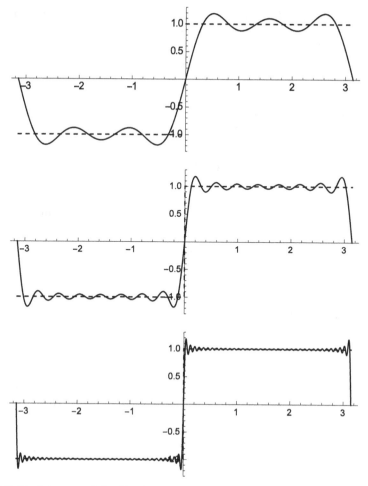

Fig. 5.7 Partial sums of the Fourier series of the square wave. Top: s_5. Middle: s_{15}. Bottom: s_{75}. The square wave itself is shown with dashed lines.

Figure 5.7 shows s_5, s_{15}, and s_{75}. It does appear from the diagram that $\|f - s_N\|_2 \to 0$ as $N \to \infty$, but we can also see *Gibbs's phenomenon* in this figure, which is that s_N does not converge *uniformly* to f. Instead, s_N always overshoots f near its points of discontinuity by an amount (about 9%) that does not decrease with N, so $\|f - s_N\|_u$ does not converge to zero. For a proof of Gibbs's phenomenon, see [DM72] or other texts on harmonic analysis. ◇

Since the normalized trigonometric system is an orthonormal basis for $L^2[-\pi, \pi]$, Theorem 5.8.1 tells us that in addition to the Fourier series expansion given in equation (5.20), the *Plancherel Equality* will hold. Writing this in terms of the Fourier coefficients a_n and b_n gives us the following result.

Theorem 5.10.4 (Plancherel's Equality). *If $f \in L^2[-\pi, \pi]$ (including any $f \in C[-\pi, \pi]$ in particular), then*

$$\|f\|_2^2 = \frac{\pi a_0^2}{2} + \pi \sum_{n=1}^{\infty} a_n^2 + \pi \sum_{n=1}^{\infty} b_n^2.$$

Proof. Applying Theorem 5.8.1(d) to the normalized trigonometric system given in equation (5.18), we see that

$$\|f\|_2^2 = \left| \left\langle f, \frac{1}{\sqrt{2\pi}} \right\rangle \right|^2 + \sum_{n=1}^{\infty} \left| \left\langle f, \frac{\cos nt}{\sqrt{\pi}} \right\rangle \right|^2 + \sum_{n=1}^{\infty} \left| \left\langle f, \frac{\sin nt}{\sqrt{\pi}} \right\rangle \right|^2$$

$$= \frac{1}{2\pi} |\langle f, 1 \rangle|^2 + \sum_{n=1}^{\infty} \frac{1}{\pi} |\langle f, \cos nt \rangle|^2 + \sum_{n=1}^{\infty} \frac{1}{\pi} |\langle f, \sin nt \rangle|^2$$

$$= \frac{\pi a_0^2}{2} + \pi \sum_{n=1}^{\infty} a_n^2 + \pi \sum_{n=1}^{\infty} b_n^2. \quad \square$$

Example 5.10.5. Let f be the square wave from Example 5.10.3. The square of the L^2-norm of f is

$$\|f\|_2^2 = \int_{-\pi}^{\pi} |f(t)|^2 \, dt = \int_{-\pi}^{\pi} 1 \, dt = 2\pi.$$

Recalling that $a_n = 0$ for every n and using the values b_n that we found in Example 5.10.3, the Plancherel Equality therefore implies that

$$2\pi = \|f\|_2^2 = \pi \sum_{n \in \mathbb{Z}} b_n^2 = \pi \sum_{\substack{n \in \mathbb{Z}, \\ n \text{ odd}}} \left| \frac{4}{\pi n} \right|^2 = \frac{16}{\pi} \sum_{\substack{n \in \mathbb{N}, \\ n \text{ odd}}} \frac{1}{n^2}.$$

Simplifying, we obtain

$$\frac{\pi^2}{8} = \sum_{\substack{n \in \mathbb{N}, \\ n \text{ odd}}} \frac{1}{n^2} = \sum_{k=1}^{\infty} \frac{1}{(2k-1)^2}. \tag{5.22}$$

Therefore

$$\sum_{n=1}^{\infty} \frac{1}{n^2} = \sum_{n=1}^{\infty} \frac{1}{(2n)^2} + \sum_{n=1}^{\infty} \frac{1}{(2n-1)^2} = \frac{1}{4} \sum_{n=1}^{\infty} \frac{1}{n^2} + \frac{\pi^2}{8}.$$

Collecting like sums and simplifying again gives us *Euler's formula*:

$$\sum_{n=1}^{\infty} \frac{1}{n^2} = \frac{\pi^2}{6}. \quad \diamond \tag{5.23}$$

Problems

5.10.6. Finish the proof that the trigonometric system is an orthogonal sequence in $C[-\pi, \pi]$ with respect to the L^2-inner product. Specifically, given any positive integers m and n, prove that:

(a) $\langle \sin mt, \cos nt \rangle = 0$,

(b) $\langle \sin mt, \sin nt \rangle = 0$ if $m \neq n$, and

(c) $\langle \cos mt, \cos nt \rangle = 0$ if $m \neq n$.

5.10.7. Prove that if $f \in C[\pi, \pi]$ is even, then $b_n = 0$ for every $n > 0$, and if $f \in C[\pi, \pi]$ is odd, then $a_n = 0$ for every $n \geq 0$.

5.10.8. Let $f \in C[-\pi, \pi]$ be given. With respect to the L^2-inner product on $C[-\pi, \pi]$, prove that the orthogonal projection of f onto the closed subspace

$$M = \text{span}\{1, \cos t, \cos 2t, \ldots, \cos Nt, \sin t, \sin 2t, \ldots, \sin Nt\}$$

is

$$p(t) = \frac{a_0}{2} + \sum_{n=1}^{N} a_n \cos nt + \sum_{n=1}^{N} b_n \sin nt.$$

5.10.9. Let $f(t) = t^2$ for $t \in [-\pi, \pi]$, and let

$$M = \text{span}\{1, \cos t, \sin t, \cos 2t, \sin 2t\}.$$

Find $\text{dist}(f, M)$, the distance from f to the subspace M with respect to the L^2-norm.

5.10.10. Given $f \in C[-\pi, \pi]$, prove that

$$\lim_{n \to \infty} \int_{-\pi}^{\pi} f(t) \cos nt \, dt = \lim_{n \to \infty} \int_{-\pi}^{\pi} f(t) \sin nt \, dt = 0.$$

5.10.11. Prove the formula in equation (5.21).

5.10.12. Let $f(t) = t^2 - \pi^2/3$ for $t \in [-\pi, \pi]$.

(a) Show that the Fourier coefficients of f are $a_0 = 0$, $a_n = 4(-1)^n/n^2$ for $n \in \mathbb{N}$, and $b_n = 0$ for $n \in \mathbb{N}$.

(b) Use the Plancherel Equality to prove that $\displaystyle\sum_{n=1}^{\infty} \frac{1}{n^4} = \frac{\pi^4}{90}$.

(c) Prove that $\displaystyle\sum_{n=1}^{\infty} \frac{1}{(2n+1)^4} = \frac{\pi^4}{96}$.

5.10.13. (Continuation of Problem 4.2.13.)

(a) Compute the Fourier coefficients a_n and b_n for the functions $f(t) = t$ and $g(t) = |t|$ on $[-\pi, \pi]$. Use this to show that the two series given in Problem 4.2.13 are the Fourier series for the functions f and g, and conclude that those series converge in L^2-norm to f and g.

(b) Use the Fourier coefficients for f to give alternative derivations of the sums stated in equations (5.22) and (5.23).

(c) Use the Fourier coefficients for g to give alternative derivations of the sums stated in parts (b) and (c) of Problem 5.10.12.

5.10.14. Let $f(t) = t(\pi - t) = \pi t - t^2$ for $t \in [0, \pi]$, and then set $f(t) = -f(-t)$ for $t \in [-\pi, 0)$, so that f is an odd function on $[-\pi, \pi]$. Compute the Fourier coefficients of f, and then use the Plancherel Equality to prove that

$$\sum_{n=1}^{\infty} \frac{1}{(2n+1)^6} = \frac{\pi^6}{960} \quad \text{and} \quad \sum_{n=1}^{\infty} \frac{1}{n^6} = \frac{\pi^6}{945}.$$

5.11 The Complex Trigonometric System

Now we will consider the complex version of the trigonometric system, which is a set of complex exponential functions. Therefore:

Throughout this section we take $\mathbb{F} = \mathbb{C}$.

We could use the facts that we have derived for the real trigonometric system to derive corresponding results for the complex system, but we instead choose to consider the complex system separately from the real system.

For simplicity of notation, it will be most convenient in this section to work with the trigonometric system in $C[0, 1]$, instead of the space $C[-\pi, \pi]$ that we used when dealing with the real trigonometric system.

Given $n \in \mathbb{Z}$ (note that n could be positive or negative), the *complex exponential of frequency n* is

$$e_n(t) = e^{2\pi i n t} = \cos(2\pi n t) + i \sin(2\pi n t), \qquad t \in \mathbb{R}.$$

The complex *trigonometric system* is

$$\mathcal{E} = \{e_n\}_{n \in \mathbb{Z}} = \{e^{2\pi i n t}\}_{n \in \mathbb{Z}}.$$

The domain of e_n is the real line, i.e., our input values t are real numbers. However, our output values $e_n(t) = e^{2\pi i n t}$ are complex numbers. In fact, $|e_n(t)| = |e^{2\pi i n t}| = 1$, so $e_n(t)$ always lies on the unit circle in $\mathbb{C}$. The graph of e_n is

$$\{(t, e^{2\pi i n t}) : t \in \mathbb{R}\} \subseteq \mathbb{R} \times \mathbb{C}.$$

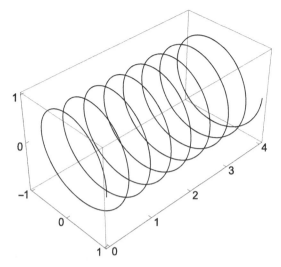

Fig. 5.8 Graph of $e_n(t) = e^{2\pi i n t}$ for $n = 2$ and $0 \le t \le 4$.

Identifying $\mathbb{R} \times \mathbb{C}$ with $\mathbb{R} \times \mathbb{R}^2 = \mathbb{R}^3$, this graph is a helix in $\mathbb{R}^3$ coiling around the t-axis, which runs down the center of the helix (see Figure 5.8). The real part of e_n is the cosine function $\cos(2\pi n t)$, while its imaginary part is the sine function $\sin(2\pi n t)$. We have $|e_n(t)| = 1$ for every t.

Recall that if n is an integer, then $e^{2\pi i n} = 1$. Therefore the complex exponential function $e_n(t) = e^{2\pi i n t}$ repeats after one unit, because

$$e_n(t+1) = e^{2\pi i n (t+1)} = e^{2\pi i n t} e^{2\pi i n} = e^{2\pi i n t} = e_n(t).$$

We therefore say that e_n is 1-*periodic*. As a consequence, once we know the values of $e_n(t)$ for t in an interval of length 1 then we know the values at all points t in $\mathbb{R}$. Hence we will usually restrict our attention just to t's that lie in the interval $[0, 1]$ instead of considering all $t \in \mathbb{R}$.

The complex exponential functions e_n are continuous and hence are Riemann integrable on $[0, 1]$. It is easy to prove that the complex trigonometric system is orthogonal, because if $m \ne n$, then we just use a u-substitution to compute that

$$\langle e_m, e_n \rangle = \int_0^1 e_m(t) \,\overline{e_n(t)} \, dt$$

$$= \int_0^1 e^{2\pi i m t} e^{-2\pi i n t} \, dt$$

$$= \int_0^1 e^{2\pi i (m-n) t} \, dt$$

$$= \left. \frac{e^{2\pi i(m-n)t}}{2\pi i(m - n)} \right|_0^1 \qquad (u = 2\pi i(m - n)t)$$

$$= \frac{e^{2\pi i(m-n)} - 1}{2\pi i(m - n)}$$

$$= \frac{1 - 1}{2\pi i(m - n)} \qquad (\text{since } e^{2\pi i(m-n)} = 1)$$

$$= 0.$$

This shows that $\{e_n\}_{n \in \mathbb{Z}}$ is an orthogonal sequence. In fact, it is *orthonormal*, because for each each $n \in \mathbb{Z}$ we have

$$\|e_n\|_2^2 = \int_0^1 |e_n(t)|^2 \, dt = \int_0^1 |e^{2\pi i n t}|^2 \, dt = \int_0^1 1^2 \, dt = 1.$$

We saw in Lemma 3.4.7 that $C[0, 1]$ is not complete with respect to the L^2-norm. Instead, it is a dense subspace of the larger Lebesgue space $L^2[0, 1]$, and we will use that fact to prove that the trigonometric system is an orthonormal basis for the Hilbert space $L^2[0, 1]$.

Theorem 5.11.1. *The trigonometric system* $\mathcal{E} = \{e^{2\pi i n t}\}_{n \in \mathbb{Z}}$ *is an orthonormal basis for* $L^2[0, 1]$.

Proof. Since we have already shown that $\mathcal{E}$ is orthonormal, Theorem 5.8.1 tells us that it remains only to prove that $\mathcal{E}$ is a *complete sequence*. That is, we must show that span($\mathcal{E}$) is dense in $L^2[0, 1]$.

Choose any function $f \in L^2[0, 1]$, and fix $\varepsilon > 0$. Since $C[0, 1]$ is dense in $L^2[0, 1]$, there exists a continuous function $g \in C[0, 1]$ such that $\|f - g\|_2 < \varepsilon$. Letting

$$C_{\mathrm{per}}[0, 1] = \{f \in C[0, 1] : f(0) = f(1)\},$$

we can use the same type of argument that was employed in the proof of Theorem 5.10.1 to obtain a function $h \in C_{\mathrm{per}}[0, 1]$ that satisfies $\|g - h\|_2 < \varepsilon$ (use equation (5.19) but with $[-\pi, \pi]$ replaced by $[0, 1]$).

Let $\mathbb{T} = \{z \in \mathbb{C} : |z| = 1\}$ be the unit circle in the complex plane. Each point $z \in \mathbb{T}$ has the form $z = e^{2\pi i t}$, where t is real, and values of t that differ by an integer amount correspond to the same point z. Therefore we can define a function $H \colon \mathbb{T} \to \mathbb{C}$ by $H(e^{2\pi i t}) = h(t)$ for $t \in [0, 1]$. The fact that h is continuous and satisfies $h(0) = h(1)$ implies that H is continuous on $\mathbb{T}$, i.e., $H \in C(\mathbb{T})$. Problem 4.7.11, which is an application of the complex version of the Stone–Weierstrass Theorem, therefore implies that there exists a complex polynomial

$$P(z) = \sum_{n=-N}^{N} c_n z^n, \qquad z \in \mathbb{T},$$

such that $\|H - P\|_u < \varepsilon$. Setting

$$p(t) = P(e^{2\pi it}) = \sum_{n=-N}^{N} c_n e^{2\pi int}, \qquad t \in [0,1],$$

gives us an element of $\text{span}(\mathcal{E})$ with $\|h - p\|_u < \varepsilon$. Consequently, the L^2-distance between h and p is

$$\|h - p\|_2^2 = \int_0^1 |h(t) - p(t)|^2 \, dt \leq \int_0^1 \|h - p\|_u^2 \, dt \leq \int_0^1 \varepsilon^2 \, dt = \varepsilon.$$

Using the Triangle Inequality to combine the above estimates, we see finally that

$$\|f - p\|_2 \leq \|f - g\|_2 + \|g - h\|_2 + \|h - p\|_2 < \varepsilon + \varepsilon + \varepsilon = 3\varepsilon.$$

Hence every $f \in L^2[0,1]$ can be approximated as closely as we like in L^2-norm by a function $p \in \text{span}(\mathcal{E})$. This shows that $\text{span}(\mathcal{E})$ is dense and therefore $\mathcal{E}$ is complete, so $\mathcal{E}$ is an orthonormal basis for $L^2[0,1]$ by Theorem 5.8.1. $\square$

Our proof of Theorem 5.11.1 relied on the Stone–Weierstrass Theorem. An alternative and more constructive proof, based on the operation of *convolution* and the notion of *approximate identities*, can be found in most texts on harmonic analysis, such as [DM72], [Ben97], [Kat04], and [Heil11, Thm. 13.23].

Given $f \in L^2[0,1]$, we usually denote the inner products of f with the complex exponentials e_n by

$$\widehat{f}(n) = \langle f, e_n \rangle = \int_0^1 f(t) \, e^{-2\pi int} \, dt, \qquad n \in \mathbb{Z}, \tag{5.24}$$

and refer to $\widehat{f}(n)$ as the *n*th *Fourier coefficient* of f (pronounced "f-hat of n"). The integral in equation (5.24) is a Lebesgue integral, but if f is Riemann integrable (for example, if $f \in C[0,1]$ or if f is piecewise continuous), then its Riemann integral and its Lebesgue integral are the same. Using this notation, the next result follows by applying Theorem 5.8.1 to the trigonometric system.

Theorem 5.11.2 (Fourier Series for $L^2[0,1]$). *For each $f \in L^2[0,1]$ we have*

$$f = \sum_{n \in \mathbb{Z}} \widehat{f}(n) \, e_n, \tag{5.25}$$

where this series converges unconditionally in the norm of $L^2[0,1]$. Furthermore, for $f \in L^2[0,1]$ the Plancherel Equality *holds:*

$$\|f\|_2^2 = \sum_{n \in \mathbb{Z}} |\widehat{f}(n)|^2, \tag{5.26}$$

and for $f, g \in L^2[0,1]$ the Parseval Equality *holds:*

$$\langle f, g \rangle = \sum_{n \in \mathbb{Z}} \widehat{f}(n)\, \overline{\widehat{g}(n)}. \qquad \Diamond$$

Equation (5.25) is called the *Fourier series representation* of f. We often write it in the form

$$f(t) = \sum_{n \in \mathbb{Z}} \widehat{f}(n)\, e^{2\pi i n t}, \tag{5.27}$$

but it is important to note that we know only that this series converges *in L^2-norm*. It need not converge *pointwise*, i.e., equation (5.27) need not hold for individual t, even if f is continuous! Indeed, establishing the convergence of Fourier series in senses other than L^2-norm can be extremely difficult. For one source where more on this topic can be found we refer the reader to [Heil11, Ch. 14].

Problems

5.11.3. Assume that α is a real number that is not an integer, and let f be the continuous, complex-valued function

$$f(t) = \frac{\pi\, e^{\pi i \alpha}}{\sin \pi \alpha}\, e^{-2\pi i \alpha t}, \qquad t \in [0,1].$$

Show that $\widehat{f}(n) = 1/(n+\alpha)$ for each $n \in \mathbb{Z}$, and use the Plancherel Equality to prove that

$$\sum_{n=-\infty}^{\infty} \frac{1}{(n+\alpha)^2} = \frac{\pi^2}{\sin^2 \pi \alpha}.$$

5.11.4. Let χ and ψ be the *box function* and the *Haar wavelet*, defined for $t \in \mathbb{R}$ by

$$\chi(t) = \begin{cases} 1, & 0 \leq t < 1, \\ 0, & \text{otherwise}, \end{cases} \qquad \text{and} \qquad \psi(t) = \begin{cases} 1, & 0 \leq t < \tfrac{1}{2}, \\ -1, & \tfrac{1}{2} \leq t < 1, \\ 0, & \text{otherwise}. \end{cases}$$

Given integers $n, k \in \mathbb{Z}$, let

$$\psi_{n,k}(t) = 2^{n/2}\psi(2^n t - k).$$

The *Haar system* on the interval $[0,1]$ is

$$\mathcal{H} = \{\chi\} \cup \{\psi_{n,k}\}_{n \geq 0,\, k=0,\dots,2^n-1}.$$

That is, the Haar system consists of the box function and all of the functions $\psi_{n,k}$ that are nonzero only within the interval $[0,1]$. Prove that the Haar system is an orthonormal sequence in $L^2[0,1]$ with respect to the L^2-inner product.

Remark: It can be shown that the Haar system is an orthonormal basis for the Lebesgue space $L^2[0,1]$ (for one proof, see [Heil11, Example 1.54]). The Haar system is the simplest example of a *wavelet orthonormal basis*.

5.11.5. For this problem we take $\mathbb{F} = \mathbb{C}$. For each $\xi \in \mathbb{R}$, define a function $e_\xi \colon \mathbb{R} \to \mathbb{C}$ by $e_\xi(t) = e^{2\pi i \xi t}$. Let $H = \operatorname{span}\{e_\xi\}_{\xi \in \mathbb{R}}$, i.e., H consists of all finite linear combinations of the functions e_ξ. Show that

$$\langle f, g \rangle = \lim_{T \to \infty} \frac{1}{2T} \int_{-T}^{T} f(t)\,\overline{g(t)}\,dt, \qquad f, g \in H,$$

defines an inner product on H, and $\{e_\xi\}_{\xi \in \mathbb{R}}$ is an uncountable orthonormal system in H. Use this to show that H is nonseparable.

Remark: H is not complete, but it lies inside of a larger complete nonseparable Hilbert space $\widetilde{H}$ that contains the important class of *almost periodic functions* (see [Kat04]).

Chapter 6
Operator Theory

Our focus in much of Chapters 2–5 was on particular vector spaces and on particular vectors in those spaces. Most of those spaces were metric, normed, or inner product spaces. Now we will concentrate on classes of *operators* that transform vectors in one space into vectors in another space. We will be especially interested in operators that map one normed space into another normed space.

6.1 Linear Operators on Normed Spaces

Since a normed space is a vector space, it has a "structure" that a generic metric space need not possess. In this section we will study *linear functions*, which preserve this vector space structure. Linearity is distinct from continuity; a continuous function preserves convergent sequences, while a linear function preserves the operations of vector addition and multiplication by scalars. We will see that we can completely characterize the linear functions that are also continuous in terms of a concept called *boundedness*. First, however, we introduce some terminology and notation.

We often refer to functions on vector spaces as *operators* or *transformations*, especially if they are linear (indeed, some authors restrict the use of these terms solely to linear functions). We will use these three terms interchangeably when dealing with vector spaces, i.e., for us the words *function*, *operator*, and *transformation* all mean the same thing.

Operators on normed spaces are often denoted by capital letters, such as A or L, or by Greek letters, such as μ or ν. Also, it is traditional, and often notationally simpler, to make the use of parentheses optional when denoting the output of a transformation. That is, we often write Ax instead of $A(x)$, although we include parentheses when there is any danger of confusion. Operators are just functions, so all of the standard terminology for functions

© Springer International Publishing AG, part of Springer Nature 2018
C. Heil, *Metrics, Norms, Inner Products, and Operator Theory*, Applied and
Numerical Harmonic Analysis, https://doi.org/10.1007/978-3-319-65322-8_6

defined in Section 1.3 applies to operators on vector spaces. Here is some additional terminology that specifically applies to operators on vector spaces.

Definition 6.1.1 (Notation for Operators). Let X and Y be vector spaces, and let $A\colon X \to Y$ be an operator.

(a) A is *linear* if $A(ax + by) = aAx + bAy$ for $x, y \in X$ and $a, b \in \mathbb{F}$.

(b) A is *antilinear* if $A(ax + by) = \overline{a}Ax + \overline{b}Ay$ for $x, y \in X$ and $a, b \in \mathbb{F}$.

(c) The *kernel* or *nullspace* of A is $\ker(A) = \{x \in X : Ax = 0\}$.

(d) The *range* of A is $\operatorname{range}(A) = A(X) = \{Ax : x \in X\}$.

(e) The *rank* of A is the vector space dimension of its range:

$$\operatorname{rank}(A) = \dim(\operatorname{range}(A)).$$

   If $\operatorname{range}(A)$ is finite-dimensional, then we say that A is a *finite-rank operator*.

(f) If $Y = \mathbb{F}$ (i.e., $A\colon X \to \mathbb{F}$ is a scalar-valued function), then A is called a *functional.* $\diamondsuit$

   The *zero operator* is the function $0\colon X \to Y$ that maps every vector in X to the zero vector in Y. The zero operator and the zero vector in a vector space are both denoted by the symbol 0, i.e., the rule for the zero operator is $0(x) = 0$ for $x \in X$. It should usually be clear from context whether the symbol 0 is meant to denote an operator or a vector.

Lemma 6.1.2. *Let X and Y be vector spaces. If $A\colon X \to Y$ is a linear operator, then the following statements hold.*

(a) $\ker(A)$ *is a subspace of X, and* $\operatorname{range}(A)$ *is a subspace of Y.*

(b) $A(0) = 0$, *and therefore* $0 \in \ker(A)$.

(c) A *is injective if and only if* $\ker(A) = \{0\}$.

(d) *If A is a linear* bijection, *then its inverse mapping* $A^{-1}\colon Y \to X$ *is also a linear bijection.*

Proof. We will prove part (c), and assign the proof of the remaining parts as Problem 6.1.5.

   (c) Assume that A is injective, and suppose that $Ax = 0$. Since A is linear, we know that we also have $A0 = 0$. But A is one-to-one, so this implies that $x = 0$. Hence $\ker(A) = \{0\}$.

   For the converse implication, suppose that $\ker(A) = 0$, and assume that $Ax = Ay$ for some $x, y \in X$. Since A is linear, this implies that $A(x - y) = Ax - Ay = 0$, and therefore $x - y \in \ker(A) = \{0\}$. Hence $x = y$, so A is injective. $\square$

Example 6.1.3. The space $C_b^1(\mathbb{R})$, which is the set of all bounded differentiable functions $f\colon \mathbb{R} \to \mathbb{F}$ whose derivative f' is both continuous and bounded, was introduced in Problem 3.5.25. Let $D\colon C_b^1(\mathbb{R}) \to C_b(\mathbb{R})$ be the derivative operator defined by $Df = f'$ for $f \in C_b^1(\mathbb{R})$. Since the derivative of a sum is the sum of the derivatives, we have

$$D(f+g) \;=\; (f+g)' \;=\; f'+g' \;=\; Df+Dg. \tag{6.1}$$

We also know that differentiation "respects" scalar multiplication, i.e.,

$$D(cf) \;=\; (cf)' \;=\; cf' \;=\; c(Df). \tag{6.2}$$

Combining equations (6.1) and (6.2), we see that D is a linear operator.

Note that it is not the functions f and g in this example that are linear but rather the operator D that transforms a function f into its derivative f'. For example, the functions $f(t) = \sin t$ and $g(t) = e^{-t^2}$ both belong to $C_b^1(\mathbb{R})$, but they are not linear. Instead, the *transformation* D is linear because it sends the input $af + bg$ to the output $af' + bg'$.

If $f(t) = c$ is any constant function, then f belongs to $C_b^1(\mathbb{R})$ and $Df = f' = 0$. Hence the kernel of D contains more than just the zero function. That is, $\ker(D) \neq \{0\}$. Since D is linear, Lemma 6.1.2 therefore implies that D is not injective. Indeed, we can see this directly because if f and g are any two constant functions, then $Df = 0 = Dg$.

Thus $\ker(D)$ contains every constant function. On the other hand, since we know that the *only* functions on the domain $\mathbb{R}$ whose derivatives are identically zero are constant functions, it follows that $\ker(D)$ is exactly the set of all constant functions. In accordance with Lemma 6.1.2, this is a subspace of $C_b^1(\mathbb{R})$; in fact, it is a one-dimensional subspace. $\Diamond$

If Y is a normed space, then we sometimes need to consider the closure of the range of A in Y. Since range(A) is a subspace of Y, Problem 3.3.12 implies that the closure of the range is a closed subspace of Y. For simplicity of notation, we denote the closure of the range by

$$\overline{\operatorname{range}}(A) \;=\; \overline{\operatorname{range}(A)}.$$

If the range of A is a closed set, i.e., if $\overline{\operatorname{range}}(A) = \operatorname{range}(A)$, then we say that A *has closed range.*

The domain X and the codomain Y of most of the operators that we will consider in this chapter will be normed spaces. Consequently, there will often be several norms in play at any given time. We typically will let the symbols $\|\cdot\|$ denote both the norm on X and the norm on Y, letting context determine the space that the norm is on. For example, if x is a vector in X, then its image Ax is a vector in Y, so in most circumstances it is safe to simply write $\|x\|$ and $\|Ax\|$, since it is clear that $\|x\|$ must denote the norm on X, while $\|Ax\|$ must denote the norm on Y. If we should need to emphasize precisely which spaces with respect to which the norm is to taken, then we will write $\|x\|_X$ and $\|Ax\|_Y$.

Problems

6.1.4. Suppose that $A \colon \mathbb{F}^n \to \mathbb{F}^m$ is a linear operator.

(a) Prove that there exists a unique $m \times n$ matrix $[a_{ij}]$ such that

$$Ax = \begin{bmatrix} a_{11} & \cdots & a_{1n} \\ \vdots & \ddots & \vdots \\ a_{m1} & \cdots & a_{mn} \end{bmatrix} \begin{bmatrix} x_1 \\ \vdots \\ x_n \end{bmatrix}, \qquad x = (x_1, \ldots, x_n) \in \mathbb{F}^n.$$

Remark: We usually identify the linear function A with the matrix $[a_{ij}]$, and simply write $A = [a_{ij}]$.

(b) The *column space* $C(A)$ of the matrix A is defined to be the span of the n columns of A. Prove that $C(A) = \text{range}(A)$, the range of the linear operator A.

6.1.5. Prove parts (a), (b), and (d) of Lemma 6.1.2.

6.1.6. Although this chapter is mostly concerned with linear operators, this problem is about an interesting nonlinear operator.

(a) Let
$$M = \{f \in C[0,1] : f(0) = 0 \text{ and } f(1) = 1\}.$$

Prove that M is a closed subset (although not a subspace) of $C[0,1]$, and therefore M is a complete metric space with respect to the uniform metric $d_u(f,g) = \|f - g\|_u$.

(b) Given $f \in M$, define Af by

$$Af(x) = \begin{cases} \frac{1}{2}f\left(\frac{x}{1-x}\right), & 0 \le x < \frac{1}{2}, \\ 1 - \frac{1}{2}f\left(\frac{1-x}{x}\right), & \frac{1}{2} \le x \le 1. \end{cases}$$

Prove that $Af \in M$, and therefore A is an operator that maps M into M.

(c) Show that

$$\|Af - Ag\|_u \le \frac{\|f - g\|_u}{2}, \qquad \text{for all } f, g \in M.$$

Consequently, A is Lipschitz on M in the sense defined in Problem 2.9.20.

(d) Use the Banach Fixed Point Theorem (Problem 2.9.21) to prove that there exists a unique function $m \in M$ such that $Am = m$. This function is called the *Minkowski question mark function* or the *slippery Devil's staircase*.

(e) Set $f_0(x) = x$, and define $f_{n+1} = Af_n$ for $n \in \mathbb{N}$. Prove that f_n converges uniformly to m (this fact was used to generate the approximation to m that appears in Figure 6.1).

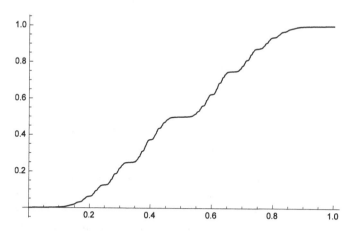

Fig. 6.1 The function f_7 from Problem 6.1.6, approximating the slippery Devil's staircase.

6.2 The Definition of a Bounded Operator

In many circumstances, it is convenient to say that a function $f \colon X \to Y$ is "bounded" on X if the range of f is a bounded subset of Y. For example, this is what we mean when we say that $f(x) = \sin x$ is a bounded function on $\mathbb{R}$, and this type of boundedness is exactly what the subscript "b" stands for in the definition of the space $C_b(X)$.

However, if X and Y are normed spaces and $A \colon X \to Y$ is a *linear* operator, then the range of A is a *subspace* of Y. No subspace other than the trivial subspace $\{0\}$ is a bounded set, so if A is not the zero operator, then range(A) is not a bounded subset of Y. Therefore, a nonzero linear operator can never be bounded in the sense that its range is a bounded subset of Y. Instead, we will formulate the idea of "boundedness" for *linear* operators by looking at the relationship between the size of $\|x\|$ and the size of $\|Ax\|$. If there is a limit to how large $\|Ax\|$ can be *in comparison to* $\|x\|$, then we will say that A *is bounded*. Here is the precise definition.

Definition 6.2.1 (Bounded Linear Operator). Let X and Y be normed spaces. We say that a linear operator $A \colon X \to Y$ is *bounded* if there exists a finite constant $K \geq 0$ such that

$$\|Ax\| \leq K \|x\|, \qquad \text{for all } x \in X. \qquad \diamond \qquad (6.3)$$

In particular, if A is bounded, then $\|Ax\| \leq K$ for every unit vector x. We give a name to the supremum of $\|Ax\|$ over the unit vectors x.

Definition 6.2.2 (Operator Norm). Let X and Y be normed spaces. If X is nontrivial, then the *operator norm* of a linear operator $A: X \to Y$ is

$$\|A\| = \sup_{\|x\|=1} \|Ax\|. \tag{6.4}$$

If $X = \{0\}$ (in which case there are no unit vectors in S), then we set $\|A\| = 0$. $\Diamond$

Note that three different uses of the symbols $\|\cdot\|$ appear in equation (6.4), with three different meanings. Each meaning is determined by context. First, since x is a vector in X, we understand that $\|x\|$ refers to the norm on the space X. Second, since Ax is a vector in Y, we implicitly know that $\|Ax\|$ means the norm on Y. Third, since A is an operator that maps X to Y, we understand that $\|A\|$ denotes the operator norm of A. On those occasions where we need to carefully distinguish among the various norms in play, we may use subscripts to indicate what space a given norm refers to. For example, we might write $\|x\|_X$, $\|Ax\|_Y$, and $\|A\|_{X \to Y}$ to distinguish between the norm on X, the norm on Y, and the operator norm. Rewriting equation (6.4) using this notation, the operator norm of A would be given by

$$\|A\|_{X \to Y} = \sup_{\|x\|_X = 1} \|Ax\|_Y.$$

Another common notation is to denote the operator norm by $\|A\|_{\mathrm{op}}$, instead of $\|A\|$.

Remark 6.2.3. We will implicitly assume in most proofs that X is nontrivial, since the case $X = \{0\}$ is usually easy to deal with. The reader should check that the statements of theorems are valid even if $X = \{0\}$. $\Diamond$

We have not yet shown that the operator norm is a norm in any sense, but to illustrate its meaning let

$$S = \{x \in X : \|x\| = 1\}$$

be the unit sphere in X. The operator norm is the supremum of the norms of the vectors in

$$A(S) = \{Ax : x \in S\} = \{Ax : \|x\| = 1\}.$$

Since $A(S)$ is the direct image of the unit sphere under A, the operator norm $\|A\|$ is in a sense the "maximum distortion of the unit sphere" by A (although the supremum in the definition of $\|A\|$ need not be achieved, and we could have $\|A\| = \infty$ *even though* $\|Ax\|$ is finite, by definition, for every $x \in S$).

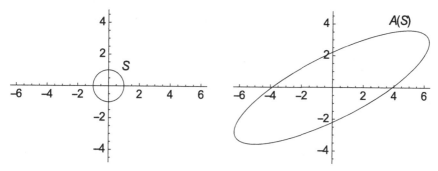

Fig. 6.2 Left: The unit circle $S = \{x \in \mathbb{R}^2 : \|x\|_2 = 1\}$. Right: The image of S under the linear operator $A: \mathbb{R}^2 \to \mathbb{R}^2$ whose matrix is given in equation (6.5). The operator norm $\|A\|$ is the length of a semimajor axis of the ellipse $A(S)$.

For example, consider $X = \mathbb{R}^n$ and $Y = \mathbb{R}^m$. Every linear operator $A: \mathbb{R}^n \to \mathbb{R}^m$ is given by an $m \times n$ matrix, which we also call A. That is, there is an $m \times n$ matrix $A = [a_{ij}]$ such that Ax (the image of x under the operator A) is simply the product of the matrix $A = [a_{ij}]$ with the vector x (see Problem 6.1.4). If we place the Euclidean norm on both $\mathbb{R}^n$ and $\mathbb{R}^m$, then S is the ordinary unit sphere in $\mathbb{R}^n$, and $A(S)$ is an ellipsoid in $\mathbb{R}^m$ (although it might be degenerate, i.e., one or more of its axes might have zero length). The ellipse $A(S)$ corresponding to $m = n = 2$ and the matrix

$$A = \begin{bmatrix} 6 & 2 \\ 2 & 3 \end{bmatrix} \tag{6.5}$$

is shown in Figure 6.2. The operator norm of A is the norm of the vector on $A(S)$ farthest from the origin, i.e., it is the length of a semimajor axis of the ellipsoid (what is that length, exactly?).

Here are some properties of bounded linear operators.

Lemma 6.2.4. *If X and Y are normed spaces and $A: X \to Y$ is a linear operator, then the following statements hold.*

(a) *A is bounded if and only if $\|A\| < \infty$.*

(b) *$\|Ax\| \leq \|A\| \, \|x\|$ for all $x \in X$.*

(c) *If A is bounded, then $\|A\|$ is the smallest constant $K \geq 0$ such that $\|Ax\| \leq K\|x\|$ for all $x \in X$.*

(d) *If $X \neq \{0\}$, then $\|A\| = \displaystyle\sup_{\|x\| \leq 1} \|Ax\| = \sup_{x \neq 0} \frac{\|Ax\|}{\|x\|}$.*

Proof. We will prove part (a), and assign the proofs of the remaining parts as Problem 6.2.5.

⇒. Assume that A is bounded. Then there exists a finite constant $K \geq 0$ such that $\|Ax\| \leq K\|x\|$ for all $x \in X$. Consequently,

$$\|A\| = \sup_{\|x\|=1} \|Ax\| \leq \sup_{\|x\|=1} K\|x\| = K < \infty.$$

$\Leftarrow$. Now suppose that $\|A\| < \infty$. Choose any nonzero vector $x \in X$, and set $y = \frac{x}{\|x\|}$. Then $\|y\| = 1$, so

$$\|A\| = \sup_{\|z\|=1} \|Az\| \geq \|Ay\| = \left\|A\left(\frac{x}{\|x\|}\right)\right\| = \frac{1}{\|x\|}\|Ax\|.$$

Rearranging, we see that $\|Ax\| \leq \|A\|\,\|x\|$ whenever $x \neq 0$. This is also true for $x = 0$, so taking $K = \|A\|$, we have $\|Ax\| \leq K\|x\|$ for every x. Since $K = \|A\|$ is a finite number, this shows that A is bounded. $\square$

Problems

6.2.5. Prove parts (b)–(c) of Lemma 6.2.4.

6.2.6. Let X and Y be normed spaces, and suppose that $A\colon X \to Y$ is a bounded linear operator. Show that if the norm on X is replaced by an equivalent norm and the norm on Y is replaced by an equivalent norm, then A remains bounded with respect to these new norms.

6.2.7. Let X be a normed space, and assume $A\colon X \to X$ is bounded and linear. Prove that if λ is an eigenvalue of A (i.e., $Ax = \lambda x$ for some nonzero $x \in X$), then $|\lambda| \leq \|A\|$.

6.2.8. Let $N \in \mathbb{N}$ be a *fixed* integer, and define an operator $A_N\colon \ell^1 \to \ell^1$ by

$$A_N x = (x_1, \dots, x_N, 0, 0, \dots), \qquad x = (x_k)_{k \in \mathbb{N}} \in \ell^1.$$

Prove the following statements.

(a) A_N is bounded and linear, and $\|A_N\| = 1$.

(b) A_N is not injective and not surjective.

(c) A_N is not an isometry, i.e., there is some $x \in \ell^1$ such that $\|Ax\|_1 \neq \|x\|_1$ (compare Definition 6.6.1).

(d) $\ker(A_N)$ is a proper closed subspace of ℓ^1.

(e) $\mathrm{range}(A_N)$ is a proper closed subspace of ℓ^1.

6.2.9. Define an operator $B\colon \ell^1 \to \ell^1$ by

$$Bx = \left(\frac{x_k}{k}\right)_{k \in \mathbb{N}} = \left(x_1, \frac{x_2}{2}, \frac{x_3}{3}, \dots\right), \qquad x = (x_k)_{k \in \mathbb{N}} \in \ell^1.$$

Prove the following statements.

(a) B is bounded and linear, and $\|B\| = 1$.

(b) B is injective (and therefore $\ker(B) = \{0\}$), but B is not surjective.

(c) B is not an isometry, i.e., there is some $x \in \ell^1$ such that $\|Bx\|_1 \neq \|x\|_1$ (compare Definition 6.6.1).

(d) range(B) is a proper dense subspace of ℓ^1, but it is *not closed*.

6.2.10. Let X be a Banach space and let Y be a normed space. Suppose that $A\colon X \to Y$ is bounded and linear, and suppose there exists a scalar $c > 0$ such that
$$\|Ax\| \geq c\,\|x\|, \qquad \text{for all } x \in X.$$

Prove that A is injective and range(A) is a closed subspace of Y. How does this relate to Problems 6.2.8 and 6.2.9?

6.3 Examples

We will give some examples of bounded and unbounded linear operators. First, however, we prove that if the *domain* of a linear operator is finite-dimensional, then the operator is automatically bounded.

Theorem 6.3.1. *Let X and Y be normed vector spaces. If X is finite-dimensional and $A\colon X \to Y$ is linear, then A is bounded.*

Proof. Because X is finite-dimensional, all norms on X are equivalent (see Theorem 3.7.2). As a consequence, A is bounded with respect to one norm on X if and only if it is bounded with respect to every other norm (this is Problem 6.2.6). Therefore we can work with any norm on X that we like, and for this proof we choose the analogue of the ℓ^1-norm on $\mathbb{F}^d$. That is, just as we did in equation (3.27), we fix a particular basis $\mathcal{B} = \{e_1, \ldots, e_d\}$ for X, write $x \in X$ as $x = \sum_{k=1}^d c_k(x)\,e_k$, and declare the norm of x to be
$$\|x\|_1 = \sum_{k=1}^d |c_k(x)|.$$

If $x \in X$, then $Ax = \sum_{k=1}^d c_k(x)\,Ae_k$, since A is linear. Therefore
$$
\begin{aligned}
\|Ax\| &= \left\| \sum_{k=1}^d c_k(x)\,Ae_k \right\| \\
&\leq \sum_{k=1}^d |c_k|\,\|Ae_k\| \qquad \text{(Triangle Inequality)} \\
&\leq \left(\sum_{k=1}^d |c_k| \right) \left(\max_{k=1,\ldots,d} \|Ae_k\| \right) = C\,\|x\|_1,
\end{aligned}
$$

where $C = \max \|Ae_k\|$. The number C is finite, since there are only finitely many vectors Ae_k. Since we have shown that $\|Ax\| \leq C \|x\|_1$ for every x, we conclude that the operator A is bounded. $\square$

Thus, all linear operators on *finite-dimensional domains* are bounded. In contrast, we will show that a linear operator on an infinite-dimensional domain can be unbounded, *even if its range is finite-dimensional!* We will illustrate this with some examples of operators whose range is the one-dimensional field of scalars $\mathbb{F}$. According to Definition 6.1.1, a function of the form $\mu \colon X \to \mathbb{F}$ is called a *functional* on X. Since $\mu(x)$ is a scalar for every x and since the norm on $\mathbb{F}$ is absolute value, the operator norm of such a functional μ is

$$\|\mu\| \;=\; \sup_{\|x\|=1} |\mu(x)|. \tag{6.6}$$

We start with an example of a *bounded* linear functional δ on the infinite-dimensional domain $C_b(\mathbb{R})$. This functional δ will map each element of $C_b(\mathbb{R})$ to a scalar. The input to δ is a continuous bounded function $f \in C_b(\mathbb{R})$, and the output is a scalar $\delta(f)$. This particular functional δ will determine which scalar to output by evaluating f at the origin.

Example 6.3.2. Consider $X = C_b(\mathbb{R})$, the space of all bounded continuous functions $f \colon \mathbb{R} \to \mathbb{F}$. Theorem 3.5.10 showed that $C_b(\mathbb{R})$ is a Banach space with respect to the uniform norm

$$\|f\|_{\mathrm{u}} \;=\; \sup_{t \in \mathbb{R}} |f(t)|.$$

Define a functional $\delta \colon C_b(\mathbb{R}) \to \mathbb{F}$ by

$$\delta(f) \;=\; f(0), \quad f \in C_b(\mathbb{R}).$$

That is, $\delta(f)$ is simply the value that f takes at the origin. For example, if $f(t) = \cos t$, then

$$\delta(f) \;=\; \cos 0 \;=\; 1;$$

if $g(t) = e^{-|t|}$, then

$$\delta(g) \;=\; e^0 \;=\; 1;$$

and if $h(t) = f(t) + g(t) = \cos t + e^{-|t|}$, then

$$\delta(h) \;=\; \cos 0 + e^0 \;=\; 2.$$

Note that δ is not injective, since $\delta(f) = \delta(g)$ even though $f \neq g$. On the other hand, δ is surjective (why?). Can you explicitly describe the kernel of δ? Is it a *closed* subspace of $C_b(\mathbb{R})$?

We will show that δ is a bounded linear functional on $C_b(\mathbb{R})$. To show linearity, choose functions $f, g \in C_b(\mathbb{R})$ and scalars $a, b \in \mathbb{F}$. Then, by definition of addition of functions,

$$\delta(af + bg) = (af + bg)(0) = af(0) + bg(0) = a\,\delta(f) + b\,\delta(g). \quad (6.7)$$

Hence δ is linear on $C_b(\mathbb{R})$.

To see that δ is bounded, fix any $f \in C_b(\mathbb{R})$. Then

$$|\delta(f)| = |f(0)| \leq \sup_{t \in \mathbb{R}} |f(t)| = \|f\|_u.$$

This shows that δ is bounded, and its operator norm satisfies

$$\|\delta\| = \sup_{\|f\|_u = 1} |\delta(f)| \leq \sup_{\|f\|_u = 1} \|f\|_u = 1.$$

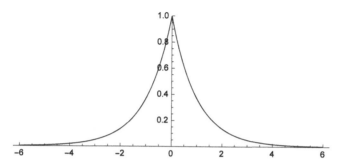

Fig. 6.3 The *two-sided exponential function* $g(t) = e^{-|t|}$.

In fact, we can determine the exact value of $\|\delta\|$. Consider the *two-sided exponential function* $g(t) = e^{-|t|}$, pictured in Figure 6.3. This function satisfies $g(0) = 1$ and $|g(t)| \leq 1$ for all t (any other function $g \in C_b(\mathbb{R})$ that has these two properties would do just as well for this argument). We have $\|g\|_u = 1$, so g is a unit vector in $C_b(\mathbb{R})$. Therefore

$$\|\delta\| = \sup_{\|f\|_u = 1} |\delta(f)| \geq |\delta(g)| = |g(0)| = 1.$$

Hence the operator norm of δ is precisely $\|\delta\| = 1$. $\diamondsuit$

The functional δ is just one of infinitely many different linear functionals that we can define on $C_b(\mathbb{R})$. For example, given $a \in \mathbb{R}$ we could define $\delta_a \colon C_b(\mathbb{R}) \to \mathbb{F}$ to be evaluation at a instead of evaluation at the origin, i.e., we could set $\delta_a(f) = f(a)$ for $f \in C_b(\mathbb{R})$. Here is a different type of bounded linear functional on $C_b(\mathbb{R})$.

Example 6.3.3. Define a functional $\mu \colon C_b(\mathbb{R}) \to \mathbb{F}$ by

$$\mu(f) = \int_{-\infty}^{\infty} f(t)\, e^{-|t|}\, dt, \qquad f \in C_b(\mathbb{R}). \quad (6.8)$$

The integral in equation (6.8) is an improper Riemann integral, defined by

$$\int_{-\infty}^{\infty} f(t)\, e^{-|t|}\, dt \;=\; \lim_{b \to \infty} \int_{-b}^{b} f(t)\, e^{-|t|}\, dt.$$

This limit exists because f is bounded and continuous and $\int_{-\infty}^{\infty} e^{-|t|}\, dt$ is finite.

The reader should verify that μ defined in this way is linear. If f is any function in $C_b(\mathbb{R})$, then, by the definition of the uniform norm, $|f(t)| \le \|f\|_u$ for each $t \in \mathbb{R}$. Consequently,

$$
\begin{aligned}
|\mu(f)| &= \left| \int_{-\infty}^{\infty} f(t)\, e^{-|t|}\, dt \right| \\
&\le \int_{-\infty}^{\infty} |f(t)|\, e^{-|t|}\, dt \\
&\le \int_{-\infty}^{\infty} \|f\|_u\, e^{-|t|}\, dt \\
&= \|f\|_u \int_{-\infty}^{\infty} e^{-|t|}\, dt \;=\; 2\,\|f\|_u.
\end{aligned}
$$

This shows that μ is bounded. Indeed, by taking the supremum over the unit vectors in $C_b(\mathbb{R})$ we see that

$$\|\mu\| \;=\; \sup_{\|f\|_u = 1} |\mu(f)| \;\le\; \sup_{\|f\|_u = 1} 2\,\|f\|_u \;=\; 2.$$

Can you determine the exact value of the operator norm of μ? $\Diamond$

Next we construct a linear functional that is bounded with respect to one norm on its domain, but unbounded when we place a different norm on the domain. The domain for this example is $X = C_b^1(\mathbb{R})$, which was also used earlier in Example 6.1.3.

Example 6.3.4. Problem 3.5.25 showed that $C_b^1(\mathbb{R})$ is a Banach space with respect to its "standard norm"

$$\|f\|_{C_b^1} \;=\; \|f\|_u + \|f'\|_u \;=\; \sup_{t \in \mathbb{R}} |f(t)| + \sup_{t \in \mathbb{R}} |f'(t)|.$$

Define a functional $\delta' \colon C_b^1(\mathbb{R}) \to \mathbb{F}$ by

$$\delta'(f) \;=\; -f'(0), \qquad f \in C_b^1(\mathbb{R}). \tag{6.9}$$

(The minus sign in the definition of δ' is not important for this example, but it is traditionally included in the definition of δ'.) An argument similar to the calculation in equation (6.7) shows that δ' is linear.

We will prove that δ' is bounded when the norm on $C_b^1(\mathbb{R})$ is $\|\cdot\|_{C_b^1}$. To see this, fix any $f \in C_b^1(\mathbb{R})$. Then

$$|\delta'(f)| = |f'(0)| \leq \|f'\|_{\mathrm{u}} \leq \|f\|_{\mathrm{u}} + \|f'\|_{\mathrm{u}} = \|f\|_{C_b^1}.$$

Hence δ' is bounded, and by taking the supremum over all unit vectors f (i.e., unit with respect to the norm $\|\cdot\|_{C_b^1}$) we see that

$$\|\delta'\| = \sup_{\|f\|_{C_b^1}=1} |\delta'(f)| \leq \sup_{\|f\|_{C_b^1}=1} \|f\|_{C_b^1} = 1.$$

In fact, we will show that the operator norm of δ' is exactly 1. To do this, consider the function $f_n(t) = \sin nt$. We have $f_n'(t) = n\cos nt$, so

$$\|f_n\|_{C_b^1} = \|f_n\|_{\mathrm{u}} + \|f_n'\|_{\mathrm{u}} = \|\sin nt\|_{\mathrm{u}} + \|n\cos nt\|_{\mathrm{u}} = 1 + n.$$

Therefore the function

$$g_n(t) = \frac{f_n(t)}{\|f_n\|_{C_b^1}} = \frac{\sin nt}{n+1}$$

satisfies $\|g_n\|_{C_b^1} = 1$, i.e., g_n is one of the unit vectors in $C_b^1(\mathbb{R})$. Since

$$\delta'(g_n) = -g_n'(0) = -\frac{n\cos 0}{n+1} = -\frac{n}{n+1},$$

we see that

$$\|\delta'\| = \sup_{\|f\|_{C_b^1}=1} |\delta'(f)| \geq \sup_{n \in \mathbb{N}} |\delta'(g_n)| = \sup_{n \in \mathbb{N}} \frac{n}{n+1} = 1.$$

Therefore $\|\delta'\| = 1$.

However, everything changes if we decide to place the uniform norm on $C_b^1(\mathbb{R})$ instead of its standard norm. To see why, let δ' be the same functional as before, i.e., δ' is defined by equation (6.9), but now assume that the norm on $C_b^1(\mathbb{R})$ is $\|\cdot\|_{\mathrm{u}}$ instead of $\|\cdot\|_{C_b^1}$. Consider again the function $f_n(x) = \sin nx$. With respect to the uniform norm we have

$$\|f_n\|_{\mathrm{u}} = 1,$$

so f_n is a unit vector with respect to $\|\cdot\|_{\mathrm{u}}$. However,

$$\delta'(f_n) = -f_n'(0) = -n\cos 0 = -n,$$

so

$$\|\delta'\| = \sup_{\|f\|_{\mathrm{u}}=1} |\delta'(f)| \geq \sup_{n \in \mathbb{N}} |\delta'(f_n)| = \sup_{n \in \mathbb{N}} n = \infty.$$

Therefore δ' is *unbounded* when the norm on $C_b^1(\mathbb{R})$ is $\|\cdot\|_u$ (even though $|\delta'(f)|$ is finite for every f in $C_b^1(\mathbb{R})$!). $\quad\Diamond$

It may seem that we cheated a bit in the preceding example, since $C_b^1(\mathbb{R})$ is not complete with respect to $\|\cdot\|_u$ (see Problem 3.5.25). It may be tempting to believe that all linear functionals on a *complete* normed space are bounded, but this is not true. Problem 6.3.8 shows that if X is *any* infinite-dimensional normed space, then there exists a linear functional $\mu\colon X \to \mathbb{F}$ that is not bounded (although we need to appeal to the Axiom of Choice to prove the existence of such functionals on arbitrary normed spaces). Boundedness is a desirable property, but we cannot take it for granted when our domain is infinite-dimensional. Although we will focus mostly on bounded operators, many important linear operators that arise in mathematics, physics, and engineering are unbounded.

Problems

6.3.5. Let M be a closed subspace of a Hilbert space H, and let P be the orthogonal projection of H onto M (see part (b) of Definition 5.6.3). Prove that P is bounded and linear, and find $\|P\|$. Is P isometric?

6.3.6. Let $A = [a_{ij}]$ be an $m \times n$ matrix, which we identify with the linear operator $A\colon \mathbb{F}^n \to \mathbb{F}^m$ that maps x to Ax. Since $\mathbb{F}^n$ is finite-dimensional, Theorem 6.3.1 implies that A is bounded with respect to any norm on $\mathbb{F}^n$. However, the exact value of the operator norm of A depends on which norms we choose to place on $\mathbb{F}^n$ and $\mathbb{F}^m$. Prove the following statements.

(a) If the norm on both $\mathbb{F}^n$ and $\mathbb{F}^m$ is $\|\cdot\|_1$, then the operator norm of A is

$$\|A\| = \max_{j=1,\ldots,n}\left\{\sum_{i=1}^{m}|a_{ij}|\right\}.$$

(b) If the norm on $\mathbb{F}^n$ and $\mathbb{F}^m$ is $\|\cdot\|_\infty$, then the operator norm of A is

$$\|A\| = \max_{i=1,\ldots,m}\left\{\sum_{j=1}^{n}|a_{ij}|\right\}.$$

(c) If the norm on $\mathbb{F}^n$ is $\|\cdot\|_1$ and the norm on $\mathbb{F}^m$ is $\|\cdot\|_\infty$, then the operator norm of A is

$$\|A\| = \max_{\substack{i=1,\ldots,m \\ j=1,\ldots,n}}|a_{ij}|.$$

(d) If the norm on $\mathbb{F}^n$ is $\|\cdot\|_\infty$ and the norm on $\mathbb{F}^m$ is $\|\cdot\|_1$, then the operator norm of A satisfies

$$\|A\| \leq \sum_{i=1}^{m} \sum_{j=1}^{n} |a_{ij}|.$$

Show by example that equality holds for some matrices but not for others.

6.3.7. Let X be the set of all continuous functions on $\mathbb{R}$ that are square-integrable, i.e.,

$$X = \left\{ f \in C(\mathbb{R}) : \int_{-\infty}^{\infty} |f(t)|^2 \, dt < \infty \right\}.$$

Define

$$\|f\|_2 = \left(\int_{-\infty}^{\infty} |f(t)|^2 \, dt \right)^{1/2},$$

and prove the following statements.

(a) $\|\cdot\|_2$ defines a norm on X.

(b) The set

$$Y = \left\{ f \in C(\mathbb{R}) : \int_{-\infty}^{\infty} |tf(t)|^2 \, dt < \infty \right\}$$

is a proper subspace of X, and $\|\cdot\|_2$ is a norm on Y.

(c) If $f \in Y$, then the function $g(t) = tf(t)$ belongs to X.

(d) The *position operator* $P \colon Y \to X$ defined by $(Pf)(t) = tf(t)$ is unbounded, i.e.,

$$\sup_{f \in Y, \, \|f\|_2 = 1} \|Pf\|_2 = \infty.$$

Observe that $\|Pf\|_2 < \infty$ for every $f \in Y$!

6.3.8. (a) Let c_{00} be the space of "finite sequences" introduced in Example 2.2.13. For this problem, the norm on c_{00} is the ℓ^1-norm. Given any $N \in \mathbb{N}$ and $x = (x_1, x_2, \ldots, x_N, 0, 0, \ldots) \in c_{00}$, define

$$Ax = (x_1, 2x_2, \ldots, Nx_N, 0, 0, \ldots).$$

Prove that $A \colon c_{00} \to c_{00}$ is linear but unbounded (even though $\|Ax\|_1 < \infty$ for every $x \in c_{00}$).

(b) Let X be any infinite-dimensional space. The Axiom of Choice implies that there exists some Hamel basis for X, say $\{x_i\}_{i \in I}$. By dividing each vector by its norm, we may assume that $\|x_i\| = 1$ for each $i \in I$. Let $J_0 = \{j_1, j_2, \ldots\}$ be any countably infinite subsequence of I. Define $\mu(x_{j_n}) = n$ for $n \in \mathbb{N}$, and set $\mu(x_i) = 0$ for $i \in I \setminus J_0$. By the definition of a Hamel basis, each nonzero vector $x \in X$ can be written uniquely as $x = \sum_{k=1}^{N} c_k x_{i_k}$ for some $N \in \mathbb{N}$, indices $i_1, \ldots, i_N \in I$, and nonzero scalars $c_1, \ldots, c_N$. Use this to extend μ to a linear functional whose domain is X, and show that $\mu \colon X \to \mathbb{F}$ is unbounded.

6.4 Equivalence of Bounded and Continuous Linear Operators

We are especially interested in linear operators that map one normed space into another normed space. We will prove that for these operators, boundedness and continuity are entirely equivalent.

Theorem 6.4.1 (Boundedness Equals Continuity). *Let X and Y be normed vector spaces. If $A\colon X \to Y$ is a linear operator, then*

$$A \text{ is bounded} \quad \Longleftrightarrow \quad A \text{ is continuous.}$$

Proof. $\Rightarrow$. Assume that A is bounded, and choose any vectors $x_n,\, x \in X$ such that $x_n \to x$. Applying linearity and Lemma 6.2.4, we see that

$$\|Ax_n - Ax\| \;=\; \|A(x_n - x)\| \;\leq\; \|A\|\,\|x_n - x\| \;\to\; 0.$$

Thus $Ax_n \to Ax$. Lemma 2.9.4 therefore implies that A is continuous.

$\Leftarrow$. Suppose that A is continuous but unbounded. Then

$$\|A\| \;=\; \sup_{\|x\|=1} \|Ax\| \;=\; \infty,$$

so for each $n \in \mathbb{N}$ there must exist a vector $x_n \in X$ such that $\|x_n\| = 1$ but $\|Ax_n\| \geq n$. Setting $y_n = x_n/n$, we see that

$$\|y_n - 0\| \;=\; \|y_n\| \;=\; \frac{\|x_n\|}{n} \;=\; \frac{1}{n} \;\to\; 0 \quad \text{as } n \to \infty.$$

Thus $y_n \to 0$. Since A is continuous and linear, it follows that $Ay_n \to A0 = 0$. By the continuity of the norm, this implies that

$$\lim_{n \to \infty} \|Ay_n\| \;=\; \|0\| \;=\; 0.$$

However, for each $n \in \mathbb{N}$ we have

$$\|Ay_n\| \;=\; \frac{1}{n}\,\|Ax_n\| \;\geq\; \frac{1}{n}\cdot n \;=\; 1,$$

which is a contradiction. Therefore A must be bounded. $\square$

In summary, if X and Y are normed spaces and $A\colon X \to Y$ is linear, then the terms "continuous" and "bounded" are entirely interchangeable when applied to A.

Problems

6.4.2. Let X and Y be normed spaces, and suppose that A, $B\colon X \to Y$ are bounded linear operators. Prove that if $Ax = Bx$ for all x in a dense set $S \subseteq X$, then $A = B$.

6.4.3. Given normed spaces X and Y, prove the following statements.

(a) If $A\colon X \to Y$ is continuous, then $\ker(A)$ is a closed subset of X (even if A is not linear).

(b) If $A\colon X \to Y$ is bounded and linear, then $\ker(A)$ is a closed subspace of X.

6.4.4. Let X and Y be normed spaces, and fix any number $0 \le M < \infty$.

(a) Suppose that $A\colon X \to Y$ is bounded and linear, and prove that $E = \{x \in X : \|Ax\| \le M\}$ is a nonempty closed subset of X.

(b) Suppose that $A_i\colon X \to Y$ is bounded and linear for every i in some (possibly uncountable) index set I. Prove that

$$
E = \left\{ x \in X : \sup_{i \in I} \|A_i x\| \le M \right\}
$$

is a nonempty closed subset of X.

6.4.5. Define an operator $D\colon C_b^1(\mathbb{R}) \to C_b(\mathbb{R})$ by $Df = f'$ for $f \in C_b^1(\mathbb{R})$, and observe that D is linear.

(a) Prove that D is continuous if we take the norm on $C_b^1(\mathbb{R})$ to be $\|\cdot\|_{C_b^1}$ and the norm on $C_b(\mathbb{R})$ to be the uniform norm.

(b) Prove that D is not continuous if we take the norm on both $C_b^1(\mathbb{R})$ and $C_b(\mathbb{R})$ to be the uniform norm.

6.5 The Space $\mathcal{B}(X, Y)$

If X and Y are normed spaces, then we collect the bounded linear operators that map X into Y to form another space that we call $\mathcal{B}(X, Y)$.

Definition 6.5.1. Let X and Y be normed vector spaces. The set of all bounded linear operators from X into Y is denoted by

$$
\mathcal{B}(X, Y) = \{A\colon X \to Y : A \text{ is bounded and linear}\}.
$$

If $X = Y$, then we write $\mathcal{B}(X) = \mathcal{B}(X, X)$. $\diamond$

Since boundedness and continuity are equivalent for linear operators on normed spaces, we can also describe $\mathcal{B}(X,Y)$ as the space of continuous linear operators from X into Y.

The following lemma (whose proof is Problem 6.5.7) shows that the operator norm is a norm on $\mathcal{B}(X,Y)$, and therefore $\mathcal{B}(X,Y)$ is a normed space with respect to the operator norm.

Lemma 6.5.2. *If X and Y are normed spaces, then the operator norm is a norm on $\mathcal{B}(X,Y)$. That is:*

(a) $0 \leq \|A\| < \infty$ *for every* $A \in \mathcal{B}(X,Y)$,

(b) $\|A\| = 0$ *if and only if* $A = 0$,

(c) $\|cA\| = |c|\,\|A\|$ *for all* $A \in \mathcal{B}(X,Y)$ *and all scalars c, and*

(d) $\|A + B\| \leq \|A\| + \|B\|$ *for all* $A, B \in \mathcal{B}(X,Y)$. $\diamondsuit$

Since the operator norm is a norm, we have a corresponding notion of convergence. That is, if $\{A_n\}_{n \in \mathbb{N}}$ is a sequence of operators in $\mathcal{B}(X,Y)$, then A_n *converges to A in operator norm* if $\|A - A_n\| \to 0$. This requires that

$$\lim_{n \to \infty} \|A - A_n\| = \lim_{n \to \infty} \left(\sup_{\|x\|=1} \|Ax - A_n x\| \right) = 0.$$

The following example shows that "pointwise convergence" of operators (i.e., $A_n x \to Ax$ for each $x \in X$) is not sufficient to imply that A_n converges to A in operator norm.

Example 6.5.3. Assume that $\{e_n\}_{n \in \mathbb{N}}$ is an orthonormal basis for a Hilbert space H. Then we know from Theorem 5.8.1 that every $x \in H$ can be written as

$$x = \sum_{n=1}^{\infty} \langle x, e_n \rangle e_n. \tag{6.10}$$

If we let $H_N = \mathrm{span}\{e_1, \dots, e_N\}$, then the orthogonal projection of H onto H_N is given by

$$P_N x = \sum_{n=1}^{N} \langle x, e_n \rangle e_n, \qquad x \in H.$$

This is the Nth partial sum of the series in equation (6.10), so we have $P_N x \to x = Ix$ as $N \to \infty$, where I is the identity operator on H. This means that $P_N x$ converges to Ix for each individual $x \in H$. This is a kind of pointwise convergence of P_N to I. But does P_N converge to I with respect to the operator norm? To determine this, we must compute the operator norm of $I - P_N$. This is given by

$$\|I - P_N\| = \sup_{\|x\|=1} \|Ix - P_N x\| = \sup_{\|x\|=1} \|x - P_N x\|.$$

Now, if $x \in H_N$, then $P_N x = x$ and $x - P_N x = 0$. However, if $x \notin H_N$, then $P_N x \neq x$. In fact, if $x = e_{N+1}$, then $P_N x = P_N e_{N+1} = 0$ (why?). Since e_{N+1} is one of the unit vectors in H, this implies that

$$\|I - P_N\| = \sup_{\|x\|=1} \|x - P_N x\| \geq \|e_{N+1} - P_N e_{N+1}\| \geq \|e_{N+1} - 0\| = 1.$$

In fact, the reader should show that $\|I - P_N\|$ is exactly equal to 1. Hence $\|I - P_N\| \nrightarrow 0$, so P_N does *not* converge to the identity operator in operator norm, even though $P_N x \to I x$ for every $x \in H$. $\diamond$

We will prove that if Y is complete, then $\mathcal{B}(X,Y)$ is complete as well. The idea of the proof is similar to previous completeness proofs that we have seen, such as those of Theorems 3.4.3 or 3.5.7. That is, given a Cauchy sequence, we first find a candidate to which the sequence appears to converge, and then prove that the sequence converges *in norm* to that candidate.

Theorem 6.5.4. *If X is a normed vector space and Y is a Banach space, then $\mathcal{B}(X,Y)$ is a Banach space with respect to the operator norm.*

Proof. Assume that $\{A_n\}_{n \in \mathbb{N}}$ is a sequence of operators in $\mathcal{B}(X,Y)$ that is Cauchy with respect to the operator norm. That is, we assume that for each $\varepsilon > 0$ there is some $N > 0$ such that

$$m, n > N \implies \|A_m - A_n\| < \varepsilon.$$

Fix any vector $x \in X$. Since

$$\|A_m x - A_n x\| \leq \|A_m - A_n\| \, \|x\|,$$

it follows that $\{A_n x\}_{n \in \mathbb{N}}$ is a Cauchy sequence of vectors in Y. Since Y is complete, this sequence must converge. That is, there exists some $y \in Y$ such that $A x_n \to y$ as $n \to \infty$. Define $Ax = y$. This gives us a candidate limit operator A, and the reader should check that this mapping A is linear.

Lemma 3.3.3 tells us that every Cauchy sequence in a normed space is bounded. So, since $\{A_n\}_{n \in \mathbb{N}}$ is Cauchy, we must have $C = \sup \|A_n\| < \infty$. If we fix $x \in X$, then $A_n x \to Ax$ by the definition of A, and therefore

$$
\begin{aligned}
\|Ax\| &= \lim_{n \to \infty} \|A_n x\| && \text{(continuity of the norm)} \\
&\leq \sup_{n \in \mathbb{N}} \|A_n x\| \\
&\leq \sup_{n \in \mathbb{N}} \|A_n\| \, \|x\| && \text{(Lemma 6.2.4)} \\
&= C \, \|x\| && \text{(definition of } C\text{)}.
\end{aligned}
$$

Taking the supremum over all unit vectors x, we see that A is bounded and $\|A\| \leq C$.

Finally, we must show that $A_n \to A$ in operator norm. Fix any $\varepsilon > 0$. Since $\{A_n\}_{n \in \mathbb{N}}$ is Cauchy, there exists an integer $N > 0$ such that

$$m, n > N \quad \Longrightarrow \quad \|A_m - A_n\| < \frac{\varepsilon}{2}.$$

Choose any vector $x \in X$ with $\|x\| = 1$. Since $A_m x \to Ax$ as $m \to \infty$, there exists some $m > N$ such that

$$\|Ax - A_m x\| < \frac{\varepsilon}{2}.$$

In fact, this inequality will hold for all large enough m, but for this proof we need only one m. Using that fixed m, given any $n > N$ we compute that

$$\begin{aligned}
\|Ax - A_n x\| &\leq \|Ax - A_m x\| + \|A_m x - A_n x\| \\
&\leq \|Ax - A_m x\| + \|A_m - A_n\| \, \|x\| \\
&< \frac{\varepsilon}{2} + \frac{\varepsilon}{2} \cdot 1 = \varepsilon.
\end{aligned}$$

Taking the supremum over all unit vectors x, it follows that

$$\|A - A_n\| = \sup_{\|x\|=1} \|Ax - A_n x\| \leq \varepsilon, \qquad \text{for all } n > N.$$

Hence $A_n \to A$ in operator norm, and therefore $\mathcal{B}(X, Y)$ is complete. $\quad \square$

Since the elements of $\mathcal{B}(X, Y)$ are functions, we know how to add elements of $\mathcal{B}(X, Y)$ and how to multiply them by scalars. However, there is also a third natural operation that we can perform on functions—we can compose them if their domains and codomains match appropriately. Given operators $A \in \mathcal{B}(X, Y)$ and $B \in \mathcal{B}(Y, Z)$, we typically denote the composition of B with A by BA. That is, $BA \colon X \to Z$ is the function whose rule is

$$(BA)(x) = B(A(x)), \qquad x \in X.$$

The next lemma (whose proof is Problem 6.5.7) shows that the composition of bounded operators is bounded.

Lemma 6.5.5. *Let X, Y, and Z be normed spaces. If $A \in \mathcal{B}(X, Y)$ and $B \in \mathcal{B}(Y, Z)$, then $BA \in \mathcal{B}(X, Z)$ and we have the following submultiplicative property of the operator norm:*

$$\|BA\| \leq \|B\| \, \|A\|. \qquad \Diamond \tag{6.11}$$

In particular, taking $X = Y$, we see that

$$A, B \in \mathcal{B}(X) \quad \Longrightarrow \quad AB, BA \in \mathcal{B}(X).$$

Therefore $\mathcal{B}(X)$ is closed under composition of operators, and composition is submultiplicative in the sense that $\|AB\| \leq \|A\|\,\|B\|$ and $\|BA\| \leq \|A\|\,\|B\|$.

Problems

6.5.6. Let m and n be positive integers. Using the identification of linear operators $A \colon \mathbb{F}^n \to \mathbb{F}^m$ with $m \times n$ matrices discussed in Problem 6.1.4, give an explicit description of $\mathcal{B}(\mathbb{F}^n, \mathbb{F}^m)$. Show further that $\mathcal{B}(\mathbb{F}^n, \mathbb{F}^m)$ is a finite-dimensional vector space, and find a Hamel basis for this space.

6.5.7. Prove Lemmas 6.5.2 and 6.5.5.

6.5.8. Let X and Y be normed spaces, and suppose that the series $\sum_{n=1}^{\infty} x_n$ converges in X. Show that if $A \in \mathcal{B}(X,Y)$, then $\sum_{n=1}^{\infty} Ax_n$ converges in Y.

6.5.9. Let X and Y be normed spaces. Suppose that $A_n \in \mathcal{B}(X,Y)$ and the series $A = \sum_{n=1}^{\infty} A_n$ converges in operator norm. Prove that for each $x \in X$ we have

$$Ax = \sum_{n=1}^{\infty} A_n x,$$

where this series converges in the norm of Y.

6.5.10. Let X be a Banach space, and let $A \in \mathcal{B}(X)$ be given. Let $A^0 = I$ be the identity map on X, and for $n > 0$ let $A^n = A \cdots A$ be the composition of A with itself n times. Prove the following statements.

(a) The series

$$e^A = \sum_{k=0}^{\infty} \frac{A^k}{k!}$$

converges absolutely in operator norm, and $\|e^A\| \leq e^{\|A\|}$.

(b) For each $x \in X$ we have

$$e^A(x) = \sum_{k=0}^{\infty} \frac{A^k x}{k!},$$

where the series on the right converges absolutely with respect to the norm of X.

(c) If $A, B \in \mathcal{B}(X)$ and $AB = BA$, then $e^A e^B = e^B e^A$.

6.5.11. This problem will show that a bounded linear operator $A \colon S \to Y$ whose domain S is a dense subspace of X and whose codomain Y is complete can be extended to the entire space X.

Let X be a normed space and let Y be a Banach space. Assume that S is a dense subspace of X, and $A\colon S \to Y$ is a bounded linear operator. Prove the following statements.

(a) If $x \in X$ and $\{x_n\}_{n\in\mathbb{N}}$ is a sequence of vectors in S such that $x_n \to x$, then $\{Ax_n\}_{n\in\mathbb{N}}$ is a Cauchy sequence in Y. Consequently, there is a vector $y \in Y$ such that $Ax_n \to y$. Moreover, this vector y is independent of the choice of sequence $\{x_n\}_{n\in\mathbb{N}}$, and therefore we can define $\widetilde{A}\colon X \to Y$ by $\widetilde{A}x = y$.

(b) $\widetilde{A}$ is bounded and linear, and its operator norm equals the operator norm of A, i.e.,

$$\|A\| = \sup_{\substack{x\in S,\\ \|x\|=1}} \|Ax\| = \sup_{\substack{x\in X,\\ \|x\|=1}} \|\widetilde{A}x\| = \|\widetilde{A}\|.$$

Further, $\widetilde{A}$ is an extension of A, i.e., $\widetilde{A}x = Ax$ for all $x \in S$.

(c) $\widetilde{A}$ is the *unique* operator in $\mathcal{B}(X,Y)$ whose restriction to S is A.

6.6 Isometries and Isomorphisms

We give the following name to operators that *preserve the norm of vectors*.

Definition 6.6.1 (Isometries). Let X and Y be normed vector spaces. An operator $A\colon X \to Y$ that satisfies

$$\|Ax\| = \|x\| \quad \text{for all } x \in X$$

is called an *isometry*, or is said to be *norm-preserving*. ◇

An isometry need not be linear. For example, if X is a normed space and we define $A\colon X \to \mathbb{F}$ by $Ax = \|x\|$, then A is an isometry, but A is nonlinear if $X \neq \{0\}$ (why?). However, we will mostly be interested in linear isometries. The next lemma gives some of their properties.

Lemma 6.6.2. *If X, Y are normed spaces and $A\colon X \to Y$ is a linear isometry, then the following statements hold.*

(a) *A is bounded, A is injective, and $\|A\| = 1$ (except when X is trivial, in which case $\|A\| = 0$).*

(b) *If X is a Banach space, then A has closed range.*

Proof. (a) If $Ax = 0$, then, since A is norm-preserving, $\|x\| = \|Ax\| = 0$. Therefore $x = 0$, so we have $\ker(A) = \{0\}$. Hence A is injective by Lemma 6.1.2. Further, if X is nontrivial (so unit vectors exist), then

$$\|A\| = \sup_{\|x\|=1} \|Ax\| = \sup_{\|x\|=1} \|x\| = 1.$$

(b) Suppose that vectors $y_n \in \text{range}(A)$ converge to a vector $y \in Y$. By the definition of the range, for each $n \in \mathbb{N}$ there is some vector $x_n \in X$ such that $Ax_n = y_n$. Since A is linear and isometric, we therefore have

$$\|x_m - x_n\| = \|A(x_m - x_n)\| = \|Ax_m - Ax_n\| = \|y_m - y_n\|. \quad (6.12)$$

But $\{y_n\}_{n \in \mathbb{N}}$ is Cauchy in Y (because it converges), so equation (6.12) implies that $\{x_n\}_{n \in \mathbb{N}}$ is Cauchy in X. Therefore, since X is complete, there is some $x \in X$ such that $x_n \to x$. Since A is bounded and therefore continuous, this implies that $Ax_n \to Ax$. By assumption we also have $Ax_n = y_n \to y$, so the uniqueness of limits implies that $y = Ax$. Thus $y \in \text{range}(A)$, so $\text{range}(A)$ is closed by Theorem 2.6.2. $\square$

While the range of an isometry is closed, the range of a general bounded linear operator need not be closed (see Problem 6.2.9 for an example, and also consider Problem 6.2.10).

We will exhibit an operator that satisfies $\|A\| = 1$ but is not an isometry.

Example 6.6.3 (Left Shift). Define $L \colon \ell^2 \to \ell^2$ by the rule

$$Lx = (x_2, x_3, \dots), \qquad x = (x_1, x_2, \dots) \in \ell^2.$$

For example, if $x = \left(1, \frac{1}{2}, \frac{1}{3}, \dots\right)$, then $Lx = \left(\frac{1}{2}, \frac{1}{3}, \frac{1}{4}, \dots\right)$. According to Problem 6.6.9, this *left-shift operator* L is bounded, linear, surjective, and satisfies $\|L\| = 1$, but it is not injective and is not an isometry. $\diamondsuit$

Although an isometry must be injective, it need not be surjective. Here is an example.

Example 6.6.4 (Right Shift). Define $R \colon \ell^2 \to \ell^2$ by

$$Rx = (0, x_1, x_2, x_3, \dots), \qquad x = (x_1, x_2, \dots) \in \ell^2.$$

According to Problem 6.6.9, this *right-shift operator* is bounded, linear, injective, and is an isometry. However, it is not surjective, since, for example, the first standard basis vector $\delta_1 = (1, 0, 0, \dots)$ does not belong to the range of R. $\diamondsuit$

Thus an isometry need not be surjective (but a *linear isometry* will always be injective). A *surjective* linear isometry $A \colon X \to Y$ has a number of remarkable properties, including the following.

- A is injective because it is a linear isometry, and it is surjective by assumption. Consequently, A is a bijection, and therefore it defines a one-to-one correspondence between the elements of X and the elements of Y. Hence A *preserves cardinality.*

- A *preserves the vector space operations* because it is linear.

- A is continuous because it is bounded. Therefore A *preserves convergent sequences*, i.e., if $x_n \to x$ in X, then $Ax_n \to Ax$ in Y.

- A *preserves the norms of vectors* because it is an isometry.

- Since A is linear and norm-preserving, $\|Ax - Ay\| = \|A(x-y)\| = \|x-y\|$ for all $x, y \in X$. Thus A *preserves distances between vectors*.

- If $y \in Y$, then $y = Ax$ for some unique $x \in X$, and therefore

$$\|A^{-1}y\| \;=\; \|x\| \;=\; \|Ax\| \;=\; \|y\|.$$

Hence A^{-1} is also a surjective linear isometry.

- Because A is continuous, the inverse image of an open set is open, and therefore A^{-1} *maps open sets to open sets*. Likewise, because A^{-1} is continuous, A *maps open sets to open sets*. Hence A *preserves the topology*.

Thus, if A is a surjective linear isometry from X to Y, then A bijectively identifies the elements of X with those of Y in a way that preserves the vector space operations, preserves the topology of the spaces, and preserves the norms of elements (and A^{-1} does the same in the reverse direction). In essence, X and Y are the "same" space with their elements renamed. We introduce some terminology for this situation.

Definition 6.6.5 (Isometric Isomorphism). Let X and Y be normed vector spaces.

(a) If $A \colon X \to Y$ is a linear isometry that is surjective, then A is called an *isometric isomorphism* of X onto Y.

(b) If there exists an isometric isomorphism $A \colon X \to Y$, then we say that X and Y are *isometrically isomorphic*, and in this case we write $X \cong Y$. $\Diamond$

The left-shift and right-shift operators L and R discussed above are not isometric isomorphisms. Here is an example of an isometric isomorphism.

Example 6.6.6. Suppose that $\{e_n\}_{n \in \mathbb{N}}$ is an orthonormal basis for an infinite-dimensional separable Hilbert space H. If x is a vector in H, then Bessel's Inequality (Theorem 5.7.1) implies that

$$\sum_{n=1}^{\infty} |\langle x, e_n \rangle|^2 \;<\; \infty.$$

Therefore the sequence of inner products $\big(\langle x, e_n \rangle\big)_{n \in \mathbb{N}}$ belongs to ℓ^2. Define $T \colon H \to \ell^2$ by

$$Tx \;=\; \big(\langle x, e_n \rangle\big)_{n \in \mathbb{N}}, \qquad x \in \ell^2.$$

The reader should check that the fact that the inner product is linear in the first variable implies that T is a linear operator.

By the Plancherel Equality,

$$\|Tx\|_2^2 = \sum_{n=1}^{\infty} |\langle x, e_n \rangle|^2 = \|x\|^2.$$

Therefore T is an isometry.

To show that T is surjective, choose any sequence $c = (c_n)_{n \in \mathbb{N}} \in \ell^2$. By Theorem 5.7.1, the series

$$x = \sum_{n=1}^{\infty} c_n x_n$$

converges, and $c_n = \langle x, e_n \rangle$ for every n. Hence $Tx = c$, so T is surjective. Thus T is an isometric isomorphism of H onto ℓ^2, so $H \cong \ell^2$.

What changes in this example if we assume only that $\{e_n\}_{n \in \mathbb{N}}$ is an orthonormal *sequence* instead of an orthonormal *basis*? $\Diamond$

Example 6.6.6 shows that *every infinite-dimensional separable Hilbert space is isometrically isomorphic to ℓ^2*. As a consequence (why?), if H and K are any two infinite-dimensional separable Hilbert spaces, then $H \cong K$. If H and K are finite-dimensional Hilbert spaces, then $H \cong K$ if and only if H and K have the same dimension (see Problem 6.6.10).

The next definition introduces a class of operators that includes the isometric isomorphisms, but is more general in the sense that the operators need not be norm-preserving.

Definition 6.6.7 (Topological Isomorphism). Let X and Y be normed spaces. If $A \colon X \to Y$ is a bijective linear operator such that both A and A^{-1} are continuous, then we say that A is a *topological isomorphism*. $\Diamond$

For example, let A be an invertible $n \times n$ matrix, and identify A with the mapping $A \colon \mathbb{F}^n \to \mathbb{F}^n$ that takes x to Ax. This is a linear map, and it is a bijection, since A is invertible. Since $\mathbb{F}^n$ is finite-dimensional, Theorem 6.3.1 implies that both A and A^{-1} are bounded, so A is a topological isomorphism.

Here is an example of a topological isomorphism on an infinite-dimensional normed space that is not an isometry.

Example 6.6.8. Let $X = Y = C_b(\mathbb{R})$, set $\phi(t) = 1 + e^{-|t|}$ for $t \in \mathbb{R}$, and define $A \colon C_b(\mathbb{R}) \to C_b(\mathbb{R})$ by

$$Af = f\phi, \qquad f \in C_b(\mathbb{R}).$$

That is, Af is simply the pointwise product of f and ϕ:

$$(Af)(t) = f(t)\left(1 + e^{-|t|}\right) = f(t) + f(t)\,e^{-|t|}, \qquad t \in \mathbb{R}.$$

Since ϕ is bounded and continuous, the product $f\phi$ is also bounded and continuous and therefore belongs to $C_b(\mathbb{R})$. Hence A does in fact map $C_b(\mathbb{R})$ into itself, and the reader should check that A is linear.

Note that ϕ is a bounded function; in fact, its uniform norm is $\|\phi\|_u = 2$. This implies that A is bounded map, because for each $f \in C_b(\mathbb{R})$ we have

$$\|Af\|_u = \|f\phi\|_u = \sup_{t \in \mathbb{R}} |f(t)\phi(t)| \leq 2 \sup_{t \in \mathbb{R}} |f(t)| = 2\|f\|_u.$$

Taking the supremum over all unit vectors f, we see that $\|A\| \leq 2$. In fact, according to Problem 6.6.11, the operator norm of A is precisely $\|A\| = 2$.

If $Af = 0$, then $f(t)\phi(t) = 0$ for every t. Since $\phi(t)$ is always nonzero, this implies that $f(t) = 0$ for every t. Hence $f = 0$, and therefore $\ker(A) = \{0\}$, i.e., the kernel of A contains only the zero function. Since A is linear, this implies that A is injective.

Now, not only is ϕ a bounded function, but it is also "bounded away from zero," because

$$1 < \phi(t) \leq 2, \qquad \text{for all } t \in \mathbb{R}. \tag{6.13}$$

Hence its reciprocal $1/\phi$ is a bounded, continuous function. Therefore, if h is a function in $C_b(\mathbb{R})$, then $f = h/\phi$ is continuous and bounded as well. This function $f \in C_b(\mathbb{R})$ satisfies $Af = f\phi = h$, so A is surjective.

Thus A is a bijection that maps $C_b(\mathbb{R})$ onto itself. The inverse function is given by $A^{-1}f = f/\phi$. Since $\phi(t) > 1$ for every t, we see that A^{-1} is bounded, because

$$\|A^{-1}f\|_u = \|f/\phi\|_u = \sup_{t \in \mathbb{R}} \left| \frac{f(t)}{\phi(t)} \right| \leq \sup_{t \in \mathbb{R}} \left| \frac{f(t)}{1} \right| = \|f\|_u.$$

This shows that $\|A^{-1}\| \leq 1$, and the reader should use Problem 6.6.11 to prove that $\|A^{-1}\| = 1$.

In summary, A is a bounded linear bijection and A^{-1} is bounded as well. Therefore A is a topological isomorphism (and so is A^{-1}). However, A is not isometric (why?), so it is not an isometric isomorphism. What happens in this example if we replace $\phi(t) = 1 + e^{-|t|}$ by $\phi(t) = e^{-|t|}$? Is A still bounded? Is it injective? Is it a topological isomorphism? $\square$

Although we will not prove it, the *Inverse Mapping Theorem* is an important result from the field of *functional analysis* that states that if X and Y are Banach spaces and $A\colon X \to Y$ is a bounded linear bijection, then A^{-1} is automatically bounded and therefore A is a topological isomorphism. Thus, when X and Y are *complete*, every continuous linear bijection is actually a topological isomorphism. For one proof of the Inverse Mapping Theorem (which is a corollary of the *Open Mapping Theorem*), see [Heil11, Thm. 2.29].

Problems

6.6.9. Prove the statements made about the left-shift and right-shift operators in Examples 6.6.3 and 6.6.4. Also show that $LR = I$, the identity operator on ℓ^2, but $RL \neq I$.

Remark: In contrast, if A and B are $n \times n$ matrices, then $AB = I$ if and only if $BA = I$.

6.6.10. Given finite-dimensional Hilbert spaces H and K, prove that $H \cong K$ if and only if $\dim(H) = \dim(K)$.

6.6.11. Given $\phi \in C_b(\mathbb{R})$, prove that

$$M_\phi f = f\phi, \qquad f \in C_b(\mathbb{R}),$$

defines a bounded linear mapping $M_\phi \colon C_b(\mathbb{R}) \to C_b(\mathbb{R})$, and its operator norm is $\|M_\phi\| = \|\phi\|_u$.

6.6.12. Given $a \in \mathbb{R}$, let $T_a \colon C_b(\mathbb{R}) \to C_b(\mathbb{R})$ be the operator that translates a function right by a units. Explicitly, if $f \in C_b(\mathbb{R})$, then $T_a f$ is the function given by $(T_a f)(t) = f(t - a)$. Prove that T_a is an isometric isomorphism. What is $(T_a)^{-1}$? Find all eigenvalues of T_a, i.e., find all scalars $\lambda \in \mathbb{F}$ for which there exists some nonzero function $f \in C_b(\mathbb{R})$ such that $T_a f = \lambda f$.

6.6.13. Let X, Y, and Z be normed spaces. Given topological isomorphisms $A \colon X \to Y$ and $B \colon Y \to Z$, prove that $BA \colon X \to Z$ is a topological isomorphism. Likewise, show that the composition of isometries is an isometry, and the composition of isometric isomorphisms is an isometric isomorphism.

6.6.14. Let X and Y be normed spaces, and assume that $A \colon X \to Y$ is a linear operator. Prove the following statements.

(a) If A is surjective and there exist constants $c, C > 0$ such that

$$c\,\|x\| \leq \|Ax\| \leq C\,\|x\|, \qquad \text{for all } x \in X,$$

then A is a topological isomorphism.

(b) If A is a topological isomorphism, then

$$\frac{\|x\|}{\|A^{-1}\|} \leq \|Ax\| \leq \|A\|\,\|x\|, \qquad \text{for all } x \in X.$$

6.6.15. Let $A \colon X \to Y$ be a topological isomorphism of a normed space X onto a normed space Y. Let $\{x_n\}_{n \in \mathbb{N}}$ be a sequence of vectors in X. Prove that $\{x_n\}_{n \in \mathbb{N}}$ is a complete sequence in X if and only if $\{Ax_n\}_{n \in \mathbb{N}}$ is a complete sequence in Y.

6.6.16. Let X be a Banach space and let Y be a normed space. Show that if there exists a topological isomorphism $A \colon X \to Y$, then Y is a Banach space.

6.6.17. Let X be a Banach space, and let $A \in \mathcal{B}(X)$ be fixed. Define $A^0 = I$ (the identity operator on X), and let $A^n = A \cdots A$ be the composition of A with itself n times.

(a) Show that if $\|A\| < 1$, then $I - A$ is a topological isomorphism and

$$(I - A)^{-1} = \sum_{n=0}^{\infty} A^n,$$

where this series (called a *Neumann series* for $(I - A)^{-1}$) converges in operator norm.

(b) Suppose that $B \in \mathcal{B}(X)$ is a topological isomorphism. Prove that if $\|A - B\| < 1/\|B^{-1}\|$, then A is also a topological isomorphism.

(c) Prove that the set of topological isomorphisms that map X onto itself is an open subset of $\mathcal{B}(X)$.

6.6.18. Let X be a normed vector space, and suppose that there exists a topological isomorphism $A \colon X \to \ell^1$. Prove that there exists a Schauder basis $\{x_n\}_{n \in \mathbb{N}}$ for X such that every vector $x \in X$ can be uniquely written as

$$x = \sum_{n=1}^{\infty} c_n(x)\, x_n, \quad \text{where} \quad \sum_{n=1}^{\infty} |c_n(x)| < \infty.$$

Remark: Such a basis is called an *absolutely convergent basis* for X.

6.6.19. This problem will show that every incomplete normed space can be naturally embedded into a larger normed space that is complete. Some knowledge of *equivalence relations* is required for this problem.

Let X be a normed space that is not complete. Let $\mathcal{C}$ be the set of all Cauchy sequences in X, and define a relation $\sim$ on $\mathcal{C}$ by declaring that if $\mathcal{X} = \{x_n\}_{n \in \mathbb{N}}$ and $\mathcal{Y} = \{y_n\}_{n \in \mathbb{N}}$ are two Cauchy sequences, then $\mathcal{X} \sim \mathcal{Y}$ if $\lim_{n \to \infty} \|x_n - y_n\| = 0$.

(a) Prove that $\sim$ is an equivalence relation on $\mathcal{C}$. That is, show that for all Cauchy sequences $\mathcal{X}$, $\mathcal{Y}$, and $\mathcal{Z}$ we have: i. (Reflexivity) $\mathcal{X} \sim \mathcal{X}$, ii. (Symmetry) If $\mathcal{X} \sim \mathcal{Y}$, then $\mathcal{Y} \sim \mathcal{X}$, and iii. (Transitivity) If $\mathcal{X} \sim \mathcal{Y}$ and $\mathcal{Y} \sim \mathcal{Z}$, then $\mathcal{X} \sim \mathcal{Z}$.

(b) Let $[\mathcal{X}] = \{\mathcal{Y} : \mathcal{Y} \sim \mathcal{X}\}$ denote the equivalence class of $\mathcal{X}$ with respect to the relation $\sim$. Let $\widetilde{X}$ be the set of all equivalence classes $[\mathcal{X}]$. Define the norm of an equivalence class $\mathcal{X} = \{x_n\}_{n \in \mathbb{N}}$ to be

$$\big\| [\mathcal{X}] \big\|_{\widetilde{X}} = \lim_{n \to \infty} \|x_n\|.$$

Prove that $\| \cdot \|_{\widetilde{X}}$ is well-defined and is a norm on $\widetilde{X}$.

(c) Given $x \in X$, let $[x]$ denote the equivalence class of the Cauchy sequence $\{x, x, x, \dots\}$. Prove that $T \colon x \mapsto [x]$ is an isometric map of X into $\widetilde{X}$.

Show also that $T(X)$ is a dense subspace of $\widetilde{X}$ (so, in the sense of identifying X with $T(X)$, we can consider X to be a subspace of $\widetilde{X}$).

(d) Prove that $\widetilde{X}$ is a Banach space with respect to $\|\cdot\|_{\widetilde{X}}$. We call $\widetilde{X}$ the *completion* of X.

(e) Prove that $\widetilde{X}$ is unique in the sense that if Y is a Banach space and $U\colon X \to Y$ is a linear isometry such that $U(X)$ is dense in Y, then there exists an isometric isomorphism $V\colon Y \to \widetilde{X}$.

6.6.20. Recall that c_{00} is an incomplete space with respect to the norm $\|\cdot\|_1$. Let $\widetilde{X}$ be the completion of c_{00} with respect to this norm, as defined in Problem 6.6.19. Prove that $\widetilde{X}$ is isometrically isomorphic to ℓ^1.

6.7 Infinite Matrices

Suppose that A is an infinite matrix, i.e.,

$$A \;=\; [a_{ij}]_{i,j\in\mathbb{N}} \;=\; \begin{bmatrix} a_{11} & a_{12} & a_{13} & \cdots \\ a_{21} & a_{22} & a_{23} & \cdots \\ a_{31} & a_{32} & a_{33} & \cdots \\ \vdots & \vdots & \vdots & \ddots \end{bmatrix},$$

where each a_{ij} is a scalar. If we think of $x = (x_k)_{k\in\mathbb{N}}$ as being a "column vector," then we can *formally define* $y = Ax$ to be the product of the matrix A with the vector x by extending the familiar definition of a matrix–vector product from finite-dimensional linear algebra. That is, we let $y = Ax$ be the vector whose ith component y_i is the ith row of A times the vector x:

$$y_i \;=\; (Ax)_i \;=\; \sum_{j=1}^{\infty} a_{ij} x_j. \tag{6.14}$$

However, this is only a "formal" definition, because we do not know whether this equation makes any sense—the series in equation (6.14) might not even converge! On the other hand, if we impose some appropriate restrictions on the infinite matrix A, then we may be able to show that the series in equation (6.14) does converge. If we can further show that for each $x \in \ell^p$ the vector $y = Ax = (y_i)_{i\in\mathbb{N}}$ belongs to ℓ^q, then we have a mapping that takes a vector $x \in \ell^p$ and outputs a vector $Ax \in \ell^q$. In this case we usually "identify" the matrix A with the mapping that sends x to Ax, and simply say that A *maps* ℓ^p *into* ℓ^q.

We will focus here on the case $q = p$. The following theorem gives a sufficient condition on an infinite matrix A that implies that A maps ℓ^p into ℓ^p for each $1 \le p \le \infty$ (in fact, this condition implies that $A\colon \ell^p \to \ell^p$ is *bounded* for each p). We call this result *Schur's Test*, because it is the discrete version of a more general theorem known by that name. In the statement of Schur's Test we write $\|A\|_{\ell^p \to \ell^p}$ to emphasize that the operator norm of A depends on the value of p.

Theorem 6.7.1 (Schur's Test). *Let $A = [a_{ij}]_{i,j \in \mathbb{N}}$ be an infinite matrix such that*

$$C_1 = \sup_{i \in \mathbb{N}} \sum_{j=1}^{\infty} |a_{ij}| < \infty, \qquad C_2 = \sup_{j \in \mathbb{N}} \sum_{i=1}^{\infty} |a_{ij}| < \infty. \tag{6.15}$$

Then the following statements hold for each $1 \le p \le \infty$.

(a) *If $x \in \ell^p$, then the series in equation (6.14) converges for each $i \in \mathbb{N}$, and the vector $y = Ax = (y_i)_{i \in \mathbb{N}}$ belongs to ℓ^p.*

(b) *A is a bounded linear mapping of ℓ^p into ℓ^p.*

(c) *The operator norm of $A \colon \ell^p \to \ell^p$ satisfies*

$$\|A\|_{\ell^p \to \ell^p} \le C_1^{1/p'} C_2^{1/p}, \tag{6.16}$$

where p' is the dual index *defined in equation (3.6), we use the convention that $1/\infty = 0$, and we take the right-hand side of equation (6.16) to be zero if A is the zero matrix.*

Proof. We will give the proof for $1 < p < \infty$, and assign the proof for the cases $p = 1$ and $p = \infty$ as Problem 6.7.8. If A is the zero matrix, then $Ax = 0$ for every x and the result is trivial, so we can assume that A is not the zero matrix. Consequently, C_1 and C_2 are strictly positive numbers.

Note first that the entries of A are bounded, since, for example,

$$|a_{ij}| \le \sum_{k=1}^{\infty} |a_{kj}| \le C_2, \qquad \text{for all } i, j \in \mathbb{N}. \tag{6.17}$$

Fix $i \in \mathbb{N}$, and let $x = (x_k)_{k \in \mathbb{N}}$ be any vector in ℓ^p. To simplify the notation, let $b_{ij} = |a_{ij}|^{1/p'}$ and $c_{ij} = |a_{ij}|^{1/p}$. Then since $\frac{1}{p} + \frac{1}{p'} = 1$, we have

$$|a_{ij}| = |a_{ij}|^{1/p'} |a_{ij}|^{1/p} = b_{ij} c_{ij}.$$

Therefore, by applying Hölder's Inequality we see that

$$\sum_{j=1}^{\infty} |a_{ij} x_j| = \sum_{j=1}^{\infty} b_{ij} \left(c_{ij} |x_j| \right) \qquad (|a_{ij}| = b_{ij} c_{ij})$$

$$\le \left(\sum_{j=1}^{\infty} b_{ij}^{p'} \right)^{1/p'} \left(\sum_{j=1}^{\infty} c_{ij}^{p} |x_j|^p \right)^{1/p} \qquad \text{(by Hölder)}$$

$$= \left(\sum_{j=1}^{\infty} |a_{ij}| \right)^{1/p'} \left(\sum_{j=1}^{\infty} |a_{ij}| \, |x_j|^p \right)^{1/p} \qquad \text{(definition of } b_{ij} \text{ and } c_{ij})$$

$$\leq C_1^{1/p'} \left(\sum_{j=1}^{\infty} |a_{ij}| \, |x_j|^p \right)^{1/p} \qquad \text{(by equation (6.15))}$$

$$\leq C_1^{1/p'} \left(\sum_{j=1}^{\infty} C_2 \, |x_j|^p \right)^{1/p} \qquad \text{(by equation (6.17))}$$

$$= C_1^{1/p'} C_2^{1/p} \, \|x\|_p \; < \; \infty.$$

This shows that the series of scalars $y_i = \sum_{j=1}^{\infty} a_{ij} x_j$ converges absolutely. Hence the vector $y = Ax = (y_i)_{i \in \mathbb{N}}$ is well-defined (i.e., it exists, because the series that define the y_i converge for every i). However, we do not yet know whether y belongs to ℓ^p.

Reusing some of the estimates made above, we see that

$$|y_i| = \left| \sum_{j=1}^{\infty} a_{ij} x_j \right| \leq \sum_{j=1}^{\infty} |a_{ij} x_j| \leq C_1^{1/p'} \left(\sum_{j=1}^{\infty} |a_{ij}| \, |x_j|^p \right)^{1/p}. \qquad (6.18)$$

Taking pth powers, we use this to estimate the ℓ^p-norm of $y = Ax$ as follows:

$$\|Ax\|_p^p = \|y\|_p^p = \sum_{i=1}^{\infty} |y_i|^p$$

$$\leq \sum_{i=1}^{\infty} C_1^{p/p'} \sum_{j=1}^{\infty} |a_{ij}| \, |x_j|^p$$

$$= C_1^{p/p'} \sum_{j=1}^{\infty} \sum_{i=1}^{\infty} |a_{ij}| \, |x_j|^p \qquad (6.19)$$

$$\leq C_1^{p/p'} \sum_{j=1}^{\infty} C_2 \, |x_j|^p$$

$$= C_1^{p/p'} C_2 \, \|x\|_p^p \; < \; \infty.$$

The interchange in the order of summation that is made at the step corresponding to equation (6.19) is justified by Problem 4.3.9 because all of the terms in the series are nonnegative (this is a discrete version of *Tonelli's Theorem*; such an interchange need not be valid if the series contains both positive and negative terms). We conclude that $Ax \in \ell^p$ whenever $x \in \ell^p$, i.e., A maps ℓ^p into ℓ^p. Further, by taking pth roots and then taking the supremum over all unit vectors in ℓ^p, we obtain a bound on the operator norm of A on ℓ^p:

$$\|A\|_{\ell^p \to \ell^p} = \sup_{\|x\|_p=1} \|Ax\|_p \leq \sup_{\|x\|_p=1} C_1^{1/p'} C_2^{1/p} \|x\|_p = C_1^{1/p'} C_2^{1/p}. \qquad \Diamond$$

An entirely analogous result holds for bi-infinite matrices $A = [a_{ij}]_{i,j\in\mathbb{Z}}$ and bi-infinite vectors $x = (x_k)_{k\in\mathbb{Z}}$; we just need to replace the index set $\mathbb{N}$ with $\mathbb{Z}$ in the results above. We will present an important corollary of Theorem 6.7.1 that holds in this bi-infinite setting. First, however, we make another "formal" definition.

Definition 6.7.2 (Convolution of Sequences). Let $x = (x_k)_{k\in\mathbb{Z}}$ and $y = (y_k)_{k\in\mathbb{Z}}$ be bi-infinite sequences of scalars. The *convolution* of x and y is formally defined to be the sequence $x * y$ whose kth component is

$$(x * y)_k = \sum_{j=-\infty}^{\infty} x_j y_{k-j}, \qquad (6.20)$$

as long as this series converges. $\Diamond$

As before, the word *formal* means that this definition need not always make sense, because the series in equation (6.20) need not always converge (give an example of x and y for which that happens!). If the series does converge for every $k \in \mathbb{Z}$, then $x * y$ exists and it is the sequence whose kth component is defined by equation (6.20). Otherwise, $x * y$ is undefined.

By rephrasing convolution in terms of an infinite matrix and applying Schur's Test, we obtain the following result. This is the discrete version of a more general theorem known as *Young's Inequality*. In this theorem, $\ell^p(\mathbb{Z})$ is the space of all bi-infinite sequences that are p-summable (if p is finite) or bounded (if $p = \infty$).

Theorem 6.7.3 (Young's Inequality). *Fix $1 \leq p \leq \infty$. If $x \in \ell^p(\mathbb{Z})$ and $y \in \ell^1(\mathbb{Z})$, then the convolution $x * y$ exists and belongs to $\ell^p(\mathbb{Z})$, and*

$$\|x * y\|_p \leq \|x\|_p \|y\|_1. \qquad (6.21)$$

Proof. Let A be the bi-infinite matrix

$$A = [y_{i-j}]_{i,j\in\mathbb{Z}}. \qquad (6.22)$$

Note that the entries of A are constant along each diagonal.

For this matrix A, the bi-infinite analogues of the numbers C_1 and C_2 defined in equation (6.15) are

$$C_1 = \sup_{i\in\mathbb{Z}} \sum_{j=-\infty}^{\infty} |y_{i-j}| \quad \text{and} \quad C_2 = \sup_{j\in\mathbb{Z}} \sum_{i=-\infty}^{\infty} |y_{i-j}|.$$

By making a change of variables, we see that

$$\sum_{j=-\infty}^{\infty} |y_{i-j}| = \sum_{j=-\infty}^{\infty} |y_j| = \|y\|_1,$$

which is a finite constant independent of i. Therefore $C_1 = \|y\|_1$, and likewise $C_2 = \|y\|_1$. Since C_1 and C_2 are both finite, Schur's Test therefore implies that Ax exists for each $x \in \ell^p$. But

$$(Ax)_i = \sum_{j=-\infty}^{\infty} y_{i-j} x_j = (x * y)_i,$$

so we conclude that the convolution $x * y = \big((x * y)_i\big)_{i \in \mathbb{Z}}$ exists. Further, Schur's Test tells us that A maps ℓ^p boundedly into itself, and

$$\begin{aligned} \|x * y\|_p &= \|Ax\|_p \\ &\leq \|A\|_{\ell^p \to \ell^p} \|x\|_p \\ &\leq C_1^{1/p'} C_2^{1/p} \|x\|_p \\ &= \|y\|_1^{1/p'} \|y\|_1^{1/p} \|x\|_p \\ &= \|y\|_1 \|x\|_p. \end{aligned}$$

This is true for every $x \in \ell^p(\mathbb{Z})$, so equation (6.21) holds. $\quad\square$

In particular, by taking $p = 1$ we see that $\ell^1(\mathbb{Z})$ *is closed under convolution.* That is, if x and y are any vectors in $\ell^1(\mathbb{Z})$, then their convolution $x * y$ exists and belongs to $\ell^1(\mathbb{Z})$ as well. Further, we have the following *submultiplicative* property of convolution:

$$\|x * y\|_1 \leq \|x\|_1 \|y\|_1, \qquad \text{for all } x, y \in \ell^1(\mathbb{Z}). \tag{6.23}$$

Remark 6.7.4. The entries of the matrix $A = [y_{i-j}]_{i,j \in \mathbb{Z}}$ defined in equation (6.22) depend only on the *difference* $i - j$ rather than on i and j individually. Such a matrix is said to be *Toeplitz.* What does a Toeplitz matrix look like? In particular, what does it look like if y belongs to c_{00}, i.e., if only finite many entries of y are nonzero? $\quad\Diamond$

Although it may not be apparent at first glance, convolution is an important operation that plays a central role in many areas, including harmonic analysis in mathematics and signal processing in engineering. Some references that include more details on convolution include [DM72], [Ben97], [Kat04], [Heil11] (and also see Problem 6.7.9).

As we have seen, an infinite matrix A can determine an operator that maps infinite sequences to infinite sequences. The next theorem defines an analogous type of operator, called an *integral operator*, that acts on continuous functions instead of sequences. Specifically, the operator L defined in this theorem maps continuous functions on $[a, b]$ to continuous functions on $[a, b]$, and it is a bounded operator on $C[a, b]$ when we place the L^p-norm $\| \cdot \|_p$ on $C[a, b]$ (for *continuous* functions, the L^∞ norm coincides with the uniform norm, i.e., $\|f\|_\infty = \|f\|_u$ for continuous functions on an interval).

Theorem 6.7.5. *Let k be a continuous function defined on the rectangle $[a, b]^2$ in $\mathbb{R}^2$, and define*

$$C_1 = \sup_{x \in [a,b]} \int_a^b |k(x, y)| \, dy, \qquad C_2 = \sup_{y \in [a,b]} \int_a^b |k(x, y)| \, dx.$$

Then the following statements hold for each $1 \le p \le \infty$.

(a) *If $f \in C[a, b]$, then*

$$Lf(x) = \int_a^b k(x, y) \, f(y) \, dy, \qquad x \in [a, b],$$

is a continuous function on $[a, b]$.

(b) *L is a bounded linear mapping of $C[a, b]$ into $C[a, b]$ with respect to the norm $\| \cdot \|_p$ on $C[a, b]$.*

(c) *The operator norm of $L \colon C[a, b] \to C[a, b]$ satisfies*

$$\|L\| \le C_1^{1/p'} C_2^{1/p}, \tag{6.24}$$

where p' is the dual index *defined in equation (3.6), we use the convention that $1/\infty = 0$, and we take the right-hand side of equation (6.24) to be zero if k is the zero function.* $\diamond$

The integral operator L defined in equation (6.24) is a generalization of the matrix–vector multiplication defined in equation (6.14). Given a sequence $x = (x_k)_{k \in \mathbb{N}}$, the product Ax is the sequence $Ax = \big((Ax)_i\big)_{i \in \mathbb{N}}$ whose ith component is

$$(Ax)_i = \sum_{j=1}^n a_{ij} \, x_j. \tag{6.25}$$

In contrast, if $f \in C[a, b]$, then Lf is the function whose value at a point x is

$$Lf(x) = \int_a^b k(x, y) \, f(y) \, dy. \tag{6.26}$$

Comparing equation (6.25) to (6.26), we see that the function values $k(x, y)$ play a role for the integral operator L that is entirely analogous to the one played by the entries a_{ij} in the matrix A. The integral that appears in equation (6.26) is a kind of "continuous sum" analogue to the series that appears in equation (6.25).

The proof of Theorem 6.7.5 is similar to the proof of Theorem 6.7.1, and is assigned as Problem 6.7.13. More generally, if we use Lebesgue integrals instead of Riemann integrals, then we can formulate a result similar to Theorem 6.7.5 but with $C[a, b]$ replaced with $L^p(E)$, where E is a measurable subset of $\mathbb{R}^d$. In particular, taking $E = \mathbb{R}^d$, we obtain the following analogue of Theorem 6.7.3 for convolution of functions (see [Heil19] for a complete discussion and proof).

Theorem 6.7.6 (Young's Inequality). *Fix $1 \le p \le \infty$. If $f \in L^p(\mathbb{R}^d)$ and $g \in L^1(\mathbb{R}^d)$, then their convolution $f * g$, defined by*

$$(f * g)(x) = \int_{\mathbb{R}^d} f(y) \, g(x - y) \, dy, \qquad (6.27)$$

is a measurable function in $L^p(\mathbb{R}^d)$ that is defined for almost every $x \in \mathbb{R}^d$, and we have $\|f * g\|_p \le \|f\|_p \, \|g\|_1$. ◇

Another example of convolution appears in Problem 6.7.15.

Problems

6.7.7. What are the infinite matrices that correspond to the left-shift and right-shift operators on ℓ^2?

6.7.8. Complete the proof of Theorem 6.7.1 for the cases $p = 1$ and $p = \infty$.

6.7.9. Let $p(t) = a_0 + a_1 t + \cdots + a_m t^m$ and $q(t) = b_0 + b_1 t + \cdots + b_n t^n$, and set $a_k = 0$ for $k > m$ and $k < 0$, and $b_k = 0$ for $k > n$ and $k < 0$. Prove that the product of the polynomials p and q is $p(t)q(t) = c_0 + c_1 t + \cdots + c_{m+n} t^{m+n}$, where

$$c_k = \sum_{j=0}^{k} a_j b_{k-j}, \qquad k = 0, \dots, m + n.$$

Explain why the sequence of coefficients $(c_0, \dots, c_{m+n})$ is the a discrete convolution of the sequence $(a_0, \dots, a_m)$ with the sequence $(b_0, \dots, b_n)$.

6.7.10. Prove equation (6.23) directly, i.e., without appealing to Theorems 6.7.1 and 6.7.3. Show further that if every component of x and y is nonnegative, then $\|x * y\|_1 = \|x\|_1 \, \|y\|_1$.

6.7.11. Prove the following properties of convolution of sequences in $\ell^1(\mathbb{Z})$, where x, y, and z are sequences in $\ell^1(\mathbb{Z})$ and c is a scalar.

(a) Commutativity: $x * y = y * x$.

(b) Associativity: $(x * y) * z = x * (y * z)$. Hint: Fubini's Theorem for series (Problem 4.3.8) is helpful.

(c) Distributive Law for Addition: $x * (y + z) = x * y + x * z$.

(d) Distributive Law for Scalar Multiplication: $c(x*y) = (cx)*y = x*(cy)$.

(e) Existence of an Identity: There is a sequence $\delta \in \ell^1(\mathbb{Z})$ such that $x * \delta = x$ for every $x \in \ell^1(\mathbb{Z})$.

6.7.12. Fix $1 \leq p \leq \infty$. Prove directly that if $x \in \ell^p(\mathbb{Z})$, and $y \in \ell^{p'}(\mathbb{Z})$, then $x * y$ exists, $x * y$ belongs to $\ell^\infty(\mathbb{Z})$, and $\|x * y\|_\infty \leq \|x\|_p \|y\|_{p'}$.

6.7.13. Prove Theorem 6.7.5.

6.7.14. Suppose that f, g are continuous functions on $\mathbb{R}$ that both have compact support, i.e., they are identically zero outside of some finite interval. Prove directly that their convolution $f * g$, defined by $(f * g)(x) = \int_{-\infty}^{\infty} f(y) g(x - y) \, dy$ for $x \in \mathbb{R}$, is continuous and compactly supported.

6.7.15. We say that a function $f \colon \mathbb{R} \to \mathbb{F}$ is 1-*periodic* if $f(x + 1) = f(x)$ for all $x \in \mathbb{R}$. Prove that if f and g are continuous 1-periodic functions, then their *convolution*, defined by

$$(f * g)(x) = \int_0^1 f(y) g(x - y) \, dy, \qquad x \in \mathbb{R},$$

is a continuous 1-periodic function. Also prove the following properties of convolution, where f, g, and h are continuous and 1-periodic, and c is a scalar.

(a) Commutativity: $f * g = g * f$. Hint: If F is continuous and 1-periodic, then $\int_0^1 F(x) \, dx = \int_a^{a+1} F(x) \, dx$ for every $a \in \mathbb{R}$.

(b) Associativity: $(f * g) * h = f * (g * h)$. Hint: If $G(x, y)$ is a continuous function of two variables defined on the square $[0, 1]^2$, then *Fubini's Theorem* implies that $\int_0^1 \int_0^1 G(x, y) \, dx \, dy = \int_0^1 \int_0^1 G(x, y) \, dy \, dx$.

(c) Distributive Law for Addition: $f * (g + h) = f * g + f * h$.

(d) Distributive Law for Scalar Multiplication: $c(f*g) = (cf)*g = f*(cg)$.

6.7.16. This problem will show that there is no identity element for the type of convolution defined in Problem 6.7.15. For simplicity of notation, we will take $\mathbb{F} = \mathbb{C}$ in this problem. We will use the notion of *Fourier coefficients*, which were defined in equation (5.24).

(a) Suppose that f and g are continuous 1-periodic functions on $\mathbb{R}$. Show that the Fourier coefficients $(f * g)^\wedge(n)$ of $f * g$ satisfy

$$(f * g)^\wedge(n) = \widehat{f}(n) \, \widehat{g}(n), \qquad n \in \mathbb{Z}.$$

Hint: Fubini's Theorem (see the hint to part (b) of Problem 6.7.15).

(b) Suppose that there were a continuous 1-periodic function e that was an identity for convolution, i.e., $f * e = f$ for every continuous 1-periodic function f. Prove that $\widehat{e}(n) = 1$ for every $n \in \mathbb{Z}$, and use this to derive a contradiction. Hint: Plancherel's Equality.

6.8 The Dual of a Hilbert Space

An operator that maps a normed space X into the field of scalars $\mathbb{F}$ is called a *functional* (see Definition 6.1.1). Typically, we use lowercase Greek letters such as λ, μ, and ν to denote functionals. The set of all bounded linear functionals on X is called the dual space of X and is denoted by X^*.

Definition 6.8.1 (Dual Space). The set of all bounded linear functionals on a normed space X is called the *dual space* of X. We denote the dual space by

$$X^* = \mathcal{B}(X, \mathbb{F}) = \{A : X \to \mathbb{F} : A \text{ is bounded and linear}\}. \qquad \Diamond$$

Because the scalar field $\mathbb{F}$ is complete, Theorem 6.5.4 implies that X^* is complete (even if X is not complete). Note that since the norm on $\mathbb{F}$ is absolute value, the operator norm of a functional $\mu \in X^*$ is given by

$$\|\mu\| = \sup_{\|x\|=1} |\mu(x)|.$$

We will prove that if H is a Hilbert space, then H^* is isometrically isomorphic to H. Specifically, we will prove that each vector in H determines a unique functional in X^*; and conversely, each functional in X^* is determined by a unique vector in H. Furthermore, the relation between these two is isometric. To see how this correspondence works, fix a vector $y \in X$, and define a functional $\mu_y : H \to \mathbb{F}$ by

$$\mu_y(x) = \langle x, y \rangle, \qquad x \in H.$$

Since the inner product is linear in the first variable, μ_y is linear:

$$\mu_y(ax + bz) = \langle ax + bz, y \rangle = a\langle x, y \rangle + b\langle z, y \rangle = a\mu_y(x) + b\mu_y(z).$$

The Cauchy–Bunyakovski–Schwarz Inequality implies that

$$|\mu_y(x)| = |\langle x, y \rangle| \leq \|x\| \|y\|, \qquad x \in H,$$

so by taking the supremum over the unit vectors in H we see that

$$\|\mu_y\| = \sup_{\|x\|=1} |\mu_y(x)| \leq \sup_{\|x\|=1} \|x\| \|y\| = \|y\|.$$

Therefore μ_y is bounded, and hence is an element of H^*. Therefore we can define a mapping $T\colon H \to H^*$ by setting

$$T(y) = \mu_y, \qquad y \in H.$$

We will prove in the next theorem that T is an isometric isomorphism. There is one twist, however, since T is not a *linear* map but rather is *antilinear*. This is actually a natural consequence of the fact that the inner product is antilinear in the second variable. While we have considered only linear maps up to now, antilinear maps are very similar, and we use terminology for antilinear maps similar to that for linear maps. In particular, an *antilinear isometric isomorphism* is a *antilinear* mapping that is both isometric and surjective. To emphasize, while μ_y is a *linear* function that sends x to $\mu_y(x)$, the mapping T is an *antilinear* function that sends y to μ_y.

Theorem 6.8.2 (Riesz Representation Theorem). *Let H be a Hilbert space. Using the notation given above, the following statements hold.*

(a) *If $y \in X$, then $\mu_y \in H^*$, and the operator norm of μ_y equals the norm of y:*

$$\|\mu_y\| = \sup_{\|x\|=1} |\mu_y(x)| = \sup_{\|x\|=1} |\langle x, y\rangle| = \|y\|.$$

(b) *If $\mu \in H^*$, then there exists a unique vector $y \in H$ such that $\mu = \mu_y$. That is, if $\mu\colon H \to \mathbb{F}$ is a bounded linear operator, then there exists a unique $y \in H$ such that*

$$\mu(x) = \langle x, y\rangle, \qquad \text{for all } x \in H.$$

(c) *The mapping $T\colon H \to H^*$ given by $T(y) = \mu_y$ is an antilinear isometric isomorphism of H onto H^*. In particular, if $y, z \in H$ and $a, b \in \mathbb{F}$, then*

$$\mu_{ay+bz} = T(ay+bz) = \bar{a}T(y) + \bar{b}T(z) = \bar{a}\mu_y + \bar{b}\mu_z.$$

Proof. (a) We have already seen that μ_y is linear and satisfies $\|\mu_y\| \le \|y\|$. If $y = 0$, then $\mu_y = 0$, and we are done with this part of the proof. On the other hand, if $y \ne 0$, then $z = y/\|y\|$ is a unit vector, and

$$|\mu_y(z)| = |\langle z, y\rangle| = \frac{\langle y, y\rangle}{\|y\|} = \frac{\|y\|^2}{\|y\|} = \|y\|.$$

Since $\|z\| = 1$, this implies that

$$\|\mu_y\| = \sup_{\|x\|=1} |\mu_y(x)| \ge |\mu_y(z)| = \|y\|.$$

Combining this with the inequality $\|\mu_y\| \le \|y\|$ proved before, we see that $\|\mu_y\| = \|y\|$.

(b) Let μ be any functional in H^*. If $\mu = 0$, then $y = 0$ is the required vector, so we may assume that μ is not the zero operator. In this case μ does not map every vector to zero, so the kernel of μ is a subspace of H that is not all of H. Hence its orthogonal complement $\ker(\mu)^\perp$ contains more than just the zero vector. Let z be any nonzero vector in $\ker(\mu)^\perp$. Then $z \notin \ker(\mu)$, so $\mu(z) \neq 0$. Therefore we can define $w = \frac{1}{\mu(z)} z$. Since μ is linear, we have

$$\mu(w) = \frac{1}{\mu(z)} \mu(z) = 1.$$

Now set

$$y = \frac{w}{\|w\|^2}.$$

Since y is a multiple of z, we have $y \in \ker(\mu)^\perp$. We will show that y is the vector that we seek.

Given any $x \in H$, we use the linearity of μ to compute that

$$\mu\big(x - \mu(x)w\big) = \mu(x) - \mu(x)\,\mu(w) = 0.$$

Hence $x - \mu(x)w \in \ker(\mu)$. But $y \in \ker(\mu)^\perp$, so $x - \mu(x)w$ and y are orthogonal. Using the fact that the inner product is linear in the first variable, we therefore compute that

$$\begin{aligned}
0 = \big\langle x - \mu(x)w,\, y \big\rangle &= \langle x, y \rangle - \mu(x)\,\langle w, y \rangle \\
&= \langle x, y \rangle - \mu(x)\,\frac{\langle w, w \rangle}{\|w\|^2} \\
&= \langle x, y \rangle - \mu(x).
\end{aligned}$$

Thus $\mu(x) = \langle x, y \rangle = \mu_y(x)$. As is true for every x, we have shown that $\mu = \mu_y$.

It remains only to show that y is unique. Suppose that we also had $\mu = \mu_z$ for some $z \in H$. Then for all $x \in H$ we would have

$$\langle x, y - z \rangle = \langle x, y \rangle - \langle x, z \rangle = \mu_y(x) - \mu_z(x) = \mu(x) - \mu(x) = 0.$$

Since this is true for every x, including $x = y - z$ in particular, it follows that $y - z = 0$. Therefore y is unique.

(c) Parts (a) and (b) show that T is both surjective and norm-preserving. Therefore, we just have to show that T is antilinear. To do this, fix any $y \in H$ and $c \in \mathbb{F}$. Then, since the inner product is antilinear in the second variable, for each $x \in H$ we have

$$\mu_{cy}(x) = \langle x, cy \rangle = \bar{c}\,\langle x, y \rangle = \bar{c}\,\mu_y(x).$$

Hence $T(cy) = \mu_{cy} = \bar{c}\,\mu_y = \bar{c}\,T_y$. A similar argument shows that

$$T(y+z) = \mu_{y+z} = \mu_y + \mu_z = T(y) + T(z),$$

so T is antilinear. $\quad\square$

Problems

6.8.3. Suppose that μ is a bounded linear functional on ℓ^2. Prove that there exists a unique sequence $y \in \ell^2$ such that

$$\mu(x) = \sum_{n=1}^{\infty} x_k \overline{y_k}, \qquad \text{for all } x = (x_k)_{k \in \mathbb{N}} \in \ell^2,$$

and a unique sequence $z \in \ell^2$ such that

$$\mu(x) = \sum_{n=1}^{\infty} x_k z_k, \qquad \text{for all } x = (x_k)_{k \in \mathbb{N}} \in \ell^2.$$

6.8.4. As usual, we let $\mathbb{F}^d$ be the set of all column vectors with d components, but for this problem we specifically take those vectors to be column vectors ($d \times 1$ matrices). The inner product on $\mathbb{F}^d$ is the ordinary dot product of vectors.

(a) Show directly that $\mu \colon \mathbb{F}^d \to \mathbb{F}$ is a linear functional on $\mathbb{F}^d$ if and only if there if exists a vector $y \in \mathbb{F}^d$ such that $\mu(x) = x \cdot y = y^H x$ for all $x \in \mathbb{F}^d$. Prove that every linear functional on $\mathbb{F}^d$ is therefore uniquely identified with a row vector y^H.

(b) Explain why the Riesz Representation Theorem for $H = \mathbb{F}^d$ can be interpreted as saying that $H^* = (\mathbb{F}^d)^*$ is (antilinearly) identified with the set of $1 \times d$ row vectors (which can itself be identified with $\mathbb{F}^d$).

6.8.5. We say that a sequence $\{y_n\}_{n \in \mathbb{N}}$ of vectors in a Hilbert space H is a *Bessel sequence* if there exists a constant $B < \infty$ such that

$$\sum_{n=1}^{\infty} |\langle x, y_n \rangle|^2 \leq B \|x\|^2, \qquad x \in H.$$

Prove the following statements.

(a) If $\{y_n\}_{n \in \mathbb{N}}$ is a Bessel sequence, then

$$Ax = \{\langle x, y_n \rangle\}_{n \in \mathbb{N}}, \qquad x \in H, \tag{6.28}$$

defines a bounded linear operator from A to ℓ^2.

(b) If $A \colon H \to \ell^2$ is a bounded linear operator, then there exists a Bessel sequence $\{y_n\}_{n \in \mathbb{N}}$ such that equation (6.28) holds.

6.9 The Dual of ℓ^p

Since ℓ^2 is Hilbert space, the Riesz Representation Theorem applies to ℓ^2, and it tells us that the dual space $(\ell^2)^*$ is (antilinearly) isometrically isomorphic to ℓ^2. Now we will try to characterize the dual of ℓ^p for p other than 2.

First we will show that every vector $y \in \ell^{p'}$ determines a bounded linear functional μ_y on ℓ^p. For convenience, and in contrast to the case $p = 2$, we will omit the complex conjugate in the definition of μ_y. This will give us the minor convenience that the isomorphism T that we create later will be linear instead of antilinear. As usual, p' denotes the dual index to p.

Theorem 6.9.1. *Fix $1 \le p \le \infty$. For each $y \in \ell^{p'}$, define $\mu_y \colon \ell^p \to \mathbb{F}$ by*

$$\mu_y(x) = \sum_{k=1}^{\infty} x_k y_k, \qquad x = (x_k)_{k \in \mathbb{N}} \in \ell^p. \tag{6.29}$$

Then $\mu_y \in (\ell^p)^*$, and the operator norm of μ_y equals the norm of y, i.e.,

$$\|\mu_y\| = \sup_{\|x\|=1} |\mu_y(x)| = \|y\|.$$

Proof. We will consider $1 < p < \infty$, and assign the proofs for the cases $p = 1$ and $p = \infty$ as Problem 6.9.6.

If $1 < p < \infty$, then we also have $1 < p' < \infty$. If $x \in \ell^p$, then, since $y \in \ell^{p'}$, we can apply Hölder's Inequality to obtain the estimate

$$|\mu_y(x)| = \left| \sum_{k=1}^{\infty} x_k y_k \right| \le \|x\|_p \|y\|_{p'}.$$

Taking the supremum over all unit vectors x in ℓ^p, it follows that μ_y is bounded, and

$$\|\mu_y\| = \sup_{\|x\|=1} |\mu_y(x)| \le \sup_{\|x\|=1} \|x\|_p \|y\|_{p'} = \|y\|_{p'}.$$

We must prove that we also have $\|\mu_y\| \ge \|y\|_{p'}$.

If y is the zero sequence, then we are done, so assume that not every component of y is zero. For each $k \in \mathbb{N}$, let α_k be a scalar such that $|\alpha_k| = 1$ and $\alpha_k y_k = |y_k|$. Set

$$z_k = \frac{\alpha_k |y_k|^{p'-1}}{\|y\|_{p'}^{p'-1}}, \qquad k \in \mathbb{N},$$

and let $z = (z_k)_{k \in \mathbb{N}}$. Then, using the fact that $(p'-1)p = p'$, we compute that

$$\|z\|_p^p = \sum_{k=1}^{\infty} \left(\frac{|y_k|^{p'-1}}{\|y\|_{p'}^{p'-1}}\right)^p = \sum_{k=1}^{\infty} \frac{|y_k|^{p'}}{\|y\|_{p'}^{p'}} = \frac{1}{\|y\|_{p'}^{p'}} \sum_{k=1}^{\infty} |y_k|^{p'} = 1.$$

Hence z is a unit vector in ℓ^p. Also,

$$|\mu_y(z)| = \left|\sum_{k=1}^{\infty} z_k y_k\right| = \sum_{k=1}^{\infty} \frac{\alpha_k |y_k|^{p'-1} y_k}{\|y\|_{p'}^{p'-1}}$$

$$= \sum_{k=1}^{\infty} \frac{|y_k|^{p'}}{\|y\|_{p'}^{p'-1}}$$

$$= \frac{\|y\|_{p'}^{p'}}{\|y\|_{p'}^{p'-1}} = \|y\|_{p'}.$$

Since z is one of the unit vectors in ℓ^p, this shows that $\|\mu_y\| \geq \|y\|_{p'}$. $\quad\square$

As a corollary of Theorem 6.9.1, we see that the mapping $T\colon \ell^{p'} \to (\ell^p)^*$ defined by

$$T(y) = \mu_y, \qquad y \in \ell^{p'},$$

is a linear isometry (but we do not know yet whether T is surjective).

Remark 6.9.2. If we wished to mimic the ℓ^2 situation more closely, we could have included complex conjugates in equation (6.29) and defined $\mu_y(x) = \sum_{k=1}^{\infty} x_k \overline{y_k}$. In this case the mapping $T(y) = \mu_y$ would be an antilinear isometry, just as it was when we considered Hilbert spaces in Section 6.8. $\quad\diamond$

Next we will prove that the mapping T is surjective *when p is finite*. Consequently $(\ell^p)^* \cong \ell^{p'}$ for $1 \leq p < \infty$, i.e., the dual space of ℓ^p is isometrically isomorphic to $\ell^{p'}$ for finite p.

Theorem 6.9.3 (Dual of ℓ^p). *Fix $1 \leq p < \infty$. Using the notation introduced before, the mapping $T\colon \ell^{p'} \to (\ell^p)^*$ given by $T(y) = \mu_y$ is surjective, and consequently T is an isometric isomorphism of $\ell^{p'}$ onto $(\ell^p)^*$.*

Proof. We proved that T is isometric in Theorem 6.9.1, so it remains only to prove that T is surjective. We will do this for $1 < p < \infty$, and assign the proof for $p = 1$ as Problem 6.9.6.

Choose any $\mu \in (\ell^p)^*$, i.e., assume that $\mu\colon \ell^p \to \mathbb{F}$ is a bounded linear functional. Define scalars

$$y_k = \mu(\delta_k), \qquad k \in \mathbb{N},$$

where δ_k is the kth standard basis vector. Then define a vector y by setting $y = (y_k)_{k\in\mathbb{N}}$. If $y_k = 0$ for every k, then μ is the zero functional (why?), and hence we have $\mu = \mu_y$ in this case. Therefore we can assume that not every y_k is zero. We will show that y belongs to $\ell^{p'}$ and $\mu = \mu_y$.

For each $k \in \mathbb{N}$, let α_k be a scalar such that $|\alpha_k| = 1$ and $\alpha_k y_k = |y_k|$. Set

$$z_k = \alpha_k |y_k|^{p'-1}, \qquad k \in \mathbb{N}.$$

Fix an integer $N \in \mathbb{N}$, and let

$$w_N = \sum_{k=1}^{N} z_k \delta_k = (z_1, \ldots, z_N, 0, 0, \ldots).$$

Then $w_N \in \ell^p$, and since μ is linear, we have

$$\mu(w_N) = \sum_{k=1}^{N} z_k \mu(\delta_k) = \sum_{k=1}^{N} z_k y_k = \sum_{k=1}^{N} \alpha_k |y_k|^{p'-1} y_k = \sum_{k=1}^{N} |y_k|^{p'}.$$

Since μ is bounded, it follows that

$$\sum_{k=1}^{N} |y_k|^{p'} = \mu(w_N) \leq \|\mu\| \, \|w_N\|_p = \|\mu\| \left(\sum_{k=1}^{N} |y_k|^{p'} \right)^{1/p}.$$

The sums on the preceding line are nonzero for all large enough N (because not every y_k is zero), so by dividing we obtain, for all large enough N,

$$\left(\sum_{k=1}^{N} |y_k|^{p'} \right)^{1/p'} = \left(\sum_{k=1}^{N} |y_k|^{p'} \right)^{1-\frac{1}{p}} = \frac{\sum_{k=1}^{N} |y_k|^{p'}}{\left(\sum_{k=1}^{N} |y_k|^{p'} \right)^{1/p}} \leq \|\mu\|.$$

Letting $N \to \infty$, we conclude that

$$\|y\|_{p'} = \left(\sum_{k=1}^{\infty} |y_k|^{p'} \right)^{1/p'} \leq \|\mu\| < \infty.$$

Hence $y \in \ell^{p'}$. Therefore μ_y, the functional determined by y, is a bounded linear functional on ℓ^p. Finally, if $x \in \ell^p$, then, since μ is linear and continuous,

$$\mu(x) = \mu\left(\sum_{k=1}^{\infty} x_k \delta_k \right) = \sum_{k=1}^{\infty} x_k \mu(\delta_k) = \sum_{k=1}^{\infty} x_k y_k = \mu_y(x).$$

Thus $\mu = \mu_y = T(y)$, so T is surjective. $\quad \square$

In contrast, the next theorem states that T is *not surjective* when $p = \infty$. The dual space $(\ell^\infty)^*$ is *not* isometrically isomorphic to ℓ^1.

Theorem 6.9.4. *Using the notation introduced above, there exists a bounded linear functional $\nu \in (\ell^\infty)^*$ such that $\nu \neq \mu_y$ for every $y \in \ell^1$. Consequently, the mapping $T \colon \ell^1 \to (\ell^\infty)^*$ defined in Lemma 6.9.1 is not surjective.* $\diamondsuit$

A typical proof of Theorem 6.9.4 employs the *Hahn–Banach Theorem* or another major theorem from functional analysis (for a sketch of how this works, see Problem 6.9.4).

Problem 6.9.7 gives another interesting characterization of a dual space. Specifically, it shows that the dual space of c_0 is isometrically isomorphic to ℓ^1, i.e., $(c_0)^* \cong \ell^1$.

Problems

6.9.5. Fix $1 \leq p \leq \infty$, and suppose that $A = [a_{ij}]_{i,j \in \mathbb{N}}$ is an infinite matrix such that each row $(a_{ij})_{j \in \mathbb{N}}$ belongs to $\ell^{p'}$. Prove that if $x \in \ell^p$, then the product Ax defined by equation (6.14) exists. Must A map ℓ^p into ℓ^q for some q?

6.9.6. (a) Complete the proof of Theorem 6.9.1 for $p = 1$ and $p = \infty$.

(b) Complete the proof of Theorem 6.9.3 for the case $p = 1$.

6.9.7. Given $y = (y_k)_{k \in \mathbb{N}} \in \ell^1$, define $\mu_y \colon c_0 \to \mathbb{F}$ by

$$\mu_y(x) = \sum_{k=1}^{\infty} x_k y_k, \qquad x = (x_k)_{k \in \mathbb{N}} \in c_0.$$

Prove that $T(y) = \mu_y$ is an isometric isomorphism that maps ℓ^1 onto c_0^*, and therefore $c_0^* \cong \ell^1$.

6.9.8. (a) Let c be the space of all convergent sequences, i.e.,

$$c = \left\{ x = (x_k)_{k \in \mathbb{N}} \in \ell^\infty : \lim_{k \to \infty} x_k \text{ exists} \right\}.$$

Show that c is a closed subspace of ℓ^∞, and therefore it is a Banach space with respect to the ℓ^∞-norm.

(b) Define $\mu \colon c \to \mathbb{F}$ by $\mu(x) = \lim_{k \to \infty} x_k$ for $x = (x_k)_{k \in \mathbb{N}} \in c$. Show that $\mu \in c^*$, i.e., μ is a bounded linear functional on c.

(c) The *Hahn–Banach Theorem* implies that there exists a bounded linear functional $\nu \in (\ell^\infty)^*$ that *extends* μ, i.e., ν is a bounded linear functional on ℓ^∞ and $\nu(x) = \mu(x)$ for all $x \in c$. Show that $\nu \neq \mu_y$ for every $y \in \ell^1$, where $\mu_y \in (\ell^\infty)^*$ is the functional defined by

$$\mu_y(x) = \sum_{k=1}^{\infty} x_k y_k, \qquad x = (x_k)_{k\in\mathbb{N}} \in \ell^{\infty}.$$

Conclude that the mapping $T(y) = \mu_y$ given in Theorem 6.9.4 is not surjective. Remark: For a discussion of the Hahn–Banach Theorem see [Heil11, Ch. 2]; for a proof of Hahn–Banach see [Con90] or [Fol99].

(d) Find a vector $\delta_0 \in c$ such that $\{\delta_n\}_{n\geq 0}$ is a Schauder basis for c.

(e) Show that $c^* \cong \ell^1$. Remark: We also have $c_0^* \cong \ell^1$ by Problem 6.9.7. Yet, even though the dual spaces of c and c_0 are isometrically isomorphic, c is not isometrically isomorphic to c_0 (for one proof, see [Heil11, Exer. 4.22]).

6.10 The Uniform Boundedness Principle

The Uniform Boundedness Principle states that a family of bounded linear operators on a Banach space that are uniformly bounded *at each individual point* must actually be *uniformly bounded in operator norm*. We will use the Baire Category Theorem to prove this result. Note that this theorem applies to *any family* of bounded linear operators, even an uncountable collection of operators.

Theorem 6.10.1 (Uniform Boundedness Principle). *Assume that X is a Banach space and let Y be a normed space. If $\{A_i\}_{i\in I}$ is a collection of operators in $\mathcal{B}(X,Y)$ that satisfies*

$$\forall x \in X, \quad \sup_{i\in I} \|A_i x\| < \infty,$$

then

$$\sup_{i\in I} \|A_i\| = \sup_{i\in I} \sup_{\|x\|=1} \|A_i x\| < \infty.$$

Proof. For each $n \in \mathbb{N}$ set

$$E_n = \left\{ x \in X : \sup_{i\in I} \|A_i x\| \leq n \right\}.$$

We have $X = \cup E_n$ by hypothesis. Furthermore, because each A_i is continuous, Problem 6.4.4 implies that every E_n is a closed subset of X. Since X is complete, we can therefore apply the Baire Category Theorem (Theorem 2.11.3) and conclude that some E_n must contain an open ball. That is, there must be some $n \in \mathbb{N}$, some point $x_0 \in X$, and some radius $r > 0$ such that

$$B_r(x_0) \subseteq E_n.$$

Choose any nonzero vector $x \in X$. We rescale and translate x to obtain a vector y that lies in the ball $B_r(x_0)$. Specifically, setting

$$y = x_0 + sx, \quad \text{where } s = \frac{r}{2\|x\|},$$

we have $y \in B_r(x_0) \subseteq E_n$. Therefore

$$
\begin{aligned}
\|A_i x\| = \left\| A_i\left(\frac{y - x_0}{s}\right) \right\| &= \frac{1}{s}\|A_i y - A_i x_0\| \\
&\leq \frac{1}{s}\left(\|A_i y\| + \|A_i x_0\|\right) \\
&\leq \frac{2\|x\|}{r}(n + n) \\
&= \frac{4n}{r}\|x\|.
\end{aligned}
$$

This is true for every nonzero $x \in X$, so by taking the supremum over all unit vectors x we see that $\|A_i\| \leq 4n/r$. And this is true for every $i \in I$, so

$$\sup_{i \in I}\|A_i\| \leq \frac{4n}{r} < \infty. \qquad \Diamond$$

Here is an important corollary of the Uniform Boundedness Principle (in some texts, the names "Uniform Boundedness Principle" and "Banach–Steinhaus Theorem" are used interchangeably).

Theorem 6.10.2 (Banach–Steinhaus Theorem). *Let X and Y be Banach spaces, and suppose that $\{A_n\}_{n \in \mathbb{N}}$ is a countable sequence of operators in $\mathcal{B}(X, Y)$ such that*

$$Ax = \lim_{n \to \infty} A_n x \quad \text{exists for each } x \in X.$$

Then $A \in \mathcal{B}(X, Y)$ and $\|A\| \leq \sup\|A_n\| \leq \infty$.

Proof. By hypothesis, $Ax = \lim_{n \to \infty} A_n x$ exists for each $x \in X$. This implies that A is linear (why?). Also, since convergent sequences are bounded, we have that $\sup_n \|A_n x\| < \infty$ for each $x \in X$. The Uniform Boundedness Principle therefore implies that $M = \sup_n \|A_n\| < \infty$. Consequently, for each $x \in X$ we have

$$
\begin{aligned}
\|Ax\| &\leq \|Ax - A_n x\| + \|A_n x\| \\
&\leq \|Ax - A_n x\| + \|A_n\|\,\|x\| \\
&\leq \|Ax - A_n x\| + M\,\|x\| \\
&\to 0 + M\,\|x\| \quad \text{as } n \to \infty.
\end{aligned}
$$

Therefore $\|Ax\| \leq M \|x\|$ for every x, so A is bounded and its operator norm satisfies $\|A\| \leq M$. $\square$

The hypotheses of the Banach–Steinhaus Theorem do *not* imply that $A_n \to A$ in operator norm (see Problem 7.1.9).

As an application of the Banach–Steinhaus Theorem we prove a kind of converse to Hölder's Inequality for sequences.

Theorem 6.10.3. *Fix $1 \leq p \leq \infty$ and let $y = (y_k)_{k \in \mathbb{N}}$ be any sequence of scalars. Then*

$$\sum_{k=1}^{\infty} x_k y_k \text{ converges for every } x = (x_k)_{k \in \mathbb{N}} \in \ell^p \quad \Longleftrightarrow \quad y \in \ell^{p'}.$$

Proof. $\Leftarrow$. This direction follows immediately from Hölder's Inequality (Theorem 3.2.2).

$\Rightarrow$. Assume that $1 < p < \infty$ (the proof for the cases $p = 1$ and $p = \infty$ are assigned as Problem 6.10.4).

Suppose that $\sum x_k y_k$ converges for all $x \in \ell^p$. Define linear functionals μ_N and μ on ℓ^p by

$$\mu(x) = \sum_{k=1}^{\infty} x_k y_k \quad \text{and} \quad \mu_N(x) = \sum_{k=1}^{N} x_k y_k. \tag{6.30}$$

Also set $C_N = \left(\sum_{k=1}^{N} |y_k|^{p'}\right)^{1/p'}$, and note that C_N is a finite constant (although it can be different for each integer N). Given $x \in \ell^p$, we compute that

$$|\mu_N(x)| = \left| \sum_{k=1}^{N} x_k y_k \right|$$

$$\leq \left(\sum_{k=1}^{N} |x_k|^p \right)^{1/p} \left(\sum_{k=1}^{N} |y_k|^{p'} \right)^{1/p'} \quad \text{(by Hölder's Inequality)}$$

$$\leq \left(\sum_{k=1}^{\infty} |x_k|^p \right)^{1/p} \left(\sum_{k=1}^{N} |y_k|^{p'} \right)^{1/p'}$$

$$= C_N \|x\|_p.$$

This shows that μ_N is bounded, i.e., $\mu_N \in (\ell^p)^*$. In fact, taking the supremum over all unit vectors x, we see that $\|\mu_N\| \leq C_N < \infty$.

Now by hypothesis, for each $x \in \ell^p$ we have $\mu_N(x) \to \mu(x)$ as $N \to \infty$. The Banach–Steinhaus Theorem therefore immediately implies that $\mu \in (\ell^p)^*$.

It remains to show that $y \in \ell^{p'}$. We saw in Theorem 6.9.3 that $(\ell^p)^*$ is isomorphic to $\ell^{p'}$. Therefore $\mu \in (\ell^p)^* \cong \ell^{p'}$, so there must exist some sequence $z \in \ell^{p'}$ such that $\mu(x) = \sum x_k z_k$ for all $x \in \ell^p$. Taking $x = \delta_k$, the kth standard basis vector, it follows that $\mu(\delta_k) = z_k$. But equation (6.30) tells us that $\mu(\delta_k) = y_k$, so $y_k = z_k$ for every k. Thus $y = z \in \ell^{p'}$. $\quad\square$

Problems

6.10.4. Prove Theorem 6.10.3 for the cases $p = 1$ and $p = \infty$.

6.10.5. Let X and Y be Banach spaces, and let S be a dense subspace of X. Suppose that we are given bounded linear operators $A_n \in \mathcal{B}(X, Y)$ such that $Ax = \lim_{n\to\infty} A_n x$ exists for each vector $x \in S$.

(a) Show that if $\sup \|A_n\| < \infty$, then A extends to a bounded map that is defined on all of X, and $Ax = \lim_{n\to\infty} A_n x$ for all $x \in X$.

(b) Show by example that the hypothesis $\sup \|A_n\| < \infty$ in part (a) is necessary.

6.10.6. Let X be a Banach space, and let S be a subset of X^*. Prove that S is bounded if and only if $\sup\{|\mu(x)| : \mu \in S\} < \infty$ for each $x \in X$.

6.10.7. Given a sequence $\{y_n\}_{n\in\mathbb{N}}$ of vectors in a Hilbert space H, prove that the following two statements are equivalent.

(a) $\sum_{n=1}^{\infty} |\langle x, y_n \rangle|^2 < \infty$ for every $x \in H$.

(b) $\{y_n\}_{n\in\mathbb{N}}$ is a *Bessel sequence* in the sense of Problem 6.8.5, i.e., there exists a constant $B < \infty$ such that $\sum_{n=1}^{\infty} |\langle x, y_n \rangle|^2 \le B \|x\|^2$ for every $x \in H$.

6.10.8. Fix $1 \le p, q \le \infty$, and let $A = [a_{ij}]_{i,j\in\mathbb{N}}$ be an infinite matrix such that:

(a) $(Ax)_i = \sum_j a_{ij} x_j$ converges for each $x \in \ell^p$ and $i \in \mathbb{N}$, and

(b) $Ax = ((Ax)_i)_{i\in\mathbb{N}} \in \ell^q$ for each $x \in \ell^p$.

Identifying the matrix A with the map $x \mapsto Ax$, prove that $A \in \mathcal{B}(\ell^p, \ell^q)$.

6.10.9. Let X, Y, Z be Banach spaces. Suppose that $B \colon X \times Y \to Z$ is bilinear, i.e., $B_x(z) = B(x, z)$ and $B^y(z) = B(z, y)$ are linear functions of z for each $x \in X$ and $y \in Y$. Prove that the following three statements are equivalent.

(a) $B_x \colon Y \to Z$ and $B^y \colon X \to Z$ are continuous for each $x \in X$ and $y \in Y$.

(b) There is a constant $C > 0$ such that

$$\|B(x, y)\| \le C \|x\| \|y\|, \qquad \text{for all } x \in X, \ y \in Y.$$

(c) B is a continuous mapping of $X \times Y$ into Z (see Problem 3.6.8 for the definition of the norm on $X \times Y$, and note that B need not be linear *on the domain* $X \times Y$).

Chapter 7
Operators on Hilbert Spaces

In this chapter we will study operators that map one Hilbert space into another. The fact that we now have an inner product to work with allows us to derive much more detailed results than we could for operators on generic normed spaces.

Throughout this chapter the letters H and K will usually denote Hilbert spaces, but some results require them to be only inner product spaces (we state the precise hypotheses in each theorem). Each of H and K has its own inner product, but typically we just write $\langle \cdot, \cdot \rangle$ for the inner product and let context determine whether this is the inner product on H or on K. For example, if x and y both belong to H, then we know that $\langle x, y \rangle$ must be the inner product from H, while if they belong to K, then $\langle x, y \rangle$ has to denote the inner product on K. If there is a chance of confusion, then we may write $\langle x, y \rangle_H$ and $\langle x, y \rangle_K$.

7.1 Unitary Operators

Recall from Definition 6.6.1 that an *isometry* is an operator A that preserves the norms of vectors (i.e., $\|Ax\| = \|x\|$ for every x). However, if H and K are inner product spaces, then in addition to considering whether a mapping $A \colon H \to K$ preserves norms, we can also ask whether it preserves inner products. We say that A is *inner product preserving* if

$$\langle Ax, Ay \rangle = \langle x, y \rangle, \qquad \text{for all } x, y \in H.$$

Taking $x = y$, we see that any operator that preserves inner products must preserve norms as well, and therefore is an isometry. The next lemma proves that the converse holds for linear operators on inner product spaces.

© Springer International Publishing AG, part of Springer Nature 2018
C. Heil, *Metrics, Norms, Inner Products, and Operator Theory*, Applied and
Numerical Harmonic Analysis, https://doi.org/10.1007/978-3-319-65322-8_7

Lemma 7.1.1. *Let H and K be inner product spaces. If $A\colon H \to K$ is a linear operator, then A is an isometry if and only if A is inner product preserving.*

Proof. $\Leftarrow$. If A preserves inner products, then $\|Ax\|^2 = \langle Ax, Ax \rangle = \langle x, x \rangle = \|x\|^2$, so A preserves norms as well.

$\Rightarrow$. Assume that A is an isometry. Given $x, y \in H$, we use the Polar Identity and the fact that A preserves norms to compute that

$$
\begin{aligned}
\|x\|^2 + 2\operatorname{Re}\langle x, y \rangle + \|y\|^2 &= \|x + y\|^2 && \text{(Polar)} \\
&= \|Ax + Ay\|^2 && \text{(isometry)} \\
&= \|Ax\|^2 + 2\operatorname{Re}\langle Ax, Ay \rangle + \|Ay\|^2 && \text{(Polar)} \\
&= \|x\|^2 + 2\operatorname{Re}\langle Ax, Ay \rangle + \|y\|^2 && \text{(isometry)}.
\end{aligned}
$$

Thus $\operatorname{Re}\langle Ax, Ay \rangle = \operatorname{Re}\langle x, y \rangle$. If $\mathbb{F} = \mathbb{R}$, then we are done. If $\mathbb{F} = \mathbb{C}$, then a similar calculation based on expanding $\|x + iy\|^2$ shows that $\operatorname{Im}\langle Ax, Ay \rangle = \operatorname{Im}\langle x, y \rangle$. $\square$

Recall from Definition 6.6.5 that an *isometric isomorphism* is a linear isometry that is a bijection (in fact, every linear isometry is automatically injective, so a linear isometry is a bijection if and only if it is surjective). We give the following name to isometric isomorphisms whose domain and codomain are Hilbert spaces.

Definition 7.1.2 (Unitary Operator). Let H and K be Hilbert spaces.

(a) If $A\colon H \to K$ is an isometric isomorphism, then we say that A is a *unitary operator*.

(b) If there exists a unitary operator $A\colon H \to K$, then we say that H and K are *unitarily isomorphic*, and in this case we write $H \cong K$. $\Diamond$

Thus, a unitary operator is a linear bijection that preserves both lengths and angles between vectors. For example, rotations and flips on Euclidean space are unitary operators. When dealing with matrices, we say that a square $d \times d$ matrix U is *unitary* if the corresponding mapping that takes $x \in \mathbb{F}^d$ to $Ux \in \mathbb{F}^d$ is a unitary operator in the sense of Definition 7.1.2. Problem 7.2.11 gives a characterization of unitary matrices.

Here is an example of a unitary operator on infinite-dimensional Hilbert spaces.

Example 7.1.3 (Analysis Operator). Let H be any separable, infinite-dimensional Hilbert space. Then H has an orthonormal basis, say $\{e_n\}_{n \in \mathbb{N}}$. If $x \in H$, then we know from the Plancherel Equality that

$$
\sum_{n=1}^{\infty} |\langle x, e_n \rangle|^2 = \|x\|^2. \tag{7.1}
$$

Therefore the sequence $\{\langle x, e_n \rangle\}_{n \in \mathbb{N}}$ belongs to the space ℓ^2, and so we can define an operator $T \colon H \to \ell^2$ by setting

$$Tx = \{\langle x, e_n \rangle\}_{n \in \mathbb{N}}.$$

That is, T maps the vector x to the sequence of inner products of x with the basis elements e_n. This operator T is called the *analysis operator* associated with the orthonormal basis $\{e_n\}_{n \in \mathbb{N}}$.

The fact that the inner product is linear in the first variable implies that T is linear. By equation (7.1), the ℓ^2-norm of the sequence Tx equals the norm of the vector x:

$$\|Tx\|_2 = \left\| \{\langle x, e_n \rangle\}_{n \in \mathbb{N}} \right\|_2 = \left(\sum_{n=1}^{\infty} |\langle x, e_n \rangle|^2 \right)^{1/2} = \|x\|.$$

Hence T is an isometry. Finally, if $c = (c_n)_{n \in \mathbb{N}}$ is any sequence in ℓ^2, then we know from Theorem 5.7.1 that the series $x = \sum c_n e_n$ converges, and that theorem also implies that $c_n = \langle x, e_n \rangle$ for every n. Therefore $Tx = c$, so T is surjective. Thus T is a surjective linear isometry, so it is a unitary operator that maps H onto ℓ^2. $\Diamond$

Next we will construct a particular family of bounded operators, and determine which operators in the family are unitary. As in Example 7.1.3, we begin with an orthonormal basis $\{e_n\}_{n \in \mathbb{N}}$ for a separable, infinite-dimensional Hilbert space H. Each $x \in H$ can be uniquely written as

$$x = \sum_{n=1}^{\infty} \langle x, e_n \rangle \, e_n, \tag{7.2}$$

and our operator will create a new vector from this by modifying the coefficients in this series.

Let $\lambda = (\lambda_n)_{n \in \mathbb{N}}$ be any sequence of scalars. The idea behind our construction is to adjust the series in equation (7.2) by rescaling each term by a factor of λ_n. That is, we would like to create a new vector, which we will call $M_\lambda x$, by defining

$$M_\lambda x = \sum_{n=1}^{\infty} \lambda_n \langle x, e_n \rangle \, e_n. \tag{7.3}$$

However, in order for this definition to make sense we have to ensure that the series in equation (7.3) converges. The following theorem shows that if the sequence λ is bounded, then the series in equation (7.3) converges for every x, and M_λ defines a bounded mapping of H into H. Furthermore, if each λ_n has unit modulus, then M_λ is unitary (for the case $\mathbb{F} = \mathbb{R}$, this amounts to requiring that λ be a sequence of ± 1's).

Theorem 7.1.4. *Let H be a separable, infinite-dimensional Hilbert space, let $\{e_n\}_{n\in\mathbb{N}}$ be an orthonormal basis for H, and let $\lambda = (\lambda_n)_{n\in\mathbb{N}}$ be a sequence of scalars.*

(a) *If λ is a bounded sequence, then the series defining $M_\lambda x$ in equation (7.3) converges for every $x \in H$, and the operator $M_\lambda\colon H \to H$ defined in this way is bounded.*

(b) *If $|\lambda_n| = 1$ for every n, then $M_\lambda\colon H \to H$ is a unitary operator.*

Proof. (a) Assume that λ is a bounded sequence, i.e., $\lambda \in \ell^\infty$. By definition of the ℓ^∞-norm, we have $|\lambda_n| \leq \|\lambda\|_\infty$ for every n. Therefore, if $x \in H$, then

$$
\begin{aligned}
\sum_{n=1}^{\infty} |\lambda_n \langle x, e_n \rangle|^2 &\leq \sum_{n=1}^{\infty} \|\lambda\|_\infty^2 \, |\langle x, e_n \rangle|^2 && \text{(since } |\lambda_n| \leq \|\lambda\|_\infty) \\
&= \|\lambda\|_\infty^2 \sum_{n=1}^{\infty} |\langle x, e_n \rangle|^2 && (\|\lambda\|_\infty^2 \text{ is a scalar}) \\
&= \|\lambda\|_\infty^2 \, \|x\|^2 && \text{(Plancherel's Equality)} \\
&< \infty.
\end{aligned}
$$

This implies, by Theorem 5.7.1, that the series $\sum \lambda_n \langle x, e_n \rangle \, e_n$ converges, and so we can indeed define $M_\lambda x$ by equation (7.3). Since the inner product is linear in the first variable and the series involved all converge, we compute that

$$
\begin{aligned}
M_\lambda(ax + by) &= \sum_{n=1}^{\infty} \lambda_n \langle ax + by, e_n \rangle \, e_n \\
&= a \sum_{n=1}^{\infty} \lambda_n \langle x, e_n \rangle \, e_n \; + \; b \sum_{n=1}^{\infty} \lambda_n \langle y, e_n \rangle \, e_n \\
&= a M_\lambda(x) \; + \; b M_\lambda(y).
\end{aligned}
$$

Thus M_λ is linear. Applying the Plancherel Equality just as we did previously, we have

$$
\|M_\lambda x\|^2 = \sum_{n=1}^{\infty} |\lambda_n \langle x, e_n \rangle|^2 \leq \|\lambda\|_\infty^2 \, \|x\|^2,
$$

so M_λ is a bounded operator. In fact, by taking the supremum of $\|M_\lambda x\|$ over the unit vectors we see that the operator norm of M_λ satisfies

$$
\|M_\lambda\| = \sup_{\|x\|=1} \|M_\lambda x\| \leq \|\lambda\|_\infty.
$$

The reader should verify (this is part of Problem 7.1.10) that we actually have equality, i.e., $\|M_\lambda\| = \|\lambda\|_\infty$.

(b) If $|\lambda_n| = 1$ for every n, then λ is a bounded sequence, so M_λ is a bounded operator by part (a). Additionally, in this case the Plancherel Equality implies that

$$\|M_\lambda x\|^2 = \sum_{n=1}^{\infty} |\lambda_n \langle x, e_n \rangle|^2 = \sum_{n=1}^{\infty} |\langle x, e_n \rangle|^2 = \|x\|^2,$$

so M_λ is isometric. To show that M_λ is surjective, fix any $x \in H$ and define

$$y = \sum_{n=1}^{\infty} \overline{\lambda_n} \langle x, e_n \rangle e_n$$

(why does this series converge?). Then

$$M_\lambda y = \sum_{n=1}^{\infty} \lambda_n \overline{\lambda_n} \langle x, e_n \rangle e_n = \sum_{n=1}^{\infty} |\lambda_n|^2 \langle x, e_n \rangle e_n = \sum_{n=1}^{\infty} \langle x, e_n \rangle e_n = x.$$

Thus M_λ is surjective, and therefore it is a unitary operator. $\square$

We sometimes refer to M_λ, or operators like it, as a "multiplication operator," or a "multiplier." To give a concrete example, take $H = \ell^2$ and assume that $\{e_n\}_{n \in \mathbb{N}}$ is the standard basis for ℓ^2. Then M_λ is simply the following componentwise multiplication:

$$M_\lambda x = (\lambda_1 x_1, \lambda_2 x_2, \dots), \qquad \text{for all } x = (x_1, x_2, \dots) \in \ell^2.$$

If $|\lambda_n| = 1$ for every n, then M_λ is a unitary operator (note that for real scalars this amounts to taking $\lambda_n = \pm 1$ for every n).

Problems

7.1.5. Let H and K be inner product spaces. Show that if two operators $A, B \colon H \to K$ satisfy $\langle Ax, y \rangle = \langle Bx, y \rangle$ for all $x, y \in H$, then $A = B$.

7.1.6. Let H and K be inner product spaces, and let $A \colon H \to K$ be a linear operator. Prove that

$$\|A\| = \sup_{\substack{\|x\|=1, \\ \|y\|=1}} |\langle Ax, y \rangle|.$$

7.1.7. (a) Prove that every separable infinite-dimensional Hilbert space is unitarily isomorphic to ℓ^2, and use this to prove that any two separable infinite-dimensional Hilbert spaces are unitarily isomorphic. That is, prove that if H and K are two separable infinite-dimensional Hilbert spaces, then $H \cong K$.

(b) Are every two finite-dimensional Hilbert spaces unitarily isomorphic? If not, fill in the blank so that the following statement is true, and provide a proof: If H and K are two finite-dimensional Hilbert spaces, then H and K are unitarily isomorphic if and only if _____.

7.1.8. Let M be a closed subspace of a Hilbert space H, and let P be the orthogonal projection of H onto M (see Definition 5.6.3). Prove the following statements.

(a) $\|x - Px\| = \operatorname{dist}(x, M)$ for every $x \in H$.

(b) P is linear and $\|Px\| \le \|x\|$ for every $x \in H$.

(c) If $M = \{0\}$, then $\|P\| = 0$, but in any other case we have $\|P\| = 1$. However, P is isometric only when $M = H$.

(d) $\ker(P) = M^\perp$ and $\operatorname{range}(P) = M$.

(e) $P^2 = P$ (an operator that equals its own square is *idempotent*.)

(f) $\langle Px, y \rangle = \langle x, Py \rangle$ for all $x, y \in H$ (hence P is *self-adjoint* in the sense of Definition 7.3.1).

(g) $I - P$ is the orthogonal projection of H onto $M^\perp$.

7.1.9. (a) Let X and Y be normed spaces. Given $A_n, A \in \mathcal{B}(X, Y)$, we say that A_n *converges to A in the strong operator topology* if $A_n x \to Ax$ for every $x \in X$. Show that if A_n converges to A in operator norm (i.e., $\|A - A_n\| \to 0$), then A_n converges to A in the strong operator topology.

(b) Define $L_n \colon \ell^1 \to \ell^1$ by

$$L_n(x) = (x_1, \dots, x_n, 0, 0, \dots), \qquad x = (x_k)_{k \in \mathbb{N}} \in \ell^1,$$

and let $I \colon \ell^1 \to \ell^1$ be the identity map. Prove that L_n is bounded and find $\|L_n\|$. Also prove that L_n converges to I in the strong operator topology, but L_n does not converge to I in operator norm.

(c) Let $\{e_n\}_{n \in \mathbb{N}}$ be an orthonormal basis for a Hilbert space H, and let P_N be the orthogonal projection of H onto $\operatorname{span}\{e_1, \dots, e_N\}$. Prove that $P_N \to I$ in the strong operator topology, but $\|I - P_N\| \not\to 0$ as $N \to \infty$.

7.1.10. Let $\lambda = (\lambda_n)_{n \in \mathbb{N}}$ be a bounded sequence of scalars, and let M_λ be the bounded operator discussed in Theorem 7.1.4. Set $\delta = \inf |\lambda_n|$, and prove the following statements.

(a) $\|M_\lambda\| = \|\lambda\|_\infty$.

(b) M_λ is injective if and only if $\lambda_n \ne 0$ for every n.

(c) M_λ is surjective if and only if $\delta > 0$.

(d) If $\delta = 0$ but $\lambda_n \ne 0$ for every n, then $\operatorname{range}(M_\lambda)$ is a proper, dense subspace of H.

(e) If $\delta > 0$, then M_λ is a topological isomorphism, but M_λ is unitary if and only if $|\lambda_n| = 1$ for every n.

7.1.11. Assume that $\{e_n\}_{n\in\mathbb{N}}$ is an orthonormal basis for a Hilbert space H, and let $\lambda = (\lambda_n)_{n\in\mathbb{N}}$ be any sequence of scalars. For those $x \in H$ *for which the following series converges*, define

$$M_\lambda x = \sum_{n=1}^{\infty} \lambda_n \langle x, e_n \rangle e_n. \tag{7.4}$$

Prove the following statements.

(a) The series in equation (7.4) converges if and only if x belongs to

$$\text{domain}(M_\lambda) = \left\{ x \in H : \sum_{n=1}^{\infty} |\lambda_n \langle x, e_n \rangle|^2 < \infty \right\}.$$

(b) If $\lambda \notin \ell^\infty$, then $\text{domain}(M_\lambda)$ is a dense but proper subspace of H, and $M_\lambda : \text{domain}(M_\lambda) \to H$ is an unbounded linear operator.

(c) $\text{domain}(M_\lambda) = H$ if and only if $\lambda \in \ell^\infty$.

7.2 Adjoints of Operators on Hilbert Spaces

Let A be an $m \times n$ matrix with scalar entries, which we write as

$$A = [a_{ij}] = \begin{bmatrix} a_{11} & a_{12} & \cdots & a_{1n} \\ a_{21} & a_{22} & \cdots & a_{2n} \\ a_{31} & a_{32} & \cdots & a_{3n} \\ \vdots & \vdots & \ddots & \vdots \\ a_{m1} & a_{m2} & \cdots & a_{mn} \end{bmatrix}. \tag{7.5}$$

The *transpose* of A is the $n \times m$ matrix A^T obtained by interchanging rows with columns in A:

$$A^\mathrm{T} = [a_{ji}] = \begin{bmatrix} a_{11} & a_{21} & a_{31} & \cdots & a_{m1} \\ a_{12} & a_{22} & a_{32} & \cdots & a_{m2} \\ \vdots & \vdots & \vdots & \ddots & \vdots \\ a_{1n} & a_{2n} & a_{3n} & \cdots & a_{mn} \end{bmatrix}.$$

The *Hermitian*, or *conjugate transpose*, of A is the $n \times m$ matrix A^H obtained by taking the complex conjugate of each entry of the transpose of A:

$$A^\mathrm{H} = \overline{A^\mathrm{T}} = [\overline{a_{ji}}] = \begin{bmatrix} \overline{a_{11}} & \overline{a_{21}} & \overline{a_{31}} & \cdots & \overline{a_{m1}} \\ \overline{a_{12}} & \overline{a_{22}} & \overline{a_{32}} & \cdots & \overline{a_{m2}} \\ \vdots & \vdots & \vdots & \ddots & \vdots \\ \overline{a_{1n}} & \overline{a_{2n}} & \overline{a_{3n}} & \cdots & \overline{a_{mn}} \end{bmatrix}. \tag{7.6}$$

Note that if $\mathbb{F} = \mathbb{R}$, then $A^{\mathrm{H}} = A^{\mathrm{T}}$.

Letting $x \cdot y$ denote the dot product of vectors, we have the following fundamental equality that relates A to A^{H}. We assign the proof of this result as Problem 7.2.10.

Theorem 7.2.1. *If A is an $m \times n$ matrix, then*

$$Ax \cdot y = x \cdot A^H y, \qquad \text{for all } x \in \mathbb{F}^n,\ y \in \mathbb{F}^m. \qquad \Diamond \qquad (7.7)$$

We will use the Riesz Representation Theorem to prove that if $A\colon H \to K$ is any bounded linear operator between Hilbert spaces H and K, then there is a bounded linear operator $A^*\colon K \to H$, called the *adjoint* of A, that is related to A in a manner similar to the way that equation (7.7) relates the matrices A and A^{H}.

Theorem 7.2.2 (Adjoints). *Let H and K be Hilbert spaces, and assume that $A\colon H \to K$ is a bounded linear operator. Then there exists a bounded linear operator $A^*\colon K \to H$ such that*

$$\langle Ax, y \rangle = \langle x, A^*y \rangle, \qquad \text{for all } x \in H,\ y \in K. \qquad (7.8)$$

Moreover, A^* is the unique mapping from K to H that satisfies equation (7.8).

Proof. Before beginning the proof, note that Ax and y both belong to K, and therefore the notation $\langle Ax, y \rangle$ in equation (7.2.2) implicitly denotes the inner product of vectors in K. Similarly, since x and A^*x belong to H, $\langle x, A^*y \rangle$ implicitly means the inner product of vectors in H.

Fix a vector $y \in K$, and define $\nu_y\colon H \to \mathbb{F}$ by

$$\nu_y(x) = \langle Ax, y \rangle, \qquad x \in H.$$

This functional ν_y is linear, because A is linear and the inner product is linear in the first variable. Applying the Cauchy–Bunyakovski–Schwarz Inequality and recalling that $\|Ax\| \le \|A\|\,\|x\|$ for every x, we compute that

$$|\nu_y(x)| = |\langle Ax, y \rangle| \le \|Ax\|\,\|y\| \le \|A\|\,\|x\|\,\|y\|.$$

Therefore ν_y is bounded. Indeed, by taking the supremum over the unit vectors in H we see that the operator norm of ν_y satisfies

$$\|\nu_y\| = \sup_{\|x\|=1} |\nu_y(x)| \le \|A\|\,\|y\| < \infty.$$

Hence $\nu_y \in H^*$. The Riesz Representation Theorem (Theorem 6.8.2) characterizes the dual space H^*. In particular, that theorem tells us that since ν_y is a bounded linear functional on H, there exists a unique vector $y^* \in H$ such that $\nu_y = \mu_{y^*}$, where μ_{y^*} is the functional defined by $\mu_{y^*}(x) = \langle x, y^* \rangle$. That is,

$$\nu_y(x) \;=\; \langle x, y^* \rangle, \qquad x \in H.$$

Let

$$A^* y \;=\; y^*.$$

Since each $y \in K$ determines a unique $y^* \in H$, we have defined a mapping $A^* \colon K \to H$. By definition, if $x \in H$ and $y \in K$, then

$$\langle Ax, y \rangle \;=\; \nu_y(x) \;=\; \langle x, y^* \rangle \;=\; \langle x, A^* y \rangle.$$

Thus equation (7.8) holds.

It remains to show that A^* is linear, bounded, and unique. Choose any vectors $y, z \in K$ and scalars $a, b \in \mathbb{F}$. Then for every $x \in H$ we have

$$
\begin{aligned}
\langle x, A^*(ay + bz) \rangle
&= \langle Ax, ay + bz \rangle && \text{(definition of } A^*\text{)} \\
&= \bar{a}\langle Ax, y \rangle + \bar{b}\langle Ax, z \rangle && \text{(antilinearity in 2nd variable)} \\
&= \bar{a}\langle x, A^* y \rangle + \bar{b}\langle x, A^* z \rangle && \text{(definition of } A^*\text{)} \\
&= \langle x, aA^* y + bA^* z \rangle && \text{(antilinearity in 2nd variable).}
\end{aligned}
$$

Since this is true for every $x \in H$, it follows that

$$A^*(ay + bz) \;=\; aA^* y + bA^* z.$$

Therefore A^* is linear.

Now we show that A^* is bounded. Given $y \in K$, we use Problem 5.2.8 to compute that

$$
\begin{aligned}
\|A^* y\| &= \sup_{\|x\|=1} |\langle x, A^* y \rangle| && \text{(by Problem 5.2.8)} \\
&= \sup_{\|x\|=1} |\langle Ax, y \rangle| && \text{(by definition of } A^*\text{)} \\
&\le \sup_{\|x\|=1} \|Ax\|\,\|y\| && \text{(CBS Inequality)} \\
&= \|A\|\,\|y\| && \text{(definition of operator norm).}
\end{aligned}
$$

Hence A^* is bounded. In fact, by taking the supremum over the unit vectors $y \in K$ we see that

$$\|A^*\| \;=\; \sup_{\|y\|=1} \|A^* y\| \;\le\; \sup_{\|y\|=1} \|A\|\,\|y\| \;=\; \|A\|. \tag{7.9}$$

Finally, we will show that A^* is unique. Suppose that $B \colon K \to H$ also satisfies $\langle Ax, y \rangle = \langle x, By \rangle$ for all $x \in H$ and $y \in K$. Then

$$\langle x, A^* y \rangle \;=\; \langle Ax, y \rangle \;=\; \langle x, By \rangle.$$

Since this holds for all $x \in H$, we have $A^*y = By$. Hence A^* and B are the same operator. $\square$

In particular, we see from equation (7.8) that for all $x \in H$ and $y \in K$ we have

$$Ax \perp y \quad \Longleftrightarrow \quad x \perp A^*y.$$

We introduce a name for the operator A^*.

Definition 7.2.3 (Adjoint Operator). Given Hilbert spaces H and K, the *adjoint* of $A \in \mathcal{B}(H, K)$ is the unique operator $A^* \in \mathcal{B}(K, H)$ that satisfies equation (7.8). $\diamond$

We will give some examples, beginning with the operator corresponding to an $m \times n$ matrix.

Example 7.2.4 (Adjoint of a Matrix). In this example our Hilbert spaces are the Euclidean spaces $\mathbb{F}^m$ and $\mathbb{F}^n$, each of whose inner product is the ordinary dot product of vectors.

Let A be the $m \times n$ matrix given in equation (7.5). If $x \in \mathbb{F}^n$, then Ax is a vector in $\mathbb{F}^m$. Thus A determines an operator, which we also call A, that sends $x \in \mathbb{F}^n$ to $Ax \in \mathbb{F}^m$. Likewise, the matrix A^H defined in equation (7.6) determines an operator that maps $y \in \mathbb{F}^m$ to $A^H y \in \mathbb{F}^n$.

By Theorem 7.2.2, there is *unique* operator $A^*: \mathbb{F}^m \to \mathbb{F}^n$ that satisfies $Ax \cdot y = x \cdot A^*y$ for all $x \in \mathbb{R}^n$ and $y \in \mathbb{R}^m$. Yet by Theorem 7.2.1 we have $Ax \cdot y = x \cdot A^H y$ for all $x \in \mathbb{R}^n$ and $y \in \mathbb{R}^m$. Hence A^H and A^* are the same operator, and therefore are determined by the same matrix, the Hermitian of A given in equation (7.6).

That is, the adjoint A^* of a matrix is precisely the Hermitian matrix $A^H = \overline{A^T}$. For the case $\mathbb{F} = \mathbb{R}$ the complex conjugate has no effect, so for a matrix with real entries the adjoint of A is simply the transpose A^T. $\diamond$

Next we compute the adjoint of the operator M_λ that was introduced in Theorem 7.1.4.

Example 7.2.5. Assume that $\{e_n\}_{n \in \mathbb{N}}$ is an orthonormal basis for a separable Hilbert space H, and let $\lambda = (\lambda_n)_{n \in \mathbb{N}}$ be a bounded sequence of scalars. Let $M_\lambda: H \to H$ be the multiplication operator defined in Theorem 7.1.4, i.e.,

$$M_\lambda x = \sum_{n=1}^{\infty} \lambda_n \langle x, e_n \rangle e_n, \qquad x \in H.$$

Since M_λ is bounded and linear, Theorem 7.2.2 implies that it has an adjoint $M_\lambda^*: H \to H$ that is also a bounded linear operator. We want to determine what this operator M_λ^* is. To do this, let $\overline{\lambda} = \left(\overline{\lambda_n} \right)_{n \in \mathbb{N}}$ be the sequence of complex conjugates of the λ_n. Then $\overline{\lambda}$ is a bounded sequence of scalars, so Theorem 7.1.4 tells us that

$$M_{\overline{\lambda}} x = \sum_{n=1}^{\infty} \overline{\lambda_n} \langle x, e_n \rangle e_n, \qquad x \in H,$$

is a bounded linear operator on H. We will show that $M_\lambda^* = M_{\overline{\lambda}}$.

Given vectors $x, y \in H$, we compute that

$$
\begin{aligned}
\langle x, M_\lambda^* y \rangle &= \langle M_\lambda x, y \rangle && \text{(definition of the adjoint)} \\[2mm]
&= \left\langle \sum_{n=1}^{\infty} \lambda_n \langle x, e_n \rangle e_n, \, y \right\rangle && \text{(definition of } M_\lambda) \\[2mm]
&= \sum_{n=1}^{\infty} \langle \lambda_n \langle x, e_n \rangle e_n, \, y \rangle && \text{(continuity of the inner product)} \\[2mm]
&= \sum_{n=1}^{\infty} \lambda_n \langle x, e_n \rangle \langle e_n, y \rangle && \text{(linearity in the first variable)} \\[2mm]
&= \sum_{n=1}^{\infty} \langle x, \overline{\lambda_n} \, \overline{\langle e_n, y \rangle} \, e_n \rangle && \text{(antilinearity in the 2nd variable)} \\[2mm]
&= \sum_{n=1}^{\infty} \langle x, \overline{\lambda_n} \langle y, e_n \rangle e_n \rangle && \text{(conjugate symmetry)} \\[2mm]
&= \left\langle x, \sum_{n=1}^{\infty} \overline{\lambda_n} \langle y, e_n \rangle e_n \right\rangle && \text{(continuity of the inner product)} \\[2mm]
&= \langle x, M_{\overline{\lambda}} y \rangle.
\end{aligned}
$$

Therefore $M_\lambda^* y = M_{\overline{\lambda}} y$ for each $y \in H$, so M_λ^* equals the multiplication operator $M_{\overline{\lambda}}$. ◇

The next theorem lays out some of the properties of adjoint operators on Hilbert spaces.

Theorem 7.2.6. *Let H, K, and L be Hilbert spaces. Given $A \in \mathcal{B}(H, K)$, the following statements hold.*

(a) *If $x \in H$ and $y \in K$, then $\langle A^* y, x \rangle = \langle y, Ax \rangle$.*

(b) *$(A^*)^* = A$ and $(cA)^* = \overline{c} A^*$ for all scalars c.*

(c) *If $B \in \mathcal{B}(H, K)$, then $(A + B)^* = A^* + B^*$.*

(d) *If $B \in \mathcal{B}(K, L)$, then $(BA)^* = A^* B^*$.*

(e) *$\|A\| = \|A^*\| = \|A^* A\|^{1/2} = \|AA^*\|^{1/2}$.*

Proof. We will prove statements (a) and (b) and part of statement (e), and assign the proofs of the remaining statements as Problem 7.2.14.

(a) If $x \in H$ and $y \in K$, then by conjugate symmetry and the definition of the adjoint we have

$$\langle A^*y, x \rangle = \overline{\langle x, A^*y \rangle} = \overline{\langle Ax, y \rangle} = \langle y, Ax \rangle.$$

(b) If $x \in H$ and $y \in K$, then $\langle (A^*)^*x, y \rangle = \langle x, A^*y \rangle = \langle Ax, y \rangle$. This implies that $(A^*)^* = A$. Likewise,

$$\langle (cA)x, y \rangle = c\langle Ax, y \rangle = c\langle x, A^*y \rangle = \langle x, \bar{c}A^*y \rangle = \langle x, (\bar{c}A^*)y \rangle,$$

so $(cA)^* = \bar{c}A^*$.

(e) We saw in equation (7.9) that $\|A^*\| \leq \|A\|$. By applying equation (7.9) to the operator A^*, we see that we also have $\|(A^*)^*\| \leq \|A^*\|$. Since part (b) established that $(A^*)^* = A$, it follows that $\|A^*\| = \|A\|$. □

The next theorem is easy to prove, but is very important. Indeed, for the case $H = \mathbb{F}^n$ and $K = \mathbb{F}^m$, where A corresponds to multiplication by an $m \times n$ matrix, Strang [Str16] calls this result the *Fundamental Theorem of Linear Algebra*.

Theorem 7.2.7. *Let H and K be Hilbert spaces. If $A \in \mathcal{B}(H,K)$, then*

$$\ker(A) = \text{range}(A^*)^\perp \quad and \quad \ker(A)^\perp = \overline{\text{range}}(A^*),$$

while

$$\ker(A^*) = \text{range}(A)^\perp \quad and \quad \ker(A^*)^\perp = \overline{\text{range}}(A).$$

Proof. Fix any vector $x \in \ker(A)$. If $z \in \text{range}(A^*)$, then $z = A^*y$ for some $y \in K$. Since $Ax = 0$, we therefore have

$$\langle x, z \rangle = \langle x, A^*y \rangle = \langle Ax, y \rangle = 0.$$

This shows that $x \in \text{range}(A^*)^\perp$, and therefore $\ker(A) \subseteq \text{range}(A^*)^\perp$.

Conversely, choose any vector $x \in \text{range}(A^*)^\perp$. Then for every $z \in H$ we have $A^*z \in \text{range}(A^*)$. Therefore $x \perp A^*z$, so

$$\langle Ax, z \rangle = \langle x, A^*z \rangle = 0.$$

Since this is true for all z, it follows that $Ax = 0$. Thus $x \in \ker(A)$, and hence $\text{range}(A^*)^\perp \subseteq \ker(A)$.

This shows that $\ker(A) = \text{range}(A^*)^\perp$. Taking complements we obtain $\ker(A)^\perp = (\text{range}(A^*)^\perp)^\perp = \overline{\text{range}}(A^*)$, and the remaining statements follow by replacing A with A^*. □

Since A is injective if and only if $\ker(A) = \{0\}$, we immediately obtain the following result.

Corollary 7.2.8. *Let H and K be Hilbert spaces. If $A \in \mathcal{B}(H,K)$, then A is injective if and only if $\text{range}(A^*)$ is dense in H.*

Proof. Applying earlier results, we have that

$$A \text{ is injective} \iff \ker(A) = \{0\}$$
$$\iff \ker(A)^{\perp} = H$$
$$\iff \overline{\text{range}(A^*)} = H$$
$$\iff \text{range}(A^*) \text{ is dense.} \quad \square$$

Here are some properties of the adjoint of a topological isomorphism, and a characterization of unitary mappings in terms of adjoints.

Theorem 7.2.9. *Let H and K be Hilbert spaces, and let $A \in \mathcal{B}(H, K)$ be given.*

(a) *If A is a topological isomorphism, then A^* is also a topological isomorphism, and $(A^{-1})^* = (A^*)^{-1}$.*

(b) *A is unitary if and only if A is a bijection and $A^* = A^{-1}$.*

Proof. (a) Suppose that A is a topological isomorphism, i.e., A is a linear bijection and both A and A^{-1} are bounded. Problem 6.6.14 therefore implies that there exists a constant $c > 0$ such that

$$\|Ax\| \geq c \|x\|, \qquad \text{for all } x \in H. \tag{7.10}$$

Fix any nonzero vector $y \in K$. Then $y = Ax$ for some $x \in H$, so

$$
\begin{aligned}
\|y\|^2 \; = \; \|Ax\|^2 \; &= \; \langle Ax, Ax \rangle && \text{(definition of the norm)} \\
&= \; \langle A^*Ax, x \rangle && \text{(definition of } A^*) \\
&\leq \; \|A^*Ax\| \, \|x\| && \text{(CBS Inequality)} \\
&\leq \; \|A^*Ax\| \, \frac{\|Ax\|}{c} && \text{(by equation (7.10))} \\
&= \; \|A^*y\| \, \frac{\|y\|}{c} && \text{(since } y = Ax).
\end{aligned}
$$

Dividing by $\|y\|$ and rearranging, we conclude that

$$\|A^*y\| \geq c \|y\|, \qquad \text{for all nonzero } y \in K. \tag{7.11}$$

This same inequality holds trivially if $y = 0$, so we have $\|A^*y\| \geq c \|y\|$ for every $y \in K$. Problem 6.2.10 therefore implies that A^* is injective and range(A^*) is a closed subspace of H. Consequently, range$(A^*) = \overline{\text{range}}(A^*)$. On the other hand, since A is injective, Corollary 7.2.8 implies that range(A^*) is dense in H. This tells us that $\overline{\text{range}}(A^*) = H$. Hence range$(A^*) = \overline{\text{range}}(A^*) = H$, so A^* is surjective.

The remainder of the proof of this part (showing that A^* and $(A^*)^{-1}$ are both bounded and $(A^*)^{-1} = (A^{-1})^*$) is assigned as Problem 7.2.15.

(b) We assign the proof of part (b) as Problem 7.2.15. □

Problems

7.2.10. (a) Show that if $x = (x_1, \ldots, x_d)$ and $y = (y_1, \ldots, y_d)$ are vectors in $\mathbb{F}^d$ and we think of x and y as being column vectors (i.e., $d \times 1$ matrices), then $x^H y = y \cdot x$.

(b) Let A be an $m \times n$ matrix, as in equation (7.5). Let $\{e_1, \ldots, e_n\}$ be the standard basis for $\mathbb{F}^n$, and let $\{f_1, \ldots, f_m\}$ be the standard basis for $\mathbb{F}^m$ (see Problem 1.11.7). Prove that $Ae_i \cdot f_j = a_{ij}$ for each $i = 1, \ldots, n$ and $j = 1, \ldots, m$.

(c) Prove Theorem 7.2.1.

7.2.11. Let $u_1, \ldots, u_d$ be the columns of a $d \times d$ matrix U, which we identify with the linear mapping $U \colon \mathbb{F}^d \to \mathbb{F}^d$ that takes x to Ux. Prove that U is unitary if and only if $\{u_1, \ldots, u_d\}$ is an orthonormal basis for $\mathbb{F}^d$.

7.2.12. Let $\|\cdot\|_2$ denote the Euclidean norm. Given an $m \times n$ matrix U, prove that the following three statements are equivalent.

(a) The columns of U are orthonormal vectors in $\mathbb{F}^m$.

(b) $\|Ux\|_2 = \|x\|_2$ for all $x \in \mathbb{F}^n$, i.e., U is isometric as an operator from $\mathbb{F}^n$ to $\mathbb{F}^m$.

(c) $U^H U = I_{n \times n}$, the $n \times n$ identity matrix.

Show further that if the statements above hold, then $n \leq m$ and UU^H is the orthogonal projection onto the column space of U (which is the span of the columns of U).

7.2.13. Let $L, R \colon \ell^2 \to \ell^2$ be the left-shift and right-shift operators from Examples 6.6.3 and 6.6.4. Prove that L and R are adjoints of each other, i.e., $L^* = R$ and $R^* = L$. Do R and L commute?

7.2.14. Prove the remaining implications in Theorem 7.2.6.

7.2.15. Complete the proof of Theorem 7.2.9.

7.2.16. Given Hilbert spaces H and K, define $T \colon \mathcal{B}(H, K) \to \mathcal{B}(K, H)$ by $T(A) = A^*$. Prove that T is an antilinear isometric isomorphism (i.e., T is antilinear, isometric, and surjective).

7.2.17. Given $A \in \mathcal{B}(H)$, where H is a Hilbert space, show that

$$\ker(A) = \ker(A^*A) \quad \text{and} \quad \overline{\text{range}(A^*A)} = \overline{\text{range}(A^*)}.$$

7.2.18. Let $\{e_n\}_{n \in \mathbb{N}}$ be an orthonormal sequence for a Hilbert space H, and let $T \colon H \to \ell^2$ be the *analysis operator* $Tx = \{\langle x, e_n \rangle\}_{n \in \mathbb{N}}$ (compare Example 7.1.3).

(a) Explicitly identify the *synthesis operator* $T^* \colon \ell^2 \to H$, i.e., find an explicit formula for T^*c for each $c \in \ell^2$.

(b) Find explicit formulas for T^*T and TT^*.

(c) When will T^*T be the identity operator?

7.2.19. Let H be a Hilbert space, and assume that $\{y_n\}_{n \in \mathbb{N}}$ is a Bessel sequence in H in the sense of Problem 5.7.5. Let $Tx = \{\langle x, y_n \rangle\}_{n \in \mathbb{N}}$ be the *analysis operator* associated with $\{y_n\}_{n \in \mathbb{N}}$. Prove that T is a bounded mapping of H into ℓ^2, and explicitly identify the *synthesis operator* T^* and the *frame operator* T^*T.

7.3 Self-Adjoint Operators

Now we take $H = K$ and focus on operators that map a Hilbert space H into itself. If $A \colon H \to H$ is bounded and linear, then its adjoint $A^* \colon H \to H$ is also bounded and linear. Therefore it is possible that A and A^* might be the same operator. We have a special name for this situation.

Definition 7.3.1 (Self-Adjoint Operator). Let H be a Hilbert space. An operator $A \in \mathcal{B}(H)$ is *self-adjoint* or *Hermitian* if $A = A^*$. Equivalently, A is self-adjoint if

$$\langle Ax, y \rangle = \langle x, Ay \rangle, \qquad \text{for all } x, y \in H. \qquad \Diamond$$

Remark 7.3.2 (Normal Operators). We will focus on self-adjoint operators, but we mention a somewhat broader class that is often useful. An operator A that *commutes* with its adjoint (but need not be equal to its adjoint) is called a *normal operator*. That is, A is normal if $AA^* = A^*A$. All self-adjoint operators are normal, but not all normal operators are self-adjoint (see Problem 7.3.18). $\Diamond$

To illustrate self-adjointness, we consider square matrices.

Example 7.3.3 (Self-Adjoint Matrices). Let $\mathbb{F} = \mathbb{C}$, and let A be an $n \times n$ complex matrix. We saw earlier that the adjoint of A is its Hermitian matrix $A^* = A^{\mathrm{H}} = \overline{A^{\mathrm{T}}}$. Hence, a matrix A is self-adjoint if and only if $A = A^{\mathrm{H}}$, i.e., A equals its Hermitian matrix. When dealing with matrices, it is customary in this case to say that A *is Hermitian*, instead of A *is self-adjoint*.

If $\mathbb{F} = \mathbb{R}$, then all of the entries of A are real, and the complex conjugate in the definition of A^{H} is irrelevant. Hence the adjoint of A is its transpose A^{T} in this case. Thus, a real matrix is self-adjoint if and only if $A = A^{\mathrm{T}}$. A matrix that equals its transpose is said to be *symmetric*. $\Diamond$

A self-adjoint operator need not be a topological isomorphism; in fact a self-adjoint operator need not be either injective or surjective. For example, the zero function $0\colon H \to H$ is self-adjoint.

Example 7.3.4. Let $\lambda = (\lambda_n)_{n\in\mathbb{N}}$ be a bounded sequence, and let M_λ be the operator defined in Theorem 7.1.4. We saw in Example 7.2.5 that the adjoint of M_λ is $M_\lambda^* = M_{\overline{\lambda}}$, where $\overline{\lambda} = (\overline{\lambda_n})_{n\in\mathbb{N}}$. We will determine when M_λ is self-adjoint.

Since $\{e_n\}_{n\in\mathbb{N}}$ is an orthonormal basis, we have $\langle e_m, e_n\rangle = \delta_{mn}$ (the Kronecker delta, which is 1 when $m = n$ and zero otherwise). Therefore

$$M_\lambda e_m = \sum_{n=1}^{\infty} \lambda_n \langle e_m, e_n\rangle\, e_n = \sum_{n=1}^{\infty} \lambda_n \delta_{mn}\, e_n = \lambda_m e_m.$$

A similar calculation shows that $M_{\overline{\lambda}} e_m = \overline{\lambda_m}\, e_m$. Consequently, if M_λ is self-adjoint, then

$$\lambda_m e_m = M_\lambda e_m = M_\lambda^* e_m = M_{\overline{\lambda}} e_m = \overline{\lambda_m}\, e_m.$$

Since $e_m \neq 0$, it follows that $\lambda_m = \overline{\lambda_m}$, which tells us that λ_m is real. This is true for every m, so we have shown that if M_λ is self-adjoint then λ is a sequence of real scalars. Conversely, if λ is a sequence of real scalars, then $\overline{\lambda} = \lambda$ and therefore $M_\lambda^* = M_{\overline{\lambda}} = M_\lambda$. Thus

M_λ *is self-adjoint if and only if each* λ_n *is real.*

If $\mathbb{F} = \mathbb{R}$, then each λ_n is real by definition, so M_λ is automatically self-adjoint in this case. However, if $\mathbb{F} = \mathbb{C}$, then there are operators M_λ that are not self-adjoint. For example, if $\lambda = (i^n)_{n\in\mathbb{N}} = (i, -1, -i, 1, \dots)$, then M_λ is not self-adjoint (but it is unitary, since $|\lambda_n| = |i^n| = 1$ for every n). $\diamond$

The following properties of self-adjoint operators are an immediate consequence of Theorem 7.2.7 and Corollary 7.2.8.

Corollary 7.3.5. *Let H be a Hilbert space. If $A \in \mathcal{B}(H)$ is self-adjoint, then the following statements hold.*

(a) $\ker(A) = \operatorname{range}(A)^\perp$.

(b) $\ker(A)^\perp = \overline{\operatorname{range}}(A)$.

(c) *A is injective if and only if* $\operatorname{range}(A)$ *is dense in H.* $\diamond$

Example 7.3.6. Let $\lambda = (\frac{1}{n})_{n\in\mathbb{N}}$, so M_λ has the form

$$M_\lambda x = \sum_{n=1}^{\infty} \frac{1}{n} \langle x, e_n\rangle\, e_n, \qquad x \in H.$$

Since λ is a real sequence, we know from Example 7.3.4 that M_λ is self-adjoint. Further, M_λ is injective (why?), so Corollary 7.3.5 implies that $\operatorname{range}(M_\lambda)$

is dense in H. However, M_λ is not surjective (for example, the sequence $y = \left(\frac{1}{n}\right)_{n \in \mathbb{N}} \in \ell^2$ does not belong to the range of M_λ). Therefore for this choice of λ we see that $\text{range}(M_\lambda)$ is a dense but proper subspace of H. $\diamondsuit$

The next theorem is one of the few results in this volume in which the choice of scalar field makes a significant difference. In particular, this theorem shows that if $\mathbb{F} = \mathbb{C}$, then a bounded operator $A \colon H \to H$ is self-adjoint if and only if the inner product $\langle Ax, x \rangle$ is real for every vector $x \in H$. The statement of this theorem becomes *false* if the scalar field is real (see Problem 7.3.12). However, as long as we are working over the complex field, this result gives us a convenient way to recognize self-adjoint operators. When we want to distinguish between a Hilbert space over the real field and one over the complex field, it is convenient to say that H is a *real Hilbert space* if it is a Hilbert space and $\mathbb{F} = \mathbb{R}$, and it is a *complex Hilbert space* if it is a Hilbert space and $\mathbb{F} = \mathbb{C}$.

Theorem 7.3.7. *Let H be a **complex** Hilbert space (i.e., we take $\mathbb{F} = \mathbb{C}$). If $A \in \mathcal{B}(H)$, then*

$$A \text{ is self-adjoint} \quad \Longleftrightarrow \quad \langle Ax, x \rangle \in \mathbb{R} \text{ for all } x \in H.$$

Proof. $\Rightarrow$. Assume that A is self-adjoint, and fix $x \in H$. Using the conjugate symmetry of the inner product, the definition of the adjoint, and the fact that $A = A^*$, we compute that

$$\overline{\langle Ax, x \rangle} = \langle x, Ax \rangle = \langle A^*x, x \rangle = \langle Ax, x \rangle.$$

Therefore $\langle Ax, x \rangle$ is real.

$\Leftarrow$. Assume that $\langle Ax, x \rangle$ is real for all $x \in H$. Given $x, y \in H$, we have

$$\langle A(x+y), x+y \rangle = \langle Ax + Ay, x+y \rangle = \langle Ax, x \rangle + \langle Ax, y \rangle + \langle Ay, x \rangle + \langle Ay, y \rangle.$$

Since the three inner products $\langle A(x+y), x+y \rangle$, $\langle Ax, x \rangle$, and $\langle Ay, y \rangle$ are all real, we conclude that $\langle Ax, y \rangle + \langle Ay, x \rangle$ is real and therefore equals its own complex conjugate. Consequently,

$$
\begin{aligned}
\langle Ax, y \rangle + \langle Ay, x \rangle &= \overline{\langle Ax, y \rangle + \langle Ay, x \rangle} && (z = \bar{z} \text{ when } z \text{ is real}) \\
&= \overline{\langle Ax, y \rangle} + \overline{\langle Ay, x \rangle} && (\overline{w + z} = \overline{w} + \overline{z}) \\
&= \langle y, Ax \rangle + \langle x, Ay \rangle && (\text{conjugate symmetry}). \quad (7.12)
\end{aligned}
$$

Similarly, after examining the equation

$$\langle A(x+iy), x+iy \rangle = \langle Ax, x \rangle - i\langle Ax, y \rangle + i\langle Ay, x \rangle + \langle Ay, y \rangle,$$

we find that

$$\langle Ax, y \rangle - \langle Ay, x \rangle = -\langle y, Ax \rangle + \langle x, Ay \rangle. \qquad (7.13)$$

Adding equations (7.12) and (7.13) gives $2\langle Ax, y\rangle = 2\langle x, Ay\rangle$. Combining this with the definition of the adjoint, we obtain

$$\langle Ax, y\rangle = \langle x, Ay\rangle = \langle A^*x, y\rangle.$$

Since this is true for all vectors x, $y \in H$, it follows that $A = A^*$. □

Theorem 7.3.7 does not hold when $\mathbb{F} = \mathbb{R}$. Indeed, when the scalar field is real we have $\langle Ax, x\rangle \in \mathbb{R}$ for every vector x, no matter whether A is self-adjoint or not (compare Problem 7.3.12).

Next we give a useful formula for computing the operator norm of a self-adjoint map. This theorem holds for both real and complex Hilbert spaces.

Theorem 7.3.8. *Let H be a Hilbert space. If $A \in \mathcal{B}(H)$ is self-adjoint, then*

$$\|A\| = \sup_{\|x\|=1} |\langle Ax, x\rangle|.$$

Proof. For convenience of notation, let $M = \sup_{\|x\|=1} |\langle Ax, x\rangle|$. If x is any vector in H, then by applying the Cauchy–Bunyakovski–Schwarz Inequality we see that

$$|\langle Ax, x\rangle| \leq \|Ax\|\,\|x\| \leq \|A\|\,\|x\|^2,$$

and consequently

$$M = \sup_{\|x\|=1} |\langle Ax, x\rangle| \leq \sup_{\|x\|=1} \|A\|\,\|x\|^2 = \|A\|.$$

To prove the opposite inequality, let x and y be unit vectors in H. Expanding the inner products, canceling terms, and using the fact that $A = A^*$, we compute that

$$\begin{aligned}
\langle A(x+y), x+y\rangle - \langle A(x-y), x-y\rangle &= 2\langle Ax, y\rangle + 2\langle Ay, x\rangle \\
&= 2\langle Ax, y\rangle + 2\langle y, Ax\rangle \\
&= 2\langle Ax, y\rangle + 2\overline{\langle Ax, y\rangle} \\
&= 4\,\mathrm{Re}\langle Ax, y\rangle.
\end{aligned}$$

Therefore

$$\begin{aligned}
4\,|\mathrm{Re}\langle Ax, y\rangle| &= |\langle A(x+y), x+y\rangle - \langle A(x-y), x-y\rangle| \\
&\leq |\langle A(x+y), x+y\rangle| + |\langle A(x-y), x-y\rangle| \\
&\leq M\,\|x+y\|^2 + M\,\|x-y\|^2 \qquad \text{(by definition of M)} \\
&= 2M\left(\|x\|^2 + \|y\|^2\right) \qquad\qquad \text{(Parallelogram Law)} \\
&= 4M.
\end{aligned}$$

Thus

$$|\mathrm{Re}\langle Ax, y\rangle| \leq M \quad \text{for all unit vectors } x, y. \tag{7.14}$$

At this point, if $\mathbb{F} = \mathbb{R}$, then we apply Problem 7.1.6 and conclude that

$$\|A\| = \sup_{\|x\|=\|y\|=1} |\langle Ax, y\rangle| = \sup_{\|x\|=\|y\|=1} |\mathrm{Re}\langle Ax, y\rangle| \leq M.$$

Therefore we are done if the scalar field is real.

On the other hand, if $\mathbb{F} = \mathbb{C}$, then we fix unit vectors $x, y \in H$ and let α be a complex scalar with unit modulus such that $\alpha \langle Ax, y\rangle = |\langle Ax, y\rangle|$. Then

$$|\langle Ax, y\rangle| = \alpha \langle Ax, y\rangle = \langle Ax, \overline{\alpha}y\rangle. \tag{7.15}$$

Since $\overline{\alpha}y$ is a unit vector and since $\langle Ax, \overline{\alpha}y\rangle$ is a real number, by combining equations (7.14) and (7.15) we obtain

$$|\langle Ax, y\rangle| = \langle Ax, \overline{\alpha}y\rangle = \mathrm{Re}\langle Ax, \overline{\alpha}y\rangle \leq M.$$

Problem 7.1.6 therefore implies that $\|A\| \leq M$. $\square$

The next corollary gives a way to recognize when an operator is the zero operator. This is very useful when we want to prove that two operators are equal, because $A = B$ if and only if $A - B = 0$ (see Problem 7.3.20 for an application).

Corollary 7.3.9. *Let $A \in \mathcal{B}(H)$, where H is a Hilbert space. If either*

(a) $\mathbb{F} = \mathbb{C}$, *or*

(b) $\mathbb{F} = \mathbb{R}$ *and A is self-adjoint,*

then

$$\langle Ax, x\rangle = 0 \text{ for every } x \in H \iff A = 0.$$

Proof. Assume that $\langle Ax, x\rangle = 0$ for every x. If $\mathbb{F} = \mathbb{R}$, then we have that A is self-adjoint by hypothesis. On the other hand, if $\mathbb{F} = \mathbb{C}$, then the hypothesis that $\langle Ax, x\rangle = 0$ implies that $\langle Ax, x\rangle$ is always real, and consequently A is self-adjoint by Theorem 7.3.7. In either case A is self-adjoint, so we can use Theorem 7.3.8 to compute the operator norm of A, and we find that

$$\|A\| = \sup_{\|x\|=1} |\langle Ax, x\rangle| = \sup_{\|x\|=1} |\langle 0, x\rangle| = 0. \quad \square$$

If $\mathbb{F} = \mathbb{R}$, then Corollary 7.3.9 fails if we omit the hypothesis that A is self-adjoint (Problem 7.3.12 asks for a counterexample).

Now we define positive and positive definite operators, which are special types of self-adjoint operators.

Definition 7.3.10 (Positive and Positive Definite Operators). Assume that $A \in \mathcal{B}(H)$ is a self-adjoint operator on a Hilbert space H.

(a) We say that A is *positive* (or *nonnegative*) if $\langle Ax, x \rangle \geq 0$ for every $x \in H$. In this case we write $A \geq 0$.

(b) We say that A is *positive definite* (or *strictly positive*) if $\langle Ax, x \rangle > 0$ for every nonzero $x \in H$. In this case we write $A > 0$.   ◇

For complex Hilbert spaces, Theorem 7.3.7 implies that the assumption of self-adjointness in Definition 7.3.10 is redundant. We state this precisely as the following lemma.

Lemma 7.3.11. *Let H be a **complex** Hilbert space (i.e., we take $\mathbb{F} = \mathbb{C}$). If $A \in \mathcal{B}(H)$ is a bounded linear operator, then the following statements hold.*

(a) *A is positive $\iff$ $\langle Ax, x \rangle \geq 0$ for every $x \in H$.*

(b) *A is positive definite $\iff$ $\langle Ax, x \rangle > 0$ for every $x \neq 0$.*   ◇

Problems

7.3.12. Give an example of a real Hilbert space H and a nonzero linear operator $A \in \mathcal{B}(H)$ such that $\langle Ax, x \rangle = 0$ for every $x \in H$. Is your operator A self-adjoint?

7.3.13. Let H be a Hilbert space, and suppose that $A, B \in \mathcal{B}(H)$ are self-adjoint. Prove that ABA and BAB are self-adjoint, but AB is self-adjoint if and only if $AB = BA$. Exhibit self-adjoint operators A and B that do not commute.

7.3.14. Let H be a Hilbert space. Show that if $A \in \mathcal{B}(H)$, then $A + A^*$, AA^*, A^*A, and $AA^* - A^*A$ are all self-adjoint, and AA^* and A^*A are positive. If $A \neq 0$, must AA^* and A^*A be positive definite?

7.3.15. Let M be a closed subspace of a Hilbert space H, and fix $P \in \mathcal{B}(H)$. Show that P is the orthogonal projection of H onto M if and only if $P^2 = P$, $P^* = P$, and range$(P) = M$.

7.3.16. Let M be a closed subspace of a separable, infinite-dimensional Hilbert space H, and let P be the orthogonal projection of H onto M. Let $\{e_n\}_{n \in \mathbb{N}}$, $\{f_m\}_{m \in I}$, and $\{g_m\}_{m \in J}$ be orthonormal bases for H, M, and $M^\perp$, respectively. Prove that

$$\sum_{n=1}^{\infty} \|Pe_n\|^2 = \sum_{n=1}^{\infty} \left(\sum_{m \in I} |\langle Pe_n, f_m \rangle|^2 + \sum_{m \in J} |\langle Pe_n, g_m \rangle|^2 \right),$$

and use this to show that $\sum \|Pe_n\|^2 = \dim(M)$.

7.3.17. Let H be a Hilbert space. Given $U \in \mathcal{B}(H)$, prove that the following three statements are equivalent.

(a) U is norm-preserving (i.e., U is an isometry).

(b) U preserves inner products.

(c) $U^*U = I$.

Show further that if the statements above hold, then U has closed range and UU^* is the orthogonal projection of H onto range(U). Moreover, U is unitary if and only if $U^* = U^{-1}$.

7.3.18. Let H be a Hilbert space. We say that an operator $A \in \mathcal{B}(H)$ is *normal* if A commutes with its adjoint, i.e., if $AA^* = A^*A$. Prove the following statements.

(a) A is normal if and only if $\|Ax\| = \|A^*x\|$ for every $x \in H$.

(b) A is unitary if and only if A is a normal isometry.

(c) If λ is any bounded sequence of scalars, then the multiplication operator M_λ introduced in Theorem 7.1.4 is normal.

(d) The right-shift and left-shift operators R and L on ℓ^2 are not normal.

(e) If $\mathbb{F} = \mathbb{C}$ and $S, T \in \mathcal{B}(H)$ are commuting self-adjoint operators, then $S + iT$ is normal.

7.3.19. Let H and K be Hilbert spaces. Given $A \in \mathcal{B}(H, K)$, prove the following statements.

(a) $A^*A \in \mathcal{B}(H)$ and $AA^* \in \mathcal{B}(K)$ are self-adjoint and positive.

(b) A^*A is positive definite $\iff$ A is injective $\iff$ A^* has dense range.

(c) AA^* is positive definite $\iff$ A^* is injective $\iff$ A has dense range.

7.3.20. We say that a sequence of vectors $\{x_n\}_{n \in \mathbb{N}}$ in a Hilbert space H is a *tight frame* for H if there exists a number $A > 0$, called the *frame bound*, such that

$$\sum_{n=1}^{\infty} |\langle x, x_n \rangle|^2 = A\|x\|^2, \qquad \text{for all } x \in H.$$

Every tight frame is a *Bessel sequence* in the sense of Problem 5.7.5, and therefore the results of that problem imply that the series

$$Sx = \sum_{n=1}^{\infty} \langle x, x_n \rangle x_n$$

converges for every $x \in H$. We refer to the mapping $S: H \to H$ as the *frame operator*. Prove that S and $S - AI$ are self-adjoint, where I is the identity operator on H, and use this to show that $S = AI$ (Hint: Corollary 7.3.9). Then prove that the following three statements are equivalent.

(a) $\|x_n\|^2 = A$ for every $n \in \mathbb{N}$.

(b) $\{x_n\}_{n\in\mathbb{N}}$ is an orthogonal (but not necessarily *orthonormal*) sequence with no zero elements.

(c) $\{x_n\}_{n\in\mathbb{N}}$ is a Schauder basis for H, i.e., for each $x \in H$ there exist unique scalars $c_n(x)$ such that $x = \sum_{n=1}^{\infty} c_n(x) x_n$, where the series converges in the norm of H.

7.3.21. We take $\mathbb{F} = \mathbb{R}$ in this problem. The *Hilbert matrix* is

$$\mathcal{H} = \begin{bmatrix} 1 & 1/2 & 1/3 & 1/4 & \cdots \\ 1/2 & 1/3 & 1/4 & 1/5 & \\ 1/3 & 1/4 & 1/5 & 1/6 & \\ 1/4 & 1/5 & 1/6 & 1/7 & \\ \vdots & & & & \ddots \end{bmatrix}.$$

Define

$$C = \begin{bmatrix} 1 & 0 & 0 & 0 & \cdots \\ 1/2 & 1/2 & 0 & 0 & \\ 1/3 & 1/3 & 1/3 & 0 & \\ 1/4 & 1/4 & 1/4 & 1/4 & \\ \vdots & & & & \ddots \end{bmatrix} \quad \text{and} \quad L = \begin{bmatrix} 1 & 1/2 & 1/3 & 1/4 & \cdots \\ 1/2 & 1/2 & 1/3 & 1/4 & \\ 1/3 & 1/3 & 1/3 & 1/4 & \\ 1/4 & 1/4 & 1/4 & 1/4 & \\ \vdots & & & & \ddots \end{bmatrix}.$$

For this problem, take as given the fact that C determines a bounded map of ℓ^2 into ℓ^2 (compare Section 6.7). Prove the following statements.

(a) $L = CC^*$, so $L \geq 0$ (i.e., L is a positive operator).

(b) $I - (I - C)(I - C)^* = \text{diag}(1, 1/2, 1/3, 1/4, \dots)$, the diagonal matrix with entries $1, 1/2, \dots$ on the diagonal.

(c) $\|(I - C)\|^2 = \|(I - C)(I - C)^*\| \leq 1$.

(d) $\|C\| \leq 2$ and $\|L\| \leq 4$.

Remark: It is a fact (though not so easy to prove) that if A, B are infinite symmetric matrices and $a_{ij} \leq b_{ij}$ for all $i, j \in \mathbb{N}$, then $\|A\| \leq \|B\|$. Consequently, $\|\mathcal{H}\| \leq \|L\| \leq 4$. It is known that the operator norm of the Hilbert matrix is precisely $\|\mathcal{H}\| = \pi$ (e.g., see [Cho83]).

7.4 Compact Operators on Hilbert Spaces

We will study the special class of *compact operators* on Hilbert spaces in this section. In order to simplify the presentation, we introduce a notation for the closed unit disk in a Hilbert space.

Notation 7.4.1. The *closed unit ball* or *closed unit disk* in H will be denoted by

$$D_H = \overline{B_1(0)} = \{x \in H : \|x\| \leq 1\}. \quad \diamondsuit$$

The closed unit disk is both closed and bounded (why?). However, D_H is compact *if and only if* H is finite-dimensional (see Problem 5.9.7).

Assume that $T: H \to K$ is a continuous linear operator, and consider $T(D_H)$, the image of the closed unit disk under T. What type of set is $T(D_H)$? If H is finite-dimensional, then D_H is a compact subset of H, and therefore $T(D_H)$ is compact, since continuous functions map compact sets to compact sets (see Lemma 2.9.2). However, if H is infinite-dimensional, then D_H is not compact, and $T(D_H)$ need not be compact either (for example, think about what happens if T is the identity operator). Since D_H is a bounded set and T is a bounded operator, we know that $T(D_H)$ is a bounded subset of K, but in general $T(D_H)$ need not be closed (see Problem 7.5.5 for an example). However, there are *some* linear operators T that map D_H into a compact subset of K. In this case the closure of $T(D_H)$ is compact. We call these the *compact operators* on H.

Definition 7.4.2 (Compact Operators). Let H and K be Hilbert spaces.

(a) A linear operator $T: H \to K$ is *compact* if the closure of $T(D_H)$ is a compact subset of K (equivalently, $T(D_H)$ is contained in a compact subset of K; see Problem 2.8.11).

(b) The set of all compact operators from H to K is denoted by

$$\mathcal{B}_0(H, K) = \{T : H \to K : T \text{ is compact}\}.$$

When $H = K$ we use the abbreviation $\mathcal{B}_0(H) = \mathcal{B}_0(H, H)$. $\quad \diamondsuit$

As we observed above, if the domain H is finite-dimensional, then every linear operator $T: H \to K$ is compact (even if K is infinite-dimensional). In contrast, Problem 7.4.8 shows that if H is infinite-dimensional, then there always exist linear operators $T: H \to K$ that are not compact.

Note that while Definition 7.4.2 does require a compact operator to be linear, it does not explicitly require continuity. However, we prove next that all compact operators are automatically continuous.

Theorem 7.4.3 (Compact Operators Are Bounded). *If H and K are Hilbert spaces, then $\mathcal{B}_0(H, K) \subseteq \mathcal{B}(H, K)$, i.e., every compact operator $T: H \to K$ is bounded.*

Proof. We will prove the contrapositive statement. Assume that $T: H \to K$ is linear but unbounded. Then the operator norm of T is infinite, i.e.,

$$\|T\| = \sup_{\|x\|=1} \|Tx\| = \infty.$$

Consequently, for each $n \in \mathbb{N}$ there must exist a vector $x_n \in H$ such that $\|x_n\| = 1$ but $\|Tx_n\| \geq n$. Hence $T(D_H)$ is an unbounded set. Since every compact set is bounded, it follows that $T(D_H)$ cannot be contained in any compact set, and therefore T is not compact. $\square$

Recall that Theorem 2.8.9 gave us several ways to view compact subsets of complete metric spaces. Using that theorem, we can give corresponding equivalent formulations of compact operators.

Theorem 7.4.4. *Let H and K be Hilbert spaces. If $T: H \to K$ is linear, then the following three statements are equivalent.*

(a) *T is compact.*

(b) *$T(D_H)$ is totally bounded.*

(c) *If $\{x_n\}_{n\in\mathbb{N}}$ is a bounded sequence in H, then $\{Tx_n\}_{n\in\mathbb{N}}$ contains a convergent subsequence in K.*

Proof. (a) $\Rightarrow$ (b). Assume that T is compact, and let $E = T(D_H)$. By hypothesis, the closure of E is compact. Theorem 2.8.9 therefore implies that $\overline{E}$ is totally bounded, i.e., for each $r > 0$, $\overline{E}$ can be covered by finitely many balls of radius r. Since E is a subset of $\overline{E}$, it is totally bounded as well.

(b) $\Rightarrow$ (a). Assume that $E = T(D_H)$ is totally bounded. We claim that $\overline{E}$ is also totally bounded. To see this, fix any $r > 0$. Then, since E is totally bounded, there exist finitely many points $x_1, \ldots, x_N$ such that

$$E \subseteq B_r(x_1) \cup \cdots \cup B_r(x_N).$$

Suppose that $y \in \overline{E}$. Then there exist points $y_n \in E$ such that $y_n \to y$. Choose n large enough that $\|y - y_n\| < r$. Since $y_n \in E$, there is some k such that $y_n \in B_r(x_k)$. Hence $\|y_n - x_k\| < r$, and therefore, by the Triangle Inequality, $\|y - x_k\| < 2r$. Hence

$$\overline{E} \subseteq B_{2r}(x_1) \cup \cdots \cup B_{2r}(x_N).$$

Thus $\overline{E}$ is totally bounded. Since it is also closed and since H is complete, Theorem 2.8.9 implies that $\overline{E}$ is compact. Therefore T is a compact operator.

(a) $\Rightarrow$ (c). Suppose that T is compact and $\{x_n\}_{n\in\mathbb{N}}$ is a bounded sequence in H. By rescaling (multiply each x_n by a fixed scalar), we may assume that the vectors x_n belong to D_H. Then for each n we have

$$Tx_n \in T(D_H) \subseteq \overline{T(D_H)}.$$

Since $\overline{T(D_H)}$ is compact, Theorem 2.8.9 implies that $\{Tx_n\}_{n\in\mathbb{N}}$ contains a subsequence that converges to an element of $\overline{T(D_H)}$.

(c) $\Rightarrow$ (a). Suppose that statement (c) holds, and let $\{y_n\}_{n\in\mathbb{N}}$ be any sequence in $\overline{T(D_H)}$. Since $T(D_H)$ is dense in $\overline{T(D_H)}$, for each $n \in \mathbb{N}$ there exists a vector $z_n \in T(D_H)$ such that

$$\|y_n - z_n\| < \frac{1}{n}.$$

Statement (c) implies that $\{z_n\}_{n\in\mathbb{N}}$ contains a convergent subsequence, say $z_{n_k} \to z$. Then $z \in \overline{T(D_H)}$, and

$$\|z - y_{n_k}\| \leq \|z - z_{n_k}\| + \|z_{n_k} - y_{n_k}\| \to 0.$$

Hence $\{y_{n_k}\}_{k\in\mathbb{N}}$ is a subsequence of $\{y_n\}_{n\in\mathbb{N}}$ that converges to $z \in \overline{T(D_H)}$. This shows that $\overline{T(D_H)}$ is sequentially compact, and therefore it is compact, since K is a normed space (see Theorem 2.8.9). $\square$

Part (c) of Theorem 7.4.4 often provides the simplest method of proving that a given operator is compact. To illustrate, we will use that reformulation to show that $\mathcal{B}_0(H, K)$ is a subspace (and not just a subset) of $\mathcal{B}(H, K)$.

Lemma 7.4.5. *Let H and K be Hilbert spaces. If operators $S, T\colon H \to K$ are compact and a, b are scalars, then $aS + bT$ is compact.*

Proof. Let $\{x_n\}_{n\in\mathbb{N}}$ be a bounded sequence of vectors in H. Since S is compact, there exists a subsequence $\{y_n\}_{n\in\mathbb{N}}$ of $\{x_n\}_{n\in\mathbb{N}}$ such that Sy_n converges. Since $\{y_n\}_{n\in\mathbb{N}}$ is bounded and T is compact, there is a subsequence $\{z_n\}_{n\in\mathbb{N}}$ of $\{y_n\}_{n\in\mathbb{N}}$ such that Tz_n converges. Consequently, Sz_n and Tz_n both converge, so $(S + T)z_n = Sz_n + Tz_n$ converges. Therefore $S + T$ is compact by Theorem 7.4.4. A similar argument establishes closure under scalar multiplication. $\square$

Next we prove that the *operator norm limit* of a sequence of compact operators is compact. Consequently, $\mathcal{B}_0(H, K)$ is a *closed* subspace of $\mathcal{B}(H, K)$.

Theorem 7.4.6 (Limits of Compact Operators). *Let H and K be Hilbert spaces, and assume that $T_n\colon H \to K$ is compact for each $n \in \mathbb{N}$. If $T \in \mathcal{B}(H, K)$ is such that $\|T - T_n\| \to 0$, then T is compact.*

Proof. We will use a Cantor diagonalization argument (see the proof of Theorem 2.8.9 for another example of this type of argument).

Suppose that $\{x_n\}_{n\in\mathbb{N}}$ is a bounded sequence in H. Since T_1 is compact, there exists a subsequence $\{x_n^{(1)}\}_{n\in\mathbb{N}}$ of $\{x_n\}_{n\in\mathbb{N}}$ such that $\{T_1 x_n^{(1)}\}_{n\in\mathbb{N}}$ converges. Then, since T_2 is compact, there exists a subsequence $\{x_n^{(2)}\}_{n\in\mathbb{N}}$ of $\{x_n^{(1)}\}_{n\in\mathbb{N}}$ such that $\{T_2 x_n^{(2)}\}_{n\in\mathbb{N}}$ converges (note that $\{T_1 x_n^{(2)}\}_{n\in\mathbb{N}}$ also converges). Continue to construct subsequences in this way.

By construction, for every m the vector $x_m^{(m)}$ belongs to the subsequence $\{x_n^{(1)}\}_{n \in \mathbb{N}}$. Likewise, if $m \geq 2$, then $x_m^{(m)}$ is an element of $\{x_n^{(2)}\}_{n \in \mathbb{N}}$. In general, if $k \in \mathbb{N}$, then $\{x_m^{(m)}\}_{m \geq k}$ is a subsequence of $\{x_n^{(k)}\}_{n \in \mathbb{N}}$. Therefore $\{T_k x_m^{(m)}\}_{m \geq k}$ converges. For simplicity of notation, let $y_m = x_m^{(m)}$. Since our original sequence $\{x_n\}_{n \in \mathbb{N}}$ is bounded, we have that $R = \sup \|y_m\| < \infty$.

Fix $\varepsilon > 0$. Then there exists some k such that $\|T - T_k\| < \varepsilon/R$. Since $\{T_k y_m\}_{m \in \mathbb{N}}$ converges, it is Cauchy in K. Therefore there is some M such that $\|T_k y_m - T_k y_n\| < \varepsilon$ for all $m, n \geq M$. Hence, if $m, n \geq M$, then

$$
\begin{aligned}
\|T y_m - T y_n\| &\leq \|T y_m - T_k y_m\| + \|T_k y_m - T_k y_n\| + \|T_k y_n - T y_n\| \\
&\leq \|T - T_k\| \, \|y_m\| + \varepsilon + \|T_k - T\| \, \|y_n\| \\
&\leq R\frac{\varepsilon}{R} + \varepsilon + R\frac{\varepsilon}{R} = 3\varepsilon.
\end{aligned}
$$

Thus $\{T y_m\}_{m \in \mathbb{N}}$ is Cauchy in K. Since K is a Hilbert space, this sequence must therefore converge. Hence T is compact. $\quad\square$

It is important in Theorem 7.4.6 that the operators T_n converge to T *in operator norm*. It is possible for compact operators T_n to satisfy $T_n x \to T x$ for every x yet T is not compact (consider Problem 7.1.9).

The next theorem (whose proof is Problem 7.4.9) shows that the composition of a compact operator with a bounded operator, in either order, is compact.

Theorem 7.4.7. *Let H, K, L be Hilbert spaces, and let $A: H \to K$ and $B: K \to L$ be linear operators. If A is compact and B is bounded, or if A is bounded and B is compact, then BA is compact.* $\diamondsuit$

In particular, if $T \in \mathcal{B}_0(H)$ and $A \in \mathcal{B}(H)$, then AT and TA both belong to $\mathcal{B}_0(H)$. Using the language of abstract algebra, this says that $\mathcal{B}_0(H)$ is a two-sided *ideal* in $\mathcal{B}(H)$.

Problems

7.4.8. Let H and K be Hilbert spaces.

(a) Prove that if H is finite-dimensional, then every linear mapping $T: H \to K$ is both compact and continuous.

(b) Assume H is infinite-dimensional. Use Problem 6.3.8 to show that there exists a linear operator $T: H \to K$ that is neither compact nor continuous (even if K is finite-dimensional).

7.4.9. Prove Theorem 7.4.7.

7.4.10. Let H and K be Hilbert spaces, and suppose that $T\colon H \to K$ is compact. Prove the following statements.

(a) $T(D_H)$ can be covered by countably many balls $B_{r_n}(x_n)$ such that $\inf r_n = 0$.

(b) Using the vectors x_n from part (a), the sequence $\{x_n\}_{n\in\mathbb{N}}$ is dense in $T(D_H)$.

(c) $\overline{\mathrm{range}}(T)$ is a separable subspace of K.

7.4.11. Let H and K be Hilbert spaces, and assume that $T\colon H \to K$ is compact and $\{e_n\}_{n\in\mathbb{N}}$ is an orthonormal sequence in H.

(a) Show that every subsequence $\{f_n\}_{n\in\mathbb{N}}$ of $\{e_n\}_{n\in\mathbb{N}}$ has a subsequence $\{g_n\}_{n\in\mathbb{N}}$ of $\{f_n\}_{n\in\mathbb{N}}$ such that $\{Tg_n\}_{n\in\mathbb{N}}$ converges to the zero vector in K.

(b) Prove that $Te_n \to 0$ as $n \to \infty$.

7.4.12. Assume H is an infinite-dimensional Hilbert space and $T\colon H \to K$ is compact and injective. Prove that $T^{-1}\colon \mathrm{range}(T) \to H$ is unbounded (and also prove that part (d) of Problem 7.1.10 gives an example of such an operator).

7.4.13. Let H and K be Hilbert spaces, and assume $T\colon H \to K$ is compact. Suppose that $\{x_n\}_{n\in\mathbb{N}}$ is a bounded sequence that *converges weakly* to $x \in H$, i.e., $\langle x_n, y\rangle \to \langle x, y\rangle$ for every $y \in H$. Prove that $Tx_n \to Tx$ in norm, i.e., $\|Tx - Tx_n\| \to 0$ as $n \to \infty$.

7.5 Finite-Rank Operators

Recall that the *rank* of a linear operator $T\colon H \to K$ is the vector space dimension of its range, and T is a *finite-rank operator* if $\mathrm{range}(T)$ is finite-dimensional. In general, a finite-rank operator need not be bounded (see Problem 6.3.8). We will denote the set of *bounded* finite-rank operators by

$$\mathcal{B}_{00}(H, K) = \{T \in \mathcal{B}(H, K) : T \text{ has finite rank}\},$$

and we let $\mathcal{B}_{00}(H) = \mathcal{B}_{00}(H, H)$.

Not every linear finite-rank operator is compact (see Example 7.5.3). However, we prove next that if a finite-rank linear operator is bounded, then it is compact.

Theorem 7.5.1 (Bounded Finite-Rank Operators Are Compact).
Let H and K be Hilbert spaces. If $T\colon H \to K$ is bounded, linear, and has finite rank, then T is compact. Hence

$$\mathcal{B}_{00}(H, K) \subseteq \mathcal{B}_0(H, K). \tag{7.16}$$

Proof. Since T is bounded, $T(D_H)$ is a bounded subset of $R = \text{range}(T)$, which is finite-dimensional. All finite-dimensional normed spaces are complete, so R is a Hilbert space. Hence $\overline{T(D_H)}$ is a closed and bounded subset of the finite-dimensional Hilbert space R. Applying Problem 3.7.5, it follows that the closure of $T(D_H)$ is a compact subset of R, and hence it is a compact subset of K (why?). This shows that T is a compact operator. $\square$

Combining Theorem 7.5.1 with Theorem 7.4.6, we obtain the following result.

Corollary 7.5.2. *Let H and K be Hilbert spaces. If $T : H \to K$ is linear and there exist bounded linear finite-rank operators T_n such that $\|T - T_n\| \to 0$, then T is compact.* $\diamond$

To illustrate, we will use Corollary 7.5.2 to prove that the multiplication operator M_λ is compact when the sequence λ belongs to c_0. This operator will not have finite rank if λ contains infinitely many nonzero components.

Example 7.5.3. Assume that $\{e_n\}_{n \in \mathbb{N}}$ is an orthonormal basis for a Hilbert space H, and fix any sequence $\lambda = (\lambda_n)_{n \in \mathbb{N}} \in c_0$ (i.e., $\lambda_n \to 0$ as $n \to \infty$). Let

$$M_\lambda x = \sum_{n=1}^\infty \lambda_n \langle x, e_n \rangle e_n, \qquad x \in H, \tag{7.17}$$

be the "multiplication operator" from Theorem 7.1.4. Since $c_0 \subseteq \ell^\infty$, that theorem tells us that M_λ is bounded. We will use Corollary 7.5.2 to show that our extra assumption that λ_n converges to zero implies that M_λ is compact.

Note that M_λ need not have finite rank. However, we will define finite-rank operators T_N that "approximate" M_λ. We do this by replacing the infinite series in equation (7.17) by a finite sum. Specifically, given $N > 0$, we define $T_N : H \to H$ by

$$T_N x = \sum_{n=1}^N \lambda_n \langle x, e_n \rangle e_n, \qquad x \in H.$$

Since $T_N x$ is always a finite linear combination of $e_1, \ldots, e_N$, we have

$$\text{range}(T_N) \subseteq \text{span}\{e_1, \ldots, e_N\}.$$

Therefore T_N has finite rank. It is also linear, and since $|\lambda_n| \le \|\lambda\|_\infty$, we have

$$\|T_N x\|^2 = \sum_{n=1}^N |\lambda_n \langle x, e_n \rangle|^2 \le \sum_{n=1}^N \|\lambda\|_\infty^2 |\langle x, e_n \rangle|^2 \le \|\lambda\|_\infty^2 \|x\|^2,$$

where the final inequality follows from Bessel's Inequality. This shows that T_N is bounded (in fact, it shows that $\|T_N\| \le \|\lambda\|_\infty$).

If we fix any particular $x \in H$, then $T_N x \to M_\lambda x$ by equation (7.17). However, we will prove more—we will prove that T_N converges *in operator norm* to M_λ. To do this, we compute that for any fixed $x \in H$ we have

$$\|M_\lambda x - T_N x\|^2 = \left\| \sum_{n=1}^{\infty} \lambda_n \langle x, e_n \rangle e_n - \sum_{n=1}^{N} \lambda_n \langle x, e_n \rangle e_n \right\|^2$$

$$= \left\| \sum_{n=N+1}^{\infty} \lambda_n \langle x, e_n \rangle e_n \right\|^2 \quad \text{(because the series converge!)}$$

$$= \sum_{n=N+1}^{\infty} |\lambda_n|^2 \, |\langle x, e_n \rangle|^2 \quad \text{(Plancherel's Equality)}$$

$$\leq \left(\sup_{n>N} |\lambda_n|^2 \right) \sum_{n=N+1}^{\infty} |\langle x, e_n \rangle|^2$$

$$\leq \left(\sup_{n>N} |\lambda_n|^2 \right) \|x\|^2 \quad \text{(Bessel's Inequality)}.$$

Taking the supremum over all unit vectors x, we obtain

$$\|M_\lambda - T_N\| = \sup_{\|x\|=1} \|(M_\lambda - T_N)x\| \leq \sup_{n>N} |\lambda_n|.$$

The reader should now check that our assumption that $\lambda_n \to 0$ implies that

$$\lim_{N \to \infty} \sup_{n>N} |\lambda_n| = 0.$$

Therefore $\|M_\lambda - T_N\| \to 0$ as $N \to \infty$. This shows that $T_N \to M_\lambda$ in operator norm, and consequently Corollary 7.5.2 implies that M_λ is compact. ◇

For example, consider $\lambda = (\frac{1}{n})_{n \in \mathbb{N}}$, i.e.,

$$M_\lambda x = \sum_{n=1}^{\infty} \frac{1}{n} \langle x, e_n \rangle e_n, \qquad x \in H.$$

Since $\lambda \in c_0$, the operator M_λ is compact. However, it does not have finite rank. In fact, every orthonormal basis vector e_k belongs to the range of M_λ, because

$$M_\lambda e_k = \sum_{n=1}^{\infty} \frac{1}{n} \langle e_k, e_n \rangle e_n = \frac{e_k}{k}.$$

Even so, Problem 7.1.10 shows that M_λ is not surjective; instead, range(M_λ) is a dense, but proper, subspace of H.

Problems

7.5.4. Let H and K be Hilbert spaces, and let $T \in \mathcal{B}(H, K)$ be given. Show that T has finite rank if and only if there exist vectors $y_1, \ldots, y_N \in H$ and $z_1, \ldots, z_N \in K$ such that

$$ Tx = \sum_{k=1}^{N} \langle x, y_k \rangle z_k, \qquad x \in H. $$

In case this holds, find T^* and show that T^* has finite rank.

7.5.5. Let $\{e_n\}_{n \in \mathbb{N}}$ be an orthonormal basis for an infinite-dimensional separable Hilbert space H. Let $\lambda = (\lambda_n)_{n \in \mathbb{N}}$ be any sequence in c_0 such that $\lambda_n \neq 0$ for every n. By Example 7.5.3, the corresponding multiplication operator $M_\lambda \colon H \to H$ is compact. Prove that $M_\lambda(D_H)$ is not a closed subset of H.

7.5.6. The *Frobenius norm* of an infinite matrix $A = [a_{ij}]_{i,j \in \mathbb{N}}$ is the ℓ^2-norm of its entries:

$$ \|A\|_2 = \left(\sum_{i=1}^{\infty} \sum_{j=1}^{\infty} |a_{ij}|^2 \right)^{1/2}. $$

Suppose that A has finite Frobenius norm, and prove the following statements.

(a) If $x \in \ell^2$, then the series that defines $(Ax)_i$ converges for each $i \in \mathbb{N}$, and $Ax = \left((Ax)_i \right)_{i \in \mathbb{N}}$ belongs to ℓ^2.

(b) If given $N \in \mathbb{N}$ we let A_N be the matrix that has the same entries as A for $i, j \leq N$ but zeros elsewhere, then $A_N \colon \ell^2 \to \ell^2$ is bounded and has finite rank.

(c) $A \colon \ell^2 \to \ell^2$ is bounded, and $A_N \to A$ in operator norm.

(d) $A \colon \ell^2 \to \ell^2$ is compact.

7.6 Eigenvalues and Invariant Subspaces

Let X be a vector space, and suppose that $A \colon X \to X$ is linear. We say that a scalar $\lambda \in \mathbb{F}$ is an *eigenvalue* of A if there exists a nonzero vector $x \in X$ such that $Ax = \lambda x$. Every such nonzero vector x is called an *eigenvector* of A. To emphasize the eigenvalue that corresponds to an eigenvector x, we sometimes say that x is a λ-*eigenvector* of A. Note that an eigenvalue λ can be zero, but an eigenvector x must be nonzero. In particular, if there is a nonzero vector $x \in \ker(A)$, then $Ax = 0 = 0x$, and therefore $\lambda = 0$ is an eigenvalue for A in this case.

Restating the definition,

$$\lambda \text{ is an eigenvalue of } A \quad \Longleftrightarrow \quad \ker(A - \lambda I) \neq \{0\}.$$

In this case we call $\ker(A - \lambda I)$ the *eigenspace* corresponding to λ, or simply the λ-*eigenspace* for short. If $\ker(A - \lambda I) \neq \{0\}$, then every nonzero vector in $\ker(A - \lambda I)$ is a λ-eigenvector for A (in particular, every nonzero vector in $\ker(A)$, if there are any, is a 0-eigenvector of A). The (geometric) *multiplicity* of an eigenvalue λ is the dimension of its λ-eigenspace $\ker(T - \lambda I)$.

If X is a normed space, then we have the following inequality relating the magnitudes of any eigenvalues of A to its operator norm.

Lemma 7.6.1. *Let X be a normed space, and let $A \colon X \to X$ be a bounded linear operator. If λ is an eigenvalue of A, then $|\lambda| \leq \|A\|$.*

Proof. Suppose that $Ax = \lambda x$, where $x \neq 0$. If we set $y = x/\|x\|$, then y is a unit vector, and $Ay = \lambda y$. Therefore

$$\|Ay\| = \|\lambda y\| = |\lambda| \, \|y\| = |\lambda|.$$

Since y is only one of the unit vectors in X, it follows that

$$\|A\| = \sup_{\|z\|=1} \|Az\| \geq \|Ay\| = |\lambda|. \qquad \square$$

We introduce a related type of subspace.

Definition 7.6.2 (Invariant Subspace). A subspace M of a vector space X is said to be *invariant* under a linear operator $A \colon X \to X$ if $A(M) \subseteq M$, where $A(M) = \{Ax : x \in M\}$ is the direct image of M under A. $\quad \Diamond$

If λ is an eigenvalue of A, then its eigenspace $M = \ker(A - \lambda I)$ is invariant under A (why?). Hence eigenspaces are special cases of invariant subspaces.

For Hilbert spaces, we have the following relationship between subspaces that are invariant under a bounded operator and those that are invariant under its adjoint (the proof is assigned as Problem 7.6.6).

Lemma 7.6.3. *Let H be a Hilbert space, and let $A \in \mathcal{B}(H)$ be a bounded linear operator. If M is a closed subspace of H that is invariant under A, then $M^{\perp}$ is invariant under A^*.* $\quad \Diamond$

We will also need the following fact about eigenvalues and eigenvectors of a self-adjoint operator.

Lemma 7.6.4. *If H is a Hilbert space and $A \in \mathcal{B}(H)$ is self-adjoint, then all eigenvalues of A are real, and eigenvectors of A corresponding to distinct eigenvalues are orthogonal.*

Proof. Suppose that λ is an eigenvalue of A, and let x be a corresponding eigenvector. Then, since $A = A^*$,

$$\lambda \|x\|^2 = \langle \lambda x, x \rangle = \langle Ax, x \rangle = \langle x, Ax \rangle = \langle x, \lambda x \rangle = \overline{\lambda} \|x\|^2.$$

Since $x \neq 0$, this implies that $\lambda = \overline{\lambda}$, and therefore λ is real.

Now suppose that $\lambda \neq \mu$ are two different eigenvalues of A, and x is a λ-eigenvector, while y is a μ-eigenvector. Since the eigenvalues are real, it follows that

$$\lambda \langle x, y \rangle = \langle \lambda x, y \rangle = \langle Ax, y \rangle = \langle x, Ay \rangle = \langle x, \mu y \rangle = \mu \langle x, y \rangle.$$

But $\lambda \neq \mu$, so we must have $\langle x, y \rangle = 0$. $\square$

Turning to compact operators, the following result shows that the multiplicity of a nonzero eigenvalue of a compact operator must be finite.

Lemma 7.6.5. *Let H be a Hilbert space. If λ is a nonzero eigenvalue of a compact operator $T : H \to H$, then the λ-eigenspace of T is finite-dimensional, i.e., $\dim(\ker(T - \lambda I)) < \infty$.*

Proof. Let $M = \ker(T - \lambda I)$. Since T is compact, it is bounded and therefore continuous. This implies that M is a closed subspace of H (see Problem 6.4.3). If M were infinite-dimensional, then it would contain an infinite orthonormal sequence $\{e_n\}_{n \in \mathbb{N}}$. By the definition of M we have $Te_n = \lambda e_n$. But then, for any $m \neq n$,

$$\|Te_m - Te_n\| = \|\lambda e_m - \lambda e_n\| = |\lambda| \|e_m - e_n\| = |\lambda| \sqrt{2},$$

the last equality following from the orthonormality of the e_n. Since $\lambda \neq 0$, this implies that $\{Te_n\}_{n \in \mathbb{N}}$ contains no Cauchy subsequences and hence has no convergent subsequences. This contradicts the fact that T is compact. $\square$

Although the multiplicity of a nonzero eigenvalue of a compact operator is finite, the multiplicity of a 0-eigenvalue can be infinite. For example, the zero operator on H is compact, yet the zero eigenspace is all of H.

Problems

7.6.6. Prove Lemma 7.6.3.

7.6.7. Let X be a normed space, let $A \in \mathcal{B}(X)$ be given, and fix $n \in \mathbb{N}$. Prove that if λ is an eigenvalue of A, then λ^n is an eigenvalue of A^n. Show further that if A is a topological isomorphism and λ is an eigenvalue of A, then $\lambda \neq 0$ and λ^{-n} is an eigenvalue of $A^{-n} = (A^{-1})^n$.

7.6.8. Show that the left-shift and right-shift operators L and R on ℓ^2 are not compact, and find all of the eigenvalues and eigenvectors of L and R.

7.6.9. Is it possible for the product of two noncompact operators to be compact? Either prove that this is not possible or exhibit two bounded but noncompact operators A and B such that $AB = 0$.

7.6.10. Let $A \in \mathcal{B}(H)$ be a positive operator on a Hilbert space H. Show that all eigenvalues of A are real and nonnegative, and if A is positive definite then all eigenvalues are real and strictly positive.

7.6.11. Show that if $U \colon H \to H$ is a unitary operator on a Hilbert space H, then $U^* = U^{-1}$, U is normal, and every eigenvalue λ of U satisfies $|\lambda| = 1$.

7.6.12. Assume $A \in \mathcal{B}(H)$ is normal, and prove the following statements.

(a) If λ is an eigenvalue of A, then $\bar{\lambda}$ is an eigenvalue of A^*.

(b) The λ-eigenspace of A is the $\bar{\lambda}$-eigenspace of A^*, i.e., $\ker(A - \lambda I) = \ker(A^* - \bar{\lambda} I)$.

(c) Eigenvectors of A corresponding to distinct eigenvalues are orthogonal.

7.6.13. Let H be a Hilbert space, and assume that $T \in \mathcal{B}(H)$ is both compact and self-adjoint. Suppose that M is a closed subspace of H that is invariant under T. Let $S \colon M \to M$ be the restriction of T to M, i.e., $Sx = Tx$ for $x \in M$. Prove that S is both compact and self-adjoint on M.

    Remark: The notation $S = T|_M$ is often used to indicate that S is the restriction of T to the domain M.

7.7 Existence of an Eigenvalue

If H is a finite-dimensional Hilbert space and $A \in \mathcal{B}(H)$, then by choosing a basis for H, we can identify A with an $n \times n$ matrix. Since every $n \times n$ matrix has at least one complex eigenvalue, we know that A must have at least one eigenvalue if $\mathbb{F} = \mathbb{C}$ (we assign the details as Problem 7.7.3). In contrast, even if $\mathbb{F} = \mathbb{C}$, a linear operator on an infinite-dimensional Hilbert space need not have any eigenvalues. For example, if we take $H = \ell^2$ and let $Rx = (0, x_1, x_2, \dots)$ be the *right-shift operator* defined in Example 6.6.4, then R has no eigenvalues at all (see Problem 7.6.8).

    The right-shift operator is not compact. However, compactness by itself is not enough to ensure that an operator will have an eigenvalue. We will prove that an operator that is *both compact and self-adjoint* must have at least one eigenvalue. In order to do this, we need the following sufficient condition for the existence of an eigenvalue of a compact operator.

Lemma 7.7.1. *Let H be a Hilbert space. If $T\colon H \to H$ is compact and $\lambda \neq 0$, then*

$$\inf_{\|x\|=1} \|Tx - \lambda x\| = 0 \quad \Longrightarrow \quad \lambda \text{ is an eigenvalue of } T.$$

Proof. Suppose that $\inf_{\|x\|=1} \|Tx - \lambda x\| = 0$. Then there exist unit vectors $x_n \in H$ such that $\|Tx_n - \lambda x_n\| \to 0$. Since T is compact, $\{Tx_n\}_{n \in \mathbb{N}}$ has a convergent subsequence. That is, there exist indices n_k and a vector $y \in H$ such that $Tx_{n_k} \to y$ as $k \to \infty$. Therefore

$$\lambda x_{n_k} = \left(\lambda x_{n_k} - Tx_{n_k}\right) + Tx_{n_k} \to 0 + y = y \quad \text{as } k \to \infty. \tag{7.18}$$

Applying the continuity of T, it follows that

$$\lambda Tx_{n_k} = T(\lambda x_{n_k}) \to Ty \quad \text{as } k \to \infty.$$

On the other hand, since $Tx_{n_k} \to y$, by multiplying by the scalar λ we also have

$$\lambda Tx_{n_k} \to \lambda y \quad \text{as } k \to \infty.$$

Therefore $Ty = \lambda y$ by the uniqueness of limits. Further, by applying the continuity of the norm to equation (7.18), we see that

$$\|y\| = \lim_{k \to \infty} \|\lambda x_{n_k}\| = |\lambda| \neq 0.$$

Thus we have both $Ty = \lambda y$ and $y \neq 0$, so y is a λ-eigenvector for T. $\quad\square$

Now we will show that compactness combined with self-adjointness implies the existence of an eigenvalue. By Lemmas 7.6.1 and 7.6.4 we know that the eigenvalues of a self-adjoint operator T are real and are bounded in absolute value by the operator norm of T. Therefore every eigenvalue λ of a self-adjoint operator T must be a real scalar in the range $-\|T\| \leq \lambda \leq \|T\|$. We will show that if T is both compact and self-adjoint, then one (or both) of these extremes must be an eigenvalue.

Theorem 7.7.2. *Let H be a Hilbert space. If $T\colon H \to H$ is compact and self-adjoint, then at least one of $\|T\|$ and $-\|T\|$ is an eigenvalue of T.*

Proof. Since T is self-adjoint, Theorem 7.3.8 tells us that

$$\|T\| = \sup_{\|x\|=1} |\langle Tx, x \rangle|.$$

Consequently, there exist unit vectors x_n such that $|\langle Tx_n, x_n \rangle| \to \|T\|$. Since T is self-adjoint, each inner product $\langle Tx_n, x_n \rangle$ is real, so either we can find a subsequence $\{\langle Ty_n, y_n \rangle\}_{n \in \mathbb{N}}$ that converges to $\|T\|$, or we can find a subsequence $\{\langle Ty_n, y_n \rangle\}_{n \in \mathbb{N}}$ that converges to $-\|T\|$. Let λ be $\|T\|$ or $-\|T\|$, as appropriate. Then $\|y_n\| = 1$ for every n and $\langle Ty_n, y_n \rangle \to \lambda$. Since both λ and $\langle Ty_n, y_n \rangle$ are real, we therefore have

$$\|Ty_n - \lambda y_n\|^2 = \|Ty_n\|^2 - 2\lambda \langle Ty_n, y_n \rangle + \lambda^2 \|y_n\|^2 \qquad \text{(Polar Identity)}$$
$$\leq \|T\|^2 \|y_n\|^2 - 2\lambda \langle Ty_n, y_n \rangle + \lambda^2 \|y_n\|^2$$
$$= \lambda^2 - 2\lambda \langle Ty_n, y_n \rangle + \lambda^2 \qquad (|\lambda| = \|T\|)$$
$$\to \lambda^2 - 2\lambda^2 + \lambda^2 = 0.$$

Consequently, Lemma 7.7.1 implies that λ is an eigenvalue of T. $\square$

The *spectrum* of an operator T is the set of all scalars λ such that $T - \lambda I$ is not invertible. In particular, each eigenvalue of T is in the spectrum, and the set of eigenvalues is called the *point spectrum* of T. For more details on the spectrum of operators, see [Con90] or [Rud91].

Problems

7.7.3. Let H be a finite-dimensional Hilbert space.

(a) Prove that if $\mathbb{F} = \mathbb{C}$ and $A \colon H \to H$ is linear, then there exists at least one scalar λ that is an eigenvalue for A.

Hint: Every $d \times d$ *matrix* has at least one eigenvalue, although that eigenvalue may be complex.

(b) Show by example that if $\mathbb{F} = \mathbb{R}$, then there exists a linear operator $A \colon H \to H$ that has no eigenvalues.

7.7.4. Let $T \colon H \to H$ be compact and fix $\lambda \neq 0$. Use Lemma 7.7.1 and Problem 6.2.10 to show that if λ is not an eigenvalue of T and $\overline{\lambda}$ is not an eigenvalue of T^*, then $T - \lambda I$ is a topological isomorphism.

7.8 The Spectral Theorem for Compact Self-Adjoint Operators

The Spectral Theorem gives us a fundamental decomposition for self-adjoint operators on a Hilbert space. The version of the Spectral Theorem that we will present asserts that if T is a *compact* self-adjoint operator on H, then there is an orthonormal sequence in H that consists of eigenvectors of T, and T has a simple representation with respect to this orthonormal sequence. This is a basic version of the Spectral Theorem. The theorem generalizes to compact normal operators with little change (essentially, eigenvalues can be complex in the normal case, but must be real for self-adjoint operators). More deeply, there is a Spectral Theorem for bounded self-adjoint operators that are not compact, and even for some that are unbounded. We refer to texts on operator theory for such extensions, e.g., see [Con00], [Rud91].

We will obtain the Spectral Theorem by iterating Theorem 7.7.2. If T is the zero operator, then the index J in the following theorem is $J = \varnothing$ (and in this case we implicitly interpret a series of the form $\sum_{n \in \varnothing}$ as being the zero vector). If $T \neq 0$ has finite rank, then the index set J will be $J = \{1, \ldots, N\}$, where $N = \dim(\text{range}(T))$, and if $T \neq 0$ does not have finite rank, then we will have $J = \mathbb{N}$. We will need to use the results of several problems in the proof, including Problems 7.4.10, 7.4.11, 7.6.13, and 7.8.5.

Theorem 7.8.1 (The Spectral Theorem for Compact Self-Adjoint Operators). *Let H be a Hilbert space, and assume that $T\colon H \to H$ is both compact and self-adjoint. Then there exist:*

(a) *countably many nonzero real numbers $(\lambda_n)_{n \in J}$, either finitely many or satisfying $\lambda_n \to 0$ if infinitely many, and*

(b) *an orthonormal basis $\{e_n\}_{n \in J}$ of $\overline{\text{range}}(T)$,*

such that

$$Tx \;=\; \sum_{n \in J} \lambda_n \left\langle x, e_n \right\rangle e_n, \qquad \text{for all } x \in H. \tag{7.19}$$

Each λ_n is an eigenvalue of T, and each e_n is a λ_n-eigenvector for T.

Proof. If $T = 0$, then we can simply take $J = \varnothing$, so assume that T is not the zero operator. Let

$$R \;=\; \overline{\text{range}}(T)$$

be the closure of the range of T. Since R is a closed subspace of H, it is itself a Hilbert space. Further, because T is compact, Problem 7.4.10 implies that R is separable. We proved in Section 5.9 that every separable Hilbert space, including R in particular, has a countable orthonormal basis.

To begin the iteration process, set $H_1 = H$ and $T_1 = T$. Theorem 7.7.2 implies that T_1 has a real eigenvalue λ_1 that satisfies $|\lambda_1| = \|T_1\| > 0$. Let e_1 be a corresponding eigenvector. Since eigenvectors are nonzero, we can rescale e_1 so that $\|e_1\| = 1$ (simply divide e_1 by its length).

Let $E_1 = \text{span}\{e_1\}$. Since e_1 is an eigenvector of T, the one-dimensional subspace E_1 is invariant under T. Therefore, by Lemma 7.6.3, its orthogonal complement

$$H_2 \;=\; E_1^{\perp} \;=\; \text{span}\{e_1\}^{\perp} \;=\; \{e_1\}^{\perp}$$

is invariant under $T^* = T$. Therefore we can create a new linear operator $T_2\colon H_2 \to H_2$ by simply restricting the domain of T to the space H_2. That is, T_2 is defined by $T_2 x = Tx$, but only for $x \in H_2$. By doing this we obtain an operator T_2 that maps H_2 into itself.

It is possible that T_2 might be the zero operator. If that is the case, then we set $J = \{1\}$ and stop the iteration. Otherwise, we apply Problem 7.6.13, which shows that $T_2\colon H_2 \to H_2$ is both compact and self-adjoint as an operator on the Hilbert space H_2. Applying Theorem 7.7.2 to T_2, we see that T_2 has an eigenvalue λ_2 such that $|\lambda_2| = \|T_2\|$. Let $e_2 \in H_2$ be a corresponding eigenvector, normalized so that $\|e_2\| = 1$. Because $H_2 = \{e_1\}^\perp$, we have $e_2 \perp e_1$. Also, because $e_2 \in H_2$, we have $Te_2 = T_2e_2 = \lambda_2e_2$, and therefore λ_2 is an eigenvalue of T (not just T_2). Further, because $T_2x = T_1x$ for all $x \in H_2$ (which is a subset of H_1), we also have

$$|\lambda_2| = \|T_2\| = \sup\{\|T_2x\| : \|x\| = 1, x \in H_2\}$$
$$\leq \sup\{\|T_1x\| : \|x\| = 1, x \in H_1\} = \|T_1\| = |\lambda_1|.$$

Now we set $E_2 = \mathrm{span}\{e_1, e_2\}$ and $H_3 = E_2^\perp = \{e_1, e_2\}^\perp$. This space is invariant under T (why?), so we can define an operator $T_3 \colon H_3 \to H_3$ by restricting T to H_3. If $T_3 = 0$, then we stop at this point. Otherwise, we continue as before to construct an eigenvalue λ_3 (which will satisfy $|\lambda_3| \leq |\lambda_2|$) and a unit eigenvector e_3 (which will be orthogonal to both e_1 and e_2). As we continue this process, there are two possibilities.

Case 1: $T_{N+1} = 0$ for some N. In this case the iteration has stopped after a finite number of steps. We have obtained eigenvalues $\lambda_1, \ldots, \lambda_N$ and corresponding orthonormal eigenvectors $e_1, \ldots, e_N$. We have defined the closed subspaces

$$E_N = \mathrm{span}\{e_1, \ldots, e_N\} \quad \text{and} \quad H_{N+1} = E_N^\perp = \{e_1, \ldots, e_N\}^\perp.$$

Since E_N and H_{N+1} are orthogonal complements in H, if we fix $x \in H$, then there exist unique vectors $p_x \in E_N$ and $q_x \in H_{N+1}$ such that $x = p_x + q_x$ (in fact, p_x is the orthogonal projection of x onto E_N, and $q_x = x - p_x$). Since $e_1, \ldots, e_N$ is an orthonormal basis for E_N, we know that p_x is given by the formula

$$p_x = \sum_{n=1}^{N} \langle x, e_n \rangle e_n.$$

Also, since $q_x \in H_{N+1}$ and $T|_{H_{N+1}} = T_{N+1} = 0$, we have

$$Tq_x = T_{N+1}q_x = 0.$$

On the other hand, we know that $Te_n = \lambda_n e_n$, so

$$Tx = Tp_x + Tq_x = \sum_{n=1}^{N} \langle x, e_n \rangle Te_n + 0 = \sum_{n=1}^{N} \lambda_n \langle x, e_n \rangle e_n. \quad (7.20)$$

This is true for every $x \in H$. Setting $J = \{1, \ldots, N\}$, it follows that equation (7.19) holds.

By equation (7.20), we always have $Tx \in \mathrm{span}\{e_1, \ldots, e_N\}$, so $\mathrm{range}(T) \subseteq \mathrm{span}\{e_1, \ldots, e_N\}$. On the other hand, the fact that $\lambda_1, \ldots, \lambda_N$ are all nonzero implies that $e_1, \ldots, e_N$ all belong to the range of T. Consequently, $\mathrm{range}(T) = \mathrm{span}\{e_1, \ldots, e_N\}$. Hence $\mathrm{range}(T)$ is closed (because it is finite-dimensional) and $\{e_1, \dots, e_N\}$ is an orthonormal basis for range(T). Therefore the proof is complete for this case.

Case 2: $T_N \neq 0$ for every N. In this case the iteration never ends. We obtain infinitely many eigenvalues $\lambda_1, \lambda_2, \dots$ (with $|\lambda_1| \geq |\lambda_2| \geq \cdots$), and corresponding orthonormal eigenvectors $e_1, e_2, \dots$. Because T is compact, Problem 7.4.11 implies that $Te_n \to 0$. Since e_n is an eigenvector and $\|e_n\| = 1$, it follows that

$$|\lambda_n| = |\lambda_n| \|e_n\| = \|\lambda_n e_n\| = \|Te_n\| \to 0 \qquad \text{as } n \to \infty.$$

Let

$$M = \overline{\text{span}}\{e_n\}_{n \in \mathbb{N}}.$$

If we choose $x \in H$, then there exist unique vectors $p_x \in M$ and $q_x \in M^\perp$ such that $x = p_x + q_x$. In fact, p_x is the orthogonal projection of x onto M, and we can write p_x explicitly as

$$p_x = \sum_{n=1}^{\infty} \langle x, e_n \rangle e_n.$$

If we fix $N > 1$, then $E_{N-1} = \text{span}\{e_1, \dots, e_{N-1}\} \subseteq M$. Therefore

$$M^\perp \subseteq E_{N-1}^\perp = H_N.$$

Since T_N is the restriction of T to H_N, it follows that

$$\|Tq_x\| = \|T_N q_x\| \leq \|T_N\| \|q_x\| = |\lambda_N| \|q_x\| \to 0 \qquad \text{as } N \to \infty.$$

Therefore $Tq_x = 0$. Consequently,

$$Tx = Tp_x + Tq_x = \sum_{n=1}^{\infty} \langle x, e_n \rangle Te_n + 0 = \sum_{n=1}^{\infty} \lambda_n \langle x, e_n \rangle e_n.$$

This is true for every $x \in H$. Setting $J = \mathbb{N}$, we see that equation (7.19) holds. It remains only to show that $\{e_n\}_{n \in \mathbb{N}}$ is an orthonormal basis for $\overline{\text{range}}(T)$, and we assign this proof as Problem 7.8.5. $\square$

We stated Theorem 7.8.1 in terms of the *nonzero* eigenvalues of T. The zero eigenspace is $\ker(T)$, and the vectors in this space are eigenvectors of T for the eigenvalue $\lambda = 0$. If H is separable, then $\ker(T)$ is separable. By combining an orthonormal basis for $\ker(T)$ with the orthonormal basis for $\overline{\text{range}}(T)$ given by Theorem 7.8.1, we obtain the following reformulation and converse of the Spectral Theorem for separable Hilbert spaces. This form states that every compact, self-adjoint operator on a separable Hilbert space is actually a multiplication operator M_λ of the form given in equation (7.3). We assign the proof of this result as Problem 7.8.7.

Corollary 7.8.2. *Assume that H is a separable Hilbert space. If H is finite-dimensional then set $J = \{1, \ldots, N\}$, where $N = \dim(H)$; otherwise, let $J = \mathbb{N}$. If $T \in \mathcal{B}(H)$, then the following two statements are equivalent.*

(a) *T is compact and self-adjoint.*

(b) *There exist a real sequence $\lambda = (\lambda_n)_{n \in J}$, with $\lambda_n \to 0$ if $J = \mathbb{N}$, and an orthonormal basis $\{e_n\}_{n \in J}$ for H such that*

$$Tx = M_\lambda x = \sum_{n \in J} \lambda_n \langle x, e_n \rangle e_n, \qquad x \in H. \qquad \Diamond \qquad (7.21)$$

Note that some (perhaps infinitely many) of the eigenvalues λ_n in equation (7.21) may be zero, whereas all of the eigenvalues λ_n in equation (7.19) are nonzero. Further, $\{e_n\}_{n \in \mathbb{N}}$ is an orthonormal basis for H in equation (7.21), while $\{e_n\}_{n \in \mathbb{N}}$ may be only an orthonormal sequence in equation (7.19).

In summary, if T is a compact, self-adjoint operator on a *separable* Hilbert space, then there exists an orthonormal basis for H that consists entirely of eigenvectors of T. For the case that H is Euclidean space $\mathbb{F}^n$, we can formulate this as the following statement (whose proof is Problem 7.8.8) about diagonalization of Hermitian matrices. Note that the inverse of a unitary matrix U satisfies $U^{-1} = U^H$.

Theorem 7.8.3 (Diagonalization of Hermitian Matrices). *If A is an $n \times n$ matrix with scalar entries, then the following two statements are equivalent.*

(a) *A is Hermitian, i.e., $A = A^H$.*

(b) *$A = U \Lambda U^{-1}$, where U is a unitary matrix and Λ is a real diagonal matrix.*

Furthermore, in this case the diagonal entries of Λ are eigenvalues of A, and the columns of U form an orthonormal basis for $\mathbb{F}^n$ consisting of eigenvectors of A. $\Diamond$

The matrix U in Theorem 7.8.3 is unitary, and hence is essentially a rotation of $\mathbb{F}^d$ (more precisely, it is a composition of rotations and flips). The diagonal matrix Λ is a stretching along the axes (although by possibly different amounts along each axis). Thus Theorem 7.8.3 says that every self-adjoint matrix on $\mathbb{F}^d$ is a composition of a rotation, stretchings along orthogonal axes, and an inverse rotation.

Example 7.8.4. Consider $m = n = 2$, with real scalars. The diagonalization of the symmetric matrix $A = \begin{bmatrix} 3/2 & 1/2 \\ 1/2 & 3/2 \end{bmatrix}$ is

$$\begin{bmatrix} 3/2 & 1/2 \\ 1/2 & 3/2 \end{bmatrix} = A = U \Lambda U^{-1} = \begin{bmatrix} 2^{-1/2} & -2^{-1/2} \\ 2^{-1/2} & 2^{-1/2} \end{bmatrix} \begin{bmatrix} 2 & 0 \\ 0 & 1 \end{bmatrix} \begin{bmatrix} 2^{-1/2} & 2^{-1/2} \\ -2^{-1/2} & 2^{-1/2} \end{bmatrix}.$$

The unitary matrix U is a rotation by an angle of $\pi/4$, so U^{-1} is a rotation by $-\pi/4$. The diagonal matrix Λ is a stretching along the coordinate axes, by factors of 2 in the first coordinate and 1 in the second. Thus we can realize $Ax = U(\Lambda(U^{-1}x))$ by rotating x by $-\pi/4$, stretching the result along the coordinate axes, and rotating that result by $\pi/4$.

We illustrate this in Figure 7.1. Let u_1 and u_2 be the columns of U. These are orthonormal eigenvectors of A, and are shown in part (a) of Figure 7.1. Part (b) of the figure shows the vectors $U^{-1}u_1$ and $U^{-1}u_2$ that result after a rotation by $-\pi/4$. Part (c) shows $\Lambda U^{-1}u_1$ and $\Lambda U^{-1}u_2$, and part (d) then rotates these vectors by $\pi/4$ to obtain $Au_1 = U\Lambda U^{-1}u_1 = 2u_1$ and $Au_2 = U\Lambda U^{-1}u_2 = u_2$.   ◇

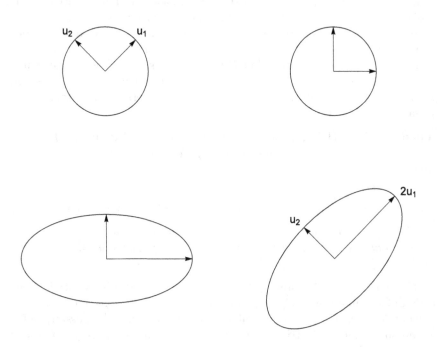

Fig. 7.1 (a) Top left: The unit circle and the orthonormal eigenvectors u_1, u_2 of A. (b) Top right: After multiplication by U^{-1} (rotation by $-\pi/4$). (c) Bottom left: After multiplication by Λ (stretching along axes). (d) Bottom right: After multiplication by U (rotation by $\pi/4$). The end result ($Au_1 = 2u_1$ and $Au_2 = u_2$) is the same as after direct multiplication by A.

Problems

7.8.5. Complete the proof of Case 2 of Theorem 7.8.1 by showing that $\{e_n\}_{n\in\mathbb{N}}$ is an orthonormal basis for $\overline{\text{range}}(T)$.

7.8.6. Verify the statements made about the diagonalization of the matrix A given in Example 7.8.4.

7.8.7. Let H be a Hilbert space, and assume that $T \in \mathcal{B}(H)$ is compact and self-adjoint.

(a) Let $\{e_n\}_{n\in J}$ be an orthonormal sequence of eigenvectors such that equation (7.19) holds, and let $\{f_m\}_{m\in I}$ be an orthonormal basis for $\ker(T)$. Prove that $\{e_n\}_{n\in J} \cup \{f_m\}_{m\in I}$ is an orthonormal basis for H.

(b) Prove Corollary 7.8.2.

7.8.8. Prove Theorem 7.8.3.

7.8.9. Let H be a Hilbert space, and assume that $T\colon H \to H$ is both compact and positive (hence all eigenvalues of T are nonnegative by Problem 7.6.10). Let equation (7.19) be the representation of T given by the Spectral Theorem. Define

$$T^{1/2}x = \sum_{n\in J} \lambda_n^{1/2} \langle x, e_n \rangle e_n, \qquad x \in H.$$

Show that $T^{1/2}$ is well-defined, compact, positive, and is a square root of T in the sense that $(T^{1/2})^2 = T$.

7.8.10. The *spectral radius* $\rho(T)$ of a square $d \times d$ matrix T is

$$\rho(T) = \max\{|\lambda| : \lambda \text{ is an eigenvalue of } T\}.$$

Let A be an $m \times n$ matrix (note that A need not be square). Show directly that if we place the ℓ^2 norm on both $\mathbb{F}^n$ and $\mathbb{F}^m$, then the operator norm of A as a mapping of $\mathbb{F}^n$ into $\mathbb{F}^m$ is $\|A\| = \rho(A^*A)^{1/2}$. Compare this to the operator norms derived in Problem 6.3.6.

7.8.11. Let vectors $v_1, \dots, v_N \in \mathbb{F}^d$ be given. The *analysis operator* for these vectors is the matrix R that has $v_1, \dots, v_N$ as rows, and the *synthesis operator* is $C = R^{\mathrm{H}}$. The *frame operator* is $S = C^{\mathrm{H}}C = RR^{\mathrm{H}}$.

(a) Give explicit formulas for Rx, Cx, and Sx, where $x \in \mathbb{F}^d$.

(b) Prove that S has an orthonormal basis of eigenvectors $\{w_1, \dots, w_d\}$ and corresponding nonnegative eigenvalues $0 \le \lambda_1 \le \cdots \le \lambda_d$.

(c) Given $x \in \mathbb{F}^d$, prove that

$$\sum_{k=1}^{N} |\langle x, v_k \rangle|^2 = \langle Sx, x \rangle = \sum_{j=1}^{d} \lambda_j |\langle x, w_j \rangle|^2.$$

(d) Prove that the following three conditions are equivalent:

    i. $\mathrm{span}\{v_1, \ldots, v_N\} = \mathbb{F}^d$,

    ii. $\lambda_1 > 0$,

    iii. there exist constants A, $B > 0$ such that

$$A\,\|x\|^2 \le \sum_{k=1}^{N} |\langle x, v_k\rangle|^2 \le B\,\|x\|^2, \qquad \text{for all } x \in \mathbb{F}^d. \qquad (7.22)$$

Remark: A set of vectors $\{v_1, \ldots, v_N\}$ that satisfies equation (7.22) is called a *frame* for $\mathbb{F}^d$.

7.9 The Singular Value Decomposition

Suppose that H and K are Hilbert spaces, and $T\colon H \to K$ is compact. In general, even if we have $H = K$, the operator T need not have any eigenvalues. However, if we compose T with its adjoint, then we obtain the self-adjoint operator $T^*T\colon H \to H$, and this operator is also compact, since the product of a compact operator and a bounded operator is compact. The Spectral Theorem implies that T^*T has eigenvalues, and in fact it tells us that there exist nonzero eigenvalues $(\mu_n)_{n \in J}$ and an orthonormal sequence of eigenvectors $\{v_n\}_{n \in J}$ such that

$$T^*Tx \;=\; \sum_{n \in J} \mu_n \langle x, v_n\rangle\, v_n, \qquad x \in H.$$

Since T^*T is a positive operator, the eigenvalues μ_n are real and strictly positive, and converge to zero if $J = \mathbb{N}$. We introduce the following terminology for this situation.

Definition 7.9.1 (Singular Numbers). Let H and K be Hilbert spaces, let $T\colon H \to K$ be compact, and let $(\mu_n)_{n \in J}$ and $\{v_n\}_{n \in J}$ be as above. The *singular numbers* or *singular values* of T are

$$s_n \;=\; \mu_n^{1/2}, \qquad n \in J,$$

taken in decreasing order:

$$s_1 \ge s_2 \ge \cdots > 0.$$

The vectors v_n are corresponding *singular vectors* of T. $\diamondsuit$

Note that according to our definition, the singular values s_n of T are nonzero (in fact, strictly positive) real scalars. If $T = 0$, then T has no singular values. However, if T is not the zero operator, then T^*T is also not the zero operator (why?), so T^*T has at least one nonzero eigenvalue, and therefore T has at least one singular value.

For a self-adjoint operator T we have $T^*T = T^2$. In this case the eigenvalues of T^*T are simply the squares of the eigenvalues of T. The singular numbers are the square roots of these eigenvalues, so this gives us the following relation between the eigenvalues and singular numbers of a compact, self-adjoint operator.

Lemma 7.9.2. *Let H be a Hilbert space. Let $T \in \mathcal{B}(H)$ be a compact, self-adjoint operator, and let $(\lambda_n)_{n \in J}$ be its nonzero eigenvalues taken in decreasing order of absolute value, i.e., $|\lambda_1| \geq |\lambda_2| \geq \cdots$. Then the singular numbers of T are*

$$s_n = |\lambda_n|, \qquad n \in J.$$

In particular, if T is positive, then $s_n = \lambda_n$ for every n. ◇

We will use the singular values of a compact operator to derive a remarkable representation of any compact operator $T \in \mathcal{B}(H, K)$.

Theorem 7.9.3 (Singular Value Decomposition). *Let H and K be Hilbert spaces. Given a compact operator $T \colon H \to K$, let $(s_n)_{n \in J}$ be the singular values for T with corresponding orthonormal singular vectors $\{v_n\}_{n \in J}$. Define*

$$u_n = s_n^{-1} T v_n, \qquad j \in J.$$

Then the following statements hold.

(a) *$\{v_n\}_{n \in J}$ is an orthonormal basis for $\overline{\mathrm{range}}(T^*)$.*

(b) *$\{u_n\}_{n \in \mathbb{N}}$ is an orthonormal basis for $\overline{\mathrm{range}}(T)$.*

(c) *$v_n = s_n^{-1} T^* u_n$ for each $n \in J$.*

(d) *For all $x \in H$ we have*

$$Tx = \sum_{n \in J} s_n \langle x, v_n \rangle u_n, \tag{7.23}$$

where this series converges in the norm of K.

Proof. (a) The Spectral Theorem (Theorem 7.8.1) implies that $\{v_n\}_{n \in J}$ is an orthonormal basis for $\overline{\mathrm{range}}(T^*T)$. However, $\overline{\mathrm{range}}(T^*T) = \overline{\mathrm{range}}(T^*)$ by Problem 7.2.17.

(b) Since v_m is an eigenvector for T^*T for the eigenvalue $\mu_m = s_m^2$, we have

$$\langle T v_m, T v_n \rangle = \langle T^* T v_m, v_n \rangle = s_m^2 \langle v_m, v_n \rangle = s_m^2 \delta_{mn}.$$

Hence $\{T v_n\}_{n \in J}$ is an orthogonal sequence. Since $\|T v_n\|^2 = \langle T v_n, T v_n \rangle = s_n^2$, it follows that $\{u_n\}_{n \in J}$ is orthonormal. Also, $u_n = s_n^{-1} T v_n \in \mathrm{range}(T)$, so $\overline{\mathrm{span}}\{u_n\}_{n \in J} \subseteq \overline{\mathrm{range}}(T)$. We will establish the opposite inclusion when we prove part (d).

(c) Since v_n is an eigenvector for T^*T for the eigenvalue s_n^2, we have

$$T^* u_n = s_n^{-1} T^* T v_n = s_n^{-1} s_n^2 v_n = s_n v_n.$$

(d) Recall from Theorem 7.2.7 that $\overline{\text{range}}(T^*)^\perp = \ker(T)$. Therefore, if we fix $x \in H$, then we have $x = p + q$, where p is the orthogonal projection of x onto $\overline{\text{range}}(T^*)$ and $q \in \overline{\text{range}}(T^*)^\perp = \ker(T)$. In fact, since $\{v_n\}_{n\in\mathbb{N}}$ is an orthonormal basis for $\overline{\text{range}}(T^*)$, we can write p explicitly as

$$p = \sum_{n\in J} \langle x, v_n \rangle\, v_n. \qquad (7.24)$$

Now, the series $\sum_{n\in J} s_n \langle x, v_n \rangle\, u_n$ converges because the s_n are bounded, the u_n are orthonormal, and $\sum |\langle x, v_n\rangle|^2 < \infty$ (by Bessel's Inequality). Since $Tq = 0$, we therefore compute that

$$
\begin{aligned}
Tx = Tp &= T\left(\sum_{n\in J} \langle x, v_n\rangle\, v_n \right) && \text{(by equation (7.24))} \\
&= \sum_{n\in J} s_n \langle x, v_n\rangle\, s_n^{-1} T v_n && \text{(T is continuous and $s_n \neq 0$)} \\
&= \sum_{n\in J} s_n \langle x, v_n\rangle\, u_n && \text{(definition of u_n).}
\end{aligned}
$$

Thus equation (7.23) holds. It follows from this that $Tx \in \overline{\text{span}}\{u_n\}_{n\in J}$, and hence $\overline{\text{range}}(T) \subseteq \overline{\text{span}}\{u_n\}_{n\in J}$. Combining this with the results obtained in part (b), we conclude that $\{u_n\}_{n\in J}$ is an orthonormal basis for $\overline{\text{range}}(T)$. $\quad\square$

The representation given in equation (7.23) is called the *Singular Value Decomposition* of T, which we often abbreviate as *SVD*. If H and K are the Euclidean spaces $\mathbb{F}^n$ and $\mathbb{F}^m$, then T corresponds to an $m \times n$ matrix. We will explore the SVD for matrices in more detail in Section 7.10.

The problems for this section will examine the following type of operator, and in particular will characterize their singular numbers.

Definition 7.9.4 (Hilbert–Schmidt Operators). Let H be a separable infinite-dimensional Hilbert space. We say that $T \in \mathcal{B}(H)$ is a *Hilbert–Schmidt operator* if there exists an orthonormal basis $\{e_n\}_{n\in\mathbb{N}}$ for H such that

$$\|T\|_{\text{HS}}^2 = \sum_{n=1}^{\infty} \|T e_n\|^2 < \infty. \qquad (7.25)$$

The space of all Hilbert–Schmidt operators on H is denoted by

$$\mathcal{B}_2(H) = \{T \in \mathcal{B}(H) : T \text{ is Hilbert–Schmidt}\}. \qquad \Diamond$$

For example, the multiplication operator M_λ introduced in Theorem 7.1.4 is Hilbert–Schmidt if and only if λ is a sequence in ℓ^2, and an infinite matrix

$A = [a_{ij}]_{i,j\in\mathbb{N}}$ determines a Hilbert–Schmidt operator on ℓ^2 if and only if its entries are square-summable (see Problem 7.9.6).

Problems

7.9.5. Let H be a Hilbert space. Prove that if $T\colon H \to H$ is a nonzero compact operator, then $\|T\| = s_1$, the first singular value of T.

7.9.6. Let H be a separable infinite-dimensional Hilbert space.

(a) Let T be a bounded linear operator on H. Prove that the quantity $\|T\|_{\mathrm{HS}}^2$ defined in equation (7.25) is independent of the choice of orthonormal basis. That is, if it is finite for one orthonormal basis, then it is finite for all and always takes the same value, and if it is infinite for one orthonormal basis, then it is infinite for all.

(b) Let M be a closed subspace of H. Prove that the orthogonal projection P of H onto M is a Hilbert–Schmidt operator if and only if M is finite-dimensional.

(c) Let M_λ be the multiplication operator defined in equation (7.3). Prove that M_λ is Hilbert–Schmidt if and only if $\lambda \in \ell^2$.

(d) Let $A = [a_{ij}]_{i,j\in\mathbb{N}}$ be an infinite matrix such that $Ax \in \ell^2$ for each $x \in \ell^2$. Prove that A is Hilbert–Schmidt if and only if its *Frobenius norm* $\|A\|_2 = \left(\sum_{i,j}|a_{ij}|^2\right)^{1/2}$ is finite (compare Problem 7.5.6).

7.9.7. Given a separable infinite-dimensional Hilbert space H, prove the following statements.

(a) The Hilbert–Schmidt norm dominates the operator norm, i.e., if T is Hilbert–Schmidt, then $\|T\| \le \|T\|_{\mathrm{HS}}$.

(b) $\|T\|_{\mathrm{HS}} = \left(\sum \|Te_n\|^2\right)^{1/2}$ defines a norm on $\mathcal{B}_2(H)$.

(c) $\mathcal{B}_2(H)$ is complete with respect to the norm $\|\cdot\|_{\mathrm{HS}}$.

7.9.8. Given a separable infinite-dimensional Hilbert space H, prove the following statements.

(a) If $T \in \mathcal{B}(H)$ is Hilbert–Schmidt, then so is T^*, and $\|T^*\|_{\mathrm{HS}} = \|T\|_{\mathrm{HS}}$.

(b) If T is Hilbert–Schmidt and $A \in \mathcal{B}(H)$, then AT and TA are Hilbert–Schmidt, $\|AT\|_{\mathrm{HS}} \le \|A\|\,\|T\|_{\mathrm{HS}}$, and $\|TA\|_{\mathrm{HS}} \le \|A\|\,\|T\|_{\mathrm{HS}}$.

(c) If T is Hilbert–Schmidt and $U \in \mathcal{B}(H)$ is unitary, then TU and UT are Hilbert–Schmidt, and $\|TU\|_{\mathrm{HS}} = \|T\|_{\mathrm{HS}} = \|UT\|_{\mathrm{HS}}$.

(d) Every bounded, finite-rank linear operator on H is Hilbert–Schmidt, and every Hilbert–Schmidt operator on H is compact.

(e) Exhibit a compact operator that is not Hilbert–Schmidt, and a Hilbert–Schmidt operator that does not have finite rank.

7.9.9. Let $T \in \mathcal{B}(H)$ be a compact operator defined on a separable infinite-dimensional Hilbert space H, and let $s = (s_n)_{n \in J}$ be the singular values of T. Prove the following statements.

(a) T is Hilbert–Schmidt if and only if $s \in \ell^2(J)$.

(b) If T is Hilbert–Schmidt, then $\|T\|_{\mathrm{HS}} = \|s\|_2 = \left(\sum s_n^2\right)^{1/2}$.

(c) If T is Hilbert–Schmidt and self-adjoint, then $\|T\|_{\mathrm{HS}} = \left(\sum \lambda_n^2\right)^{1/2}$, where $(\lambda_n)_{n \in J}$ is the sequence of nonzero eigenvalues of T.

7.9.10. Let $T \in \mathcal{B}(H)$ be a compact operator defined on a separable infinite-dimensional Hilbert space H.

(a) Prove that T^*T is a compact positive operator.

(b) Let $|T| = (T^*T)^{1/2}$ be the square root of T^*T, as constructed in Problem 7.8.9. We call $|T|$ the *absolute value* of T. Prove that the nonzero eigenvalues of $|T|$ coincide with the singular numbers of T.

(c) Show that $\overline{\mathrm{range}}(|T|) = \overline{\mathrm{range}}(T^*)$ and $\ker(|T|) = \ker(T)$.

7.9.11. A compact operator $T \in \mathcal{B}(H)$ is *trace-class* if its singular numbers $(s_n)_{n \in J}$ are summable, i.e., $\sum s_n < \infty$. The space of *trace-class operators* on H is denoted by $\mathcal{B}_1(H)$. Note that since $\ell^1 \subseteq \ell^2$, every trace-class operator is Hilbert–Schmidt, i.e., $\mathcal{B}_1(H) \subseteq \mathcal{B}_2(H)$.

Let T be a compact operator on H, and let $|T|$ be its absolute value, as defined in Problem 7.9.10. Given any orthonormal basis $\{e_k\}_{k \in \mathbb{N}}$ for H, prove that

$$\sum_{k=1}^{\infty} \langle |T|e_k, e_k \rangle = \sum_{n \in J} s_n.$$

Note that these quantities might be infinite. Prove that the following three statements are equivalent.

(a) T is trace-class.

(b) $\sum \langle |T|e_k, e_k \rangle < \infty$ for some orthonormal basis $\{e_k\}_{k \in \mathbb{N}}$ for H.

(c) $|T|^{1/2}$ is Hilbert–Schmidt.

7.10 The Singular Value Decomposition of Matrices

We will apply Theorem 7.9.3 with $H = \mathbb{F}^n$ and $K = \mathbb{F}^m$ to obtain the Singular Value Decomposition (SVD) of an $m \times n$ matrix A. First, however, we review some terminology and facts.

As usual we identify the matrix A with the linear mapping that takes $x \in \mathbb{F}^n$ to $Ax \in \mathbb{F}^m$. For a matrix, the range of A equals the span of the columns of A. We therefore call this the *column space* of A, and denote it by

$$C(A) = \text{range}(A) = \{Ax : x \in \mathbb{F}^n\}.$$

The kernel of a matrix A is usually called the *nullspace* of A. We will write this as

$$N(A) = \ker(A) = \{x \in \mathbb{F}^n : Ax = 0\}.$$

By the Fundamental Theorem of Linear Algebra (compare Theorem 7.2.7),

$$C(A^H) \text{ and } N(A) \text{ are orthogonal complements in } \mathbb{F}^n,$$

while

$$C(A) \text{ and } N(A^H) \text{ are orthogonal complements in } \mathbb{F}^m.$$

If a matrix A has real entries, then we often say that $C(A^H) = C(A^T)$ is the *row space* of A.

The *rank* of A is the dimension of its column space,

$$\text{rank}(A) = \dim(C(A)).$$

As a consequence of the singular value decomposition we will see that $C(A)$ and $C(A^H)$ have the same dimension (hence we sometimes say that the *column rank* of A equals its *row rank*). For matrices it is easy to give a direct proof that $\dim(C(A)) = \dim(C(A^H))$, because if we use Gaussian elimination to compute the reduced echelon form R of A, then the number of columns of R that have a pivot is exactly the same as the number of rows of R that have a pivot, and this number is the rank of A (for details, see [Str16] or other texts on linear algebra).

One more notion that we need is the idea of a diagonal matrix. It is clear what this should mean for square matrices, but we will also need to apply this terminology to nonsquare matrices. If $A = [a_{ij}]$ is an $m \times n$ matrix, then we say that A is *diagonal* if $a_{ij} = 0$ whenever $i \neq j$. If $m \leq n$, then a diagonal matrix can have at most m nonzero entries, while if $n \leq m$, then it can have at most n nonzero entries.

Now we prove the SVD for linear operators on Euclidean spaces.

Theorem 7.10.1 (Singular Value Decomposition of Matrices). *If A is an $m \times n$ matrix with scalar entries, then there exist a unitary $m \times m$ matrix U, a unitary $n \times n$ matrix V, and an $m \times n$ diagonal matrix Σ such that*

$$A = U\Sigma V^H = U\Sigma V^{-1}.$$

Proof. We will give the proof for the square case $m = n$, and assign the proof for $m \neq n$ as Problem 7.10.5.

We will apply Theorem 7.9.3 to A. The range of A is the column space $C(A)$, and we let $r = \text{rank}(A) = \dim(C(A))$ be the rank of A. Using this notation, the index set J in Theorem 7.9.3 is $J = \{1, \ldots, r\}$. That theorem tells us that A has r nonzero singular values $s_1, \ldots, s_r$, corresponding vectors

$v_1, \ldots, v_r$ that form an orthonormal basis for $C(A^H)$, and vectors $u_1, \ldots, u_r$ that form an orthonormal basis for $C(A)$, so that

$$Ax = \sum_{k=1}^{r} s_k \, (x \cdot v_k) \, u_k, \qquad \text{for all } x \in \mathbb{F}^n. \tag{7.26}$$

Extend the nonzero singular values by setting $s_{r+1} = \cdots = s_n = 0$. The orthogonal complement of $C(A^H)$ is $N(A)$, the nullspace of A. Therefore if we let $\{v_{r+1}, \ldots, v_n\}$ be an orthonormal basis for $N(A)$, then $\{v_1, \ldots, v_n\}$ is an orthonormal basis for $\mathbb{F}^n$. Likewise, if we choose an orthonormal basis $\{u_{r+1}, \ldots, u_n\}$ for $N(A^H)$, then we obtain an orthonormal basis $\{u_1, \ldots, u_n\}$ for $\mathbb{F}^n$. Let V be the matrix that has $v_1, \ldots, v_n$ as columns, let U be the matrix that has $u_1, \ldots, u_n$ as columns, and let Σ be the matrix that has $s_1, \ldots, s_n$ as its diagonal entries and zeros elsewhere. Then U and V are unitary, while Σ is diagonal. Finally, rewriting equation (7.26) in matrix terms, we obtain

$$Ax = \sum_{k=1}^{r} s_k \, (x \cdot v_k) \, u_k = \sum_{k=1}^{n} s_k \, (x \cdot v_k) \, u_k$$

$$= \begin{bmatrix} | & & | \\ u_1 & \cdots & u_n \\ | & & | \end{bmatrix} \begin{bmatrix} s_1 \, (x \cdot v_1) \\ \vdots \\ s_n \, (x \cdot v_n) \end{bmatrix}$$

$$= \begin{bmatrix} | & & | \\ u_1 & \cdots & u_n \\ | & & | \end{bmatrix} \begin{bmatrix} s_1 & & \\ & \ddots & \\ & & s_n \end{bmatrix} \begin{bmatrix} x \cdot v_1 \\ \vdots \\ x \cdot v_n \end{bmatrix}$$

$$= \begin{bmatrix} | & & | \\ u_1 & \cdots & u_n \\ | & & | \end{bmatrix} \begin{bmatrix} s_1 & & \\ & \ddots & \\ & & s_n \end{bmatrix} \begin{bmatrix} - & v_1^H & - \\ & \vdots & \\ - & v_n^H & - \end{bmatrix} x$$

$$= U \Sigma V^H x. \qquad \square$$

In summary, Theorem 7.10.1 implies that every matrix A is a composition of a unitary matrix V^H (essentially a rotation of $\mathbb{F}^n$), followed by a diagonal matrix Σ (which stretches along orthogonal axes), followed by a unitary matrix U (essentially a rotation of $\mathbb{F}^m$). In the next example we compute the SVD of a nonsymmetric matrix; compare this to diagonalization of a symmetric matrix presented in Example 7.8.4.

Example 7.10.2. Consider $m = n = 2$, with real scalars. The singular value decomposition of the nonsymmetric matrix $A = \begin{bmatrix} 1/2 & 3/2 \\ -3/2 & -1/2 \end{bmatrix}$ is

$$\begin{bmatrix} 1/2 & 3/2 \\ -3/2 & -1/2 \end{bmatrix} = A = U\Sigma V^T = \begin{bmatrix} 2^{-1/2} & 2^{-1/2} \\ -2^{-1/2} & 2^{-1/2} \end{bmatrix} \begin{bmatrix} 2 & 0 \\ 0 & 1 \end{bmatrix} \begin{bmatrix} 2^{-1/2} & 2^{-1/2} \\ -2^{-1/2} & 2^{-1/2} \end{bmatrix}.$$

That is, U is a rotation by $-\pi/4$, the diagonal matrix Σ stretches along the coordinate axes, and V is a rotation by $\pi/4$. Thus we can realize $Ax = U(\Sigma(V^Tx)))$ by first rotating x by $-\pi/4$ (the action of V^T on x), then stretching that result along the coordinate axes, and finally rotating that result by $-\pi/4$. Note that $V^T = V^{-1}$ is not an inverse rotation to U, in contrast to the situation in Example 7.8.4. In the general SVD, U and V can be two different unitary matrices.

We illustrate this in Figure 7.2. Let v_1 and v_2 be the columns of V. These are the singular vectors of A, and are shown in part (a) of Figure 7.2. Part (b) of the figure shows the vectors V^Tv_1 and V^Tv_2 that result after a rotation by $-\pi/4$. Part (c) shows ΣV^Tv_1 and ΣV^Tv_2, and part (d) then rotates these vectors by $-\pi/4$ to obtain $Av_1 = U\Sigma V^Tv_1 = 2u_1$ and $Av_2 = U\Sigma V^Tv_2 = u_2$, where u_1, u_2 are the columns of U. As can be seen in the figure, v_1 and v_2 are not eigenvectors of A. $\quad\Diamond$

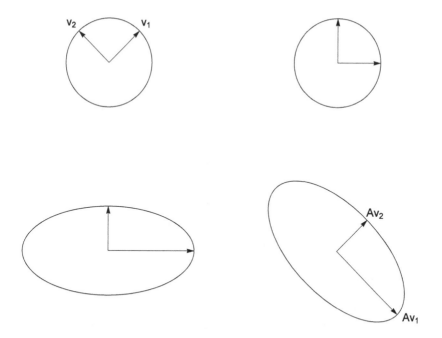

Fig. 7.2 (a) Top left: The unit circle and the orthonormal singular vectors v_1, v_2 of A. (b) Top right: After multiplication by V^T (rotation by $-\pi/4$). (c) Bottom left: After multiplication by Σ (stretching along axes). (d) Bottom right: After multiplication by U (rotation by $-\pi/4$). The end result is the same as direct multiplication by A.

Looking again at Theorem 7.10.1, we see that A maps $C(A^H)$ bijectively onto $C(A)$, with the orthonormal vectors $v_1, \ldots, v_r$ that form a basis for $C(A^H)$ mapping to the orthogonal vectors $s_1 u_1, \ldots, s_r u_r$ that form a basis for $C(A)$. Likewise, A^H maps $C(A)$ bijectively onto $C(A^H)$, and the orthonormal vectors $u_1, \ldots, u_r$ are mapped by A^H to the orthogonal vectors $s_1 v_1, \ldots, s_r v_r$. We see again that the singular vectors are eigenvectors of $A^H A$, because $A^H A v_k = s_k^2 v_k$.

Now, A maps $C(A^H)$ bijectively onto $C(A)$ and A^H maps $C(A)$ bijectively back to $C(A^H)$, but A^H is not an inverse of A on this space. On the other hand, we can see how to invert A on $C(A)$, and by extending this map from $C(A)$ to all of $\mathbb{F}^m$ we obtain a matrix that inverts A *on the column space* $C(A)$. It is not an inverse for all vectors, but only for those that belong to the column space. Using the same notation as before, we can exhibit this matrix explicitly as follows.

Theorem 7.10.3. *Let A be an $m \times n$ matrix. Let $\Sigma^\dagger$ be the $n \times m$ diagonal matrix whose first r diagonal entries are $s_1^{-1}, \ldots, s_r^{-1}$, and set*

$$A^\dagger = V \Sigma^\dagger U^H. \qquad (7.27)$$

Then

$$A^\dagger y = \sum_{k=1}^{r} s_k^{-1} (y \cdot u_k) v_k, \qquad \text{for all } y \in \mathbb{F}^m, \qquad (7.28)$$

and the following statements hold.

(a) *$AA^\dagger$ is the orthogonal projection of $\mathbb{F}^m$ onto $C(A)$. In particular,*

$$AA^\dagger y = y, \qquad \text{for all } y \in C(A).$$

(b) *$A^\dagger A$ is the orthogonal projection of $\mathbb{F}^n$ onto $C(A^H)$. In particular,*

$$A^\dagger A x = x, \qquad \text{for all } x \in C(A^H).$$

Proof. Choose any vector $y \in \mathbb{F}^m$. When we write out $A^\dagger y = V \Sigma^\dagger U^H y$, we obtain equation (7.28). Further, since $A v_k = s_k u_k$, we have

$$AA^\dagger y = A(A^\dagger y) = \sum_{k=1}^{r} s_k^{-1} (y \cdot u_k) A v_k = \sum_{k=1}^{r} (y \cdot u_k) u_k.$$

Since $\{u_1, \ldots, u_r\}$ is an orthonormal basis for $C(A)$, this tells us that $AA^\dagger y$ is the orthogonal projection of y onto $C(A)$.

Now choose any vector $x \in \mathbb{F}^n$. Since $A^H u_k = s_k v_k$, we see that

$$A^\dagger A x = \sum_{k=1}^{r} s_k^{-1} (Ax \cdot u_k) v_k$$

$$= \sum_{k=1}^{r} s_k^{-1} (x \cdot A^H u_k) v_k$$

$$= \sum_{k=1}^{r} s_k^{-1} (x \cdot s_k v_k) v_k$$

$$= \sum_{k=1}^{r} (x \cdot v_k) v_k,$$

which is the orthogonal projection of x onto $C(A^H)$. $\square$

We give the following name to the matrix $A^\dagger$.

Definition 7.10.4 (Pseudoinverse). Given an $m \times n$ matrix A, the $n \times m$ matrix $A^\dagger$ constructed in Theorem 7.10.3 is called the *Moore–Penrose pseudo-inverse*, or simply the *pseudoinverse*, of A. $\diamond$

More generally, if H and K are any Hilbert spaces, then it is possible to construct a pseudoinverse for any bounded operator $A \in \mathcal{B}(H, K)$ that has closed range. For one proof, see [Heil11, Thm. 2.33].

Problems

7.10.5. What changes must be made to the proof of Theorem 7.10.1 if $m < n$ or $n < m$?

7.10.6. Use the SVD to give another solution to Problem 7.8.10.

7.10.7. Verify the statements made about the SVD of the matrix A in Example 7.10.2.

7.10.8. Let H and K be Hilbert spaces, and suppose that $T \colon H \to K$ is a bounded linear finite-rank operator. Explain how to construct a pseudoin-verse of T that is analogous to the pseudoinverse of a matrix constructed in Theorem 7.10.3.

7.10.9. Show that if A is a $d \times d$ matrix with $\operatorname{rank}(A) = d$, then A is invertible and $A^\dagger = A^{-1}$.

7.10.10. Suppose that $A = uv^H$, where u and v are nonzero vectors in $\mathbb{F}^n$. Show that $\operatorname{rank}(A) = 1$, and find the eigenvalues, eigenvectors, singular values, singular vectors, and pseudoinverse of A.

Index of Symbols

Sets

Symbol	Description
$\varnothing$	Empty set
$B_r(x)$	Open ball of radius r centered at x
$\mathbb{C}$	Complex plane
$\mathbb{F}$	Choice of $\mathbb{R}$ or $\mathbb{C}$
$\mathcal{M}$	Set of monic monomials, $\{1, x, x^2, \dots\}$
$\mathbb{N}$	Natural numbers, $\{1, 2, 3, \dots\}$
$\mathcal{P}$	Set of all polynomials
$\mathbb{Q}$	Rational numbers, $\{m/n : m, n \in \mathbb{Z},\, n \neq 0\}$
$\mathbb{R}$	Real line
$\mathbb{Z}$	Integers, $\{\dots, -1, 0, 1, \dots\}$

Operations on Sets

Symbol	Description
$A^{\mathrm{C}} = X \backslash A$	Complement of a set $A \subseteq X$
A°	Interior of a set A
$\overline{A}$	Closure of a set A
∂A	Boundary of a set A
$A \times B$	Cartesian product of A and B
$A \backslash B = A \cap B^{\mathrm{C}}$	Relative complement of A in B
$\mathrm{diam}(A)$	Diameter of a set A
$\mathrm{dim}(X)$	Dimension of a vector space X

© Springer International Publishing AG, part of Springer Nature 2018
C. Heil, *Metrics, Norms, Inner Products, and Operator Theory*, Applied and
Numerical Harmonic Analysis, https://doi.org/10.1007/978-3-319-65322-8

$\mathrm{dist}(x, A)$	Distance from a point to a set
$\mathrm{dist}(A, B)$	Distance between two sets
$\inf(S)$	Infimum of a set of real numbers S
$\mathrm{span}(A)$	Finite linear span of a set A
$\overline{\mathrm{span}}(A)$	Closure of the finite linear span of A
$\sup(S)$	Supremum of a set of real numbers S
$\bigcup_{i \in I} X_i$	Union of sets X_i
$\bigcap_{i \in I} X_i$	Intersection of sets X_i

Sequences

Symbol	Description
$\{x_n\}_{n \in \mathbb{N}}$	A sequence of points $x_1, x_2, \ldots$
$(x_k)_{k \in \mathbb{N}}$	A sequence of scalars $x_1, x_2, \ldots$
δ_n	nth standard basis vector

Functions, Operators, Matrices

Symbol	Description
$f : A \to B$	A function from A to B
$f(A)$	Image of A under f
$f^{-1}(B)$	Inverse image of B under f
f^{-1}	Inverse function (if f is bijective)
$\mathrm{range}(f)$	Range of f
$\overline{\mathrm{range}}(f)$	Closure of the range of f
$\ker(A)$	kernel (or nullspace) of an operator A
A^*	Adjoint of an operator A
A^{T}	Transpose of a matrix A
$A^{\mathrm{H}} = \overline{A^{\mathrm{T}}}$	Hermitian, or conjugate transpose, of a matrix A
$A^{\dagger}$	Pseudoinverse of a matrix A
$C(A)$	Column space of a matrix A $(= \mathrm{range}(A))$
$N(A)$	Nullspace of a matrix A $(= \ker(A))$

Vector Spaces

Symbol	Description
c_{00}	"Finite sequences"
c_0	Sequences vanishing at infinity
$C(X)$	Continuous functions on X
$C_b(X)$	Bounded continuous functions on X
$C[a,b]$	Continuous functions on $[a,b]$
$C_0(\mathbb{R})$	Continuous functions vanishing at infinity
$C_c(\mathbb{R})$	Continuous functions with compact support
$C_b^1(I)$	Differentiable functions with continuous derivative
$\mathcal{F}_b(X)$	Bounded functions on X
ℓ^1	Summable sequences
ℓ^2	Square-summable sequences
ℓ^p	p-summable sequences (if $p < \infty$)
ℓ^∞	Bounded sequences
$L^p(E)$	Lebesgue space
$\mathrm{Lip}(I)$	Lipschitz functions on I

Metrics, Norms, and Inner Products

Symbol	Description
$A^\perp$	Orthogonal complement of a set A
$\mathrm{d}(\cdot,\cdot)$	Generic metric
$\langle \cdot,\cdot \rangle$	Generic inner product
$\|\cdot\|$	Generic norm
$\|f\|_u$	Uniform norm of a function f
$\|f\|_1$	L^1-norm of a function f
$\|f\|_2$	L^2-norm of a function f
$\|f\|_p$	L^p-norm of a function f
$x \perp y$	Orthogonal vectors
$\|x\|_1$	ℓ^1-norm of a sequence x
$\|x\|_2$	ℓ^2-norm of a sequence x
$\|x\|_p$	ℓ^p-norm of a sequence x
$\|x\|_\infty$	ℓ^∞-norm of a sequence x

Miscellaneous Symbols

Symbol	Description
□	End of proof
◇	End of Remark, Example, or Exercise
◇	End of Theorem whose proof is omitted

References

[Ask75]   R. Askey, *Orthogonal Polynomials and Special Functions*, SIAM, Philadelphia, 1975.

[Axl15]   S. Axler, *Linear Algebra Done Right*, Third Edition, Springer, Cham, 2015.

[Bar76]   R. G. Bartle, *The Elements of Real Analysis*, Second Edition, Wiley, New York, 1976.

[Ben97]   J. J. Benedetto, *Harmonic Analysis and Applications*, CRC Press, Boca Raton, FL, 1997.

[Cho83]   M. N. Choi, Tricks or treats with the Hilbert matrix. *Amer. Math. Monthly*, **90** (1983), pp. 301–312.

[Chr16]   O. Christensen, *An Introduction to Frames and Riesz Bases*, Second Edition, Birkhäuser, Boston, 2016.

[Con90]   J. B. Conway, *A Course in Functional Analysis*, Second Edition, Springer-Verlag, New York, 1990.

[Con00]   J. B. Conway, *A Course in Operator Theory*, American Mathematical Society, Providence, RI, 2000.

[DM72]   H. Dym and H. P. McKean, *Fourier Series and Integrals*, Academic Press, New York–London, 1972.

[Enf73]   P. Enflo, A counterexample to the approximation problem in Banach spaces, *Acta Math.*, **130** (1973), pp. 309–317.

[Fol99]   G. B. Folland, *Real Analysis*, Second Edition, Wiley, New York, 1999.

[GG01]   I. Gohberg and S. Goldberg, *Basic Operator Theory*, Birkhäuser, Boston, 2001 (reprint of the 1981 original).

[HKLW07]   D. Han, K. Kornelson, D. Larson, and E. Weber, *Frames for Undergraduates*, American Mathematical Society, Providence, RI, 2007.

[HHW18]   J. Haas, C. Heil, and M. D. Weir, *Thomas' Calculus*, 14th Edition, Pearson, Boston, 2018.

[Heil06]   C. Heil, Linear independence of finite Gabor systems, in: *Harmonic Analysis and Applications,* Birkhäuser, Boston (2006), pp. 171–206.

[Heil11]   C. Heil, *A Basis Theory Primer*, Expanded Edition, Birkhäuser, Boston, 2011.

[Heil19]   C. Heil, *Introduction to Real Analysis*, Birkhäuser, Boston, to appear.

[HS15]   C. Heil and D. Speegle, The HRT Conjecture and the Zero Divisor Conjecture for the Heisenberg group, in: *Excursions in Harmonic Analysis, Volume 3*, Birkhäuser, Boston (2015), pp. 159–176.

[Kap77]   I. Kaplansky, *Set Theory and Metric Spaces*, Second Edition, AMS Chelsea Publishing, New York, 1977.

[Kat04]   Y. Katznelson, *An Introduction to Harmonic Analysis*, Third Edition, Cambridge University Press, Cambridge, UK, 2004.

© Springer International Publishing AG, part of Springer Nature 2018

C. Heil, *Metrics, Norms, Inner Products, and Operator Theory*, Applied and Numerical Harmonic Analysis, https://doi.org/10.1007/978-3-319-65322-8

[Kre78]   E. Kreyszig, *Introductory Functional Analysis with Applications*, Wiley, New York, 1978.

[Mun75]   J. R. Munkres, *Topology: A First Course*, Prentice-Hall, Englewood Cliffs, NJ, 1975.

[Rud76]   W. Rudin, *Principles of Mathematical Analysis*, Third Edition. McGraw-Hill, New York, 1976.

[Rud87]   W. Rudin, *Real and Complex Analysis*, Third Edition. McGraw-Hill, New York, 1987.

[Rud91]   W. Rudin, *Functional Analysis*, Second Edition. McGraw-Hill, New York, 1991.

[ST76]   I. M. Singer and J. A. Thorpe, *Lecture Notes on Elementary Topology and Geometry*, Springer-Verlag, New York–Heidelberg, 1976 (reprint of the 1967 edition).

[SS05]   E. M. Stein and R. Shakarchi, *Real Analysis*, Princeton University Press, Princeton, NJ, 2005.

[Str16]   G. Strang, *Introduction to Linear Algebra*, Fifth Edition, Wellesley–Cambridge Press, Wellesley, MA 2016.

[Sze75]   G. Szegö, *Orthogonal Polynomials*, Fourth edition, American Mathematical Society, Providence, R.I., 1975.

[Wal02]   D. F. Walnut, *An Introduction to Wavelet Analysis*, Birkhäuser Boston, Boston, MA, 2002.

[WZ77]   R. L. Wheeden and A. Zygmund, *Measure and Integral*, Marcel Dekker, New York–Basel, 1977.

Index

© Springer International Publishing AG, part of Springer Nature 2018
C. Heil, *Metrics, Norms, Inner Products, and Operator Theory*, Applied and
Numerical Harmonic Analysis, https://doi.org/10.1007/978-3-319-65322-8

Applied and Numerical Harmonic Analysis
(81 volumes)

© Springer International Publishing AG, part of Springer Nature 2018
C. Heil, *Metrics, Norms, Inner Products, and Operator Theory*, Applied and Numerical Harmonic Analysis, https://doi.org/10.1007/978-3-319-65322-8

42. Cabrelli and J.L. Torrea: *Recent Developments in Real and Harmonic Analysis* (ISBN 978-0-8176-4531-1)
43. M.V. Wickerhauser: *Mathematics for Multimedia* (ISBN 978-0-8176-4879-4)
44. B. Forster, P. Massopust, O. Christensen, K. Gröchenig, D. Labate, P. Vandergheynst, G. Weiss, and Y. Wiaux: *Four Short Courses on Harmonic Analysis* (ISBN 978-0-8176-4890-9)
45. O. Christensen: *Functions, Spaces, and Expansions* (ISBN 978-0-8176-4979-1)
46. J. Barral and S. Seuret: *Recent Developments in Fractals and Related Fields* (ISBN 978-0-8176-4887-9)
47. O. Calin, D.-C. Chang, and K. Furutani, and C. Iwasaki: *Heat Kernels for Elliptic and Sub-elliptic Operators* (ISBN 978-0-8176-4994-4)
48. C. Heil: *A Basis Theory Primer* (ISBN 978-0-8176-4686-8)
49. J.R. Klauder: *A Modern Approach to Functional Integration* (ISBN 978-0-8176-4790-2)
50. J. Cohen and A.I. Zayed: *Wavelets and Multiscale Analysis* (ISBN 978-0-8176-8094-7)
51. Joyner and J.-L. Kim: *Selected Unsolved Problems in Coding Theory* (ISBN 978-0-8176-8255-2)
52. G.S. Chirikjian: *Stochastic Models, Information Theory, and Lie Groups, Volume 2* (ISBN 978-0-8176-4943-2)
53. J.A. Hogan and J.D. Lakey: *Duration and Bandwidth Limiting* (ISBN 978-0-8176-8306-1)
54. Kutyniok and D. Labate: *Shearlets* (ISBN 978-0-8176-8315-3)
55. P.G. Casazza and P. Kutyniok: *Finite Frames* (ISBN 978-0-8176-8372-6)
56. V. Michel: *Lectures on Constructive Approximation* (ISBN 978-0-8176-8402-0)
57. D. Mitrea, I. Mitrea, M. Mitrea, and S. Monniaux: *Groupoid Metrization Theory* (ISBN 978-0-8176-8396-2)
58. T.D. Andrews, R. Balan, J.J. Benedetto, W. Czaja, and K.A. Okoudjou: *Excursions in Harmonic Analysis, Volume 1* (ISBN 978-0-8176-8375-7)
59. T.D. Andrews, R. Balan, J.J. Benedetto, W. Czaja, and K.A. Okoudjou: *Excursions in Harmonic Analysis, Volume 2* (ISBN 978-0-8176-8378-8)
60. D.V. Cruz-Uribe and A. Fiorenza: *Variable Lebesgue Spaces* (ISBN 978-3-0348-0547-6)
61. W. Freeden and M. Gutting: *Special Functions of Mathematical (Geo-)Physics* (ISBN 978-3-0348-0562-9)
62. A. I. Saichev and W.A. Woyczyński: *Distributions in the Physical and Engineering Sciences, Volume 2: Linear and Nonlinear Dynamics of Continuous Media* (ISBN 978-0-8176-3942-6)
63. S. Foucart and H. Rauhut: *A Mathematical Introduction to Compressive Sensing* (ISBN 978-0-8176-4947-0)
64. Herman and J. Frank: *Computational Methods for Three-Dimensional Microscopy Reconstruction* (ISBN 978-1-4614-9520-8)

65. Paprotny and M. Thess: *Realtime Data Mining: Self-Learning Techniques for Recommendation Engines* (ISBN 978-3-319-01320-6)

66. Zayed and G. Schmeisser: *New Perspectives on Approximation and Sampling Theory: Festschrift in Honor of Paul Butzer's 85th Birthday* (ISBN 978-3-319-08800-6)

67. R. Balan, M. Begue, J. Benedetto, W. Czaja, and K.A Okoudjou: *Excursions in Harmonic Analysis, Volume 3* (ISBN 978-3-319-13229-7)

68. Boche, R. Calderbank, G. Kutyniok, J. Vybiral: *Compressed Sensing and its Applications* (ISBN 978-3-319-16041-2)

69. S. Dahlke, F. De Mari, P. Grohs, and D. Labate: *Harmonic and Applied Analysis: From Groups to Signals* (ISBN 978-3-319-18862-1)

70. Aldroubi, *New Trends in Applied Harmonic Analysis* (ISBN 978-3-319-27871-1)

71. M. Ruzhansky: *Methods of Fourier Analysis and Approximation Theory* (ISBN 978-3-319-27465-2)

72. G. Pfander: *Sampling Theory, a Renaissance* (ISBN 978-3-319-19748-7)

73. R. Balan, M. Begue, J. Benedetto, W. Czaja, and K.A Okoudjou: *Excursions in Harmonic Analysis, Volume 4* (ISBN 978-3-319-20187-0)

74. O. Christensen: *An Introduction to Frames and Riesz Bases, Second Edition* (ISBN 978-3-319-25611-5)

75. E. Prestini: *The Evolution of Applied Harmonic Analysis: Models of the Real World, Second Edition* (ISBN 978-1-4899-7987-2)

76. J.H. Davis: *Methods of Applied Mathematics with a Software Overview, Second Edition* (ISBN 978-3-319-43369-1)

77. M. Gilman, E. M. Smith, S. M. Tsynkov: *Transionospheric Synthetic Aperture Imaging* (ISBN 978-3-319-52125-1)

78. S. Chanillo, B. Franchi, G. Lu, C. Perez, E.T. Sawyer: *Harmonic Analysis, Partial Differential Equations and Applications* (ISBN 978-3-319-52741-3)

79. R. Balan, J. Benedetto, W. Czaja, M. Dellatorre, and K.A Okoudjou: *Excursions in Harmonic Analysis, Volume 5* (ISBN 978-3-319-54710-7)

80. Pesenson, Q.T. Le Gia, A. Mayeli, H. Mhaskar, D.X. Zhou: *Frames and Other Bases in Abstract and Function Spaces: Novel Methods in Harmonic Analysis, Volume 1* (ISBN 978-3-319-55549-2)

81. Pesenson, Q.T. Le Gia, A. Mayeli, H. Mhaskar, D.X. Zhou: *Recent Applications of Harmonic Analysis to Function Spaces, Differential Equations, and Data Science: Novel Methods in Harmonic Analysis, Volume 2* (ISBN 978-3-319-55555-3)

82. F. Weisz: *Convergence and Summability of Fourier Transforms and Hardy Spaces* (ISBN 978-3-319-56813-3)

83. Heil: *Metrics, Norms, Inner Products, and Operator Theory* (ISBN 978-3-319-65321-1)

84. S. Waldron: *An Introduction to Finite Tight Frames: Theory and Applications.* (ISBN: 978-0-8176-4814-5)

85. Joyner and C.G. Melles: *Adventures in Graph Theory: A Bridge to Advanced Mathematics.* (ISBN: 978-3-319-68381-2)

For an up-to-date list of ANHA titles, please visit **http://www.springer.com/series/4968**

Printed in the United States
By Bookmasters